TRENDS IN OPTICAL MATERIALS RESEARCH

TRENDS IN OPTICAL MATERIALS RESEARCH

OWEN G. GARDENS
EDITOR

Nova Science Publishers, Inc.
New York

NOTICE TO THE READER

LIBRARY OF CONGRESS CATALOGING-IN-PUBLICATION DATA

Trends in optical materials research / Owen G. Gardens, editor.
p. cm.
Includes index.
ISBN 1-60021-290-5
1. Optical materials. 2. Nonlinear optics--Materials. I. Gardens, Owen G.
QC374.T74 2006
621.36--dc22
2006018373

Published by Nova Science Publishers, Inc. ✢ New York

Contents

Preface		**vii**
Chapter 1	Optical Properties of Metallophthalocyanines, Alq_3 and Perylene Derivatives Investigated by Photoluminescence, Absorption, Reflection and Photomodulated Spectroscopy *Michał Wojdyła, Mateusz Rębarz, Paweł Dalasiński, Wacław Bała, Zbigniew Łukasiak, Beata Derkowska and Bouchta Sahraoui*	**1**
Chapter 2	Photoluminescence and Electroluminescence Studies on the Morphology of Organic Molecules in Thin Solid Films *Taiju Tsuboi*	**55**
Chapter 3	Understanding the Upconversion Luminescence Properties of Rare-Earth Doped Oxide Nanocrystals *Amitava Patra*	**91**
Chapter 4	Photo-Physics Involved in the Excitation of Pyrylium Salts Under Ultra-Short Pulsed Illumination *I. Polyzos, G. Tsigaridas, M. Fakis, V. Giannetas and P. Persephonis*	**113**
Chapter 5	Study of the Optical and Thermal Properties of Phosphate Glasses *Eric Tong Yih Lee and Elizabeth M. Taylor*	**137**
Chapter 6	Azo Polymer Syntheses, Electrostatic Layer-by-Layer Self-Assembly and Photoinduced Surface Modulation Properties *Yaning He and Xiaogong Wang*	**185**
Chapter 7	Stimulated Emission in Photonic Crystal *Xu Xingsheng and Zhang Daozhong*	**219**
Index		**257**

PREFACE

This new book is the newest research on the physical properties of optical materials used in all types of lasers and optical systems. The scope includes the most important optical materials, including crystals, glasses, polymers, metals, liquids, and gases. The properties detailed include both linear and nonlinear optical properties, mechanical properties, thermal properties together with many additional special properties, such as electro-, magneto-, and elasto-optic properties.

Chapter 1 - Thin layers of metallophthalocyanines (CuPc, ZnPc, MgPc, CoPc), tris(8-hydroxyquinoline) aluminium (Alq3) and perylene derivatives: 3,4,9,10-perylene-tetracarboxylic-dianhydride (PTCDA) and 3,4,9,10-tetra(heptyloxycarbonyl) perylene (PeC7) deposited on glass, quartz and (001) silicon substrates were investigated by photoluminescence (PL), photoluminescence excitation (PLE), reflection (R), absorption (A) and sensitive modulated photoabsorption (PA)/photoreflection (PR) spectroscopy. These investigations were made in different temperatures. We have studied the optical parameters, the electronic structures and the mechanisms of energy transfer in these structures. Almost all samples exhibit strong luminescence in the temperature range from 13 K to 320 K and we also observed the thermal quenching of photoluminescence. We found that the strong temperature dependence of the PL spectra of thin films of Alq3 and perylene derivatives (PTCDA, PeC7) can be attributed to the reduced exciton-phonon interaction with decreasing temperature. Moreover, the PL spectra of thin films measured at room temperature were compared with results obtained for corresponding solutions.

The absorption spectra of all studied metallophthalocyanines (MPcs) show intense B (Soret) band in the UV region and Q band in the visible region. The absorption coefficient α and the extinction coefficient k were directly found from the transmission and reflection measurements and treats with the help of multioscillator Lorentz model modified by the Lorentz-Lorenz formula. We found that the position and intensity of the Q and B bands depend on the metallic substitution. Photomodulated spectra allowed us to enhance weak features, which are not clearly visible in usual absorption or reflectance spectroscopy. The energy of observed transitions and their broadening parameters were determined by fitting Aspnes formula to the experimental PA and PR data. The spectral distribution of complex refractive index and dielectric constants (real ε_1 and imaginary ε_2 part) was also found from reflection and transmission spectra.

The low temperature photoluminescence of metallophthalocyanines (ZnPc, MgPc) thin layers have been studied for the first time. The dependence of photoluminescence on temperature is connected with phonon interaction or/and the changes of the molecular symmetry of phthalocyanines.

Chapter 2 - Photoluminescence, excitation and absorption spectra are investigated for thin films lightly and heavily doped with TPD, $Ir(ppy)_3$, Alq_3, PtOEP and mCP and for these neat films at various temperatures between 10 and 290 K. The photoluminescence spectra are compared with the electroluminescence spectra of organic light emitting diodes with emitting layers of $Ir(ppy)_3$ and PtOEP. Aggregates such as dimer and trimer are observed in the neat films of $Ir(ppy)_3$ and Alq_3. On the other hand, dimer and trimer are formed in TPD, PtOEP- and mCP-doped films and these neat films, together with monomer. In the case of PtOEP, large size aggregates which consist of many PtOEP molecules are also formed even in lightly doped films. Such a difference between the group of $Ir(ppy)_3$ and Alq_3 and the group of TPD, PtOEP and mCP is caused by difference of competition and balance among solid state solvation effect, dipole-dipole interactions between dopant molecules, and intermolecular covalent bonding. It is suggested that molecular geometrical structure gives rise to the difference.

Chapter 3 - Here, we report the role of dopant and co-dopant concentration on the modification of crystal structure and the upconversion properties (tuning and efficiency) of rare-earth doped oxide nanocrystals. We have synthesized rare-earth doped oxide nanocrystals using non-ionic surfactant-stabilized reverse emulsions. The non-ionic surfactant in the Span series, i.e. sorbitan monooleate (Span 80) of hydrophilic-lipophilic balance (HLB) values of 4.3 has been used for the stabilization of emulsions. Bright green (550 nm) and red (670 nm) upconversion emission are observed at room temperature with 980 nm excitation from the $^4S_{3/2}$ and $^4F_{9/2}$ levels of ZrO_2:Er^{3+} nanocrystals. The total luminescence values are 11960, 7132 and 302 Cd/m^2 for 1000°C heated 0.05,0.25 and 1.0 mol% Er doped ZrO_2 nanocrystals, using 850 mW of cw excitation power source. The effects of co-dopants (Yb^{3+}, Tm^{3+} and Y^{3+}) in the ZrO_2: Er matrix on the upconverted emission are also reported. The emission spectra and the pump intensity dependence of the luminescence intensities are used to understand the excitation mechanisms,i.e. excited-state absorption (ESA) or energy transfer upconversion (ETU) processes. It is observed that the upconversion intensity increased as the excitation power increased under lower excitation power, but saturated at higher excitation power at high concentration of dopant ions.

Chapter 4 - Pyrylium salts constitute a new class of very promising materials for applications in the field of photonics. In the current article we present the photo-physics involved in the excitation of a group of representative pyrylium salts. This group consists of molecules with the same basic chemical structure differing only to a specific substituent. The two-photon absorption (TPA) properties of these molecules are investigated experimentally. Large TPA cross-section values are reported. Furthermore, it is shown that under ultra-short pulse irradiation these molecules are photobleached efficiently. The photobleaching process is realized via two, diverse, simultaneously present, mechanisms. It is proved experimentally that these mechanisms are step-wise multi-photon processes. In most of the cases more than two-photons are involved in the photobleaching process. Excited state absorption is found to play a major role in this process. Furthermore, the effect of substituent change on the photobleaching properties is investigated. It is shown that photobleaching depends strongly on the specific substituent. Finally, the decay dynamics of these molecules are studied through femtosecond fluorescence time-resolved spectroscopy.

Chapter 5 - Lightwave devices such as gratings and filters have a drawback where the central wavelength of the devices is sensitive to changes in temperature, which compromises their performance. This problem is particularly acute in systems that employ Wavelength Division Multiplexing (WDM), where the stability of wavelength with regards to temperature is essential. An athermal waveguide, which is inherently insensitive to temperature, is an ideal choice to solve this problem instead of using a passive temperature control unit, which could limit the optimum performance of a WDM system.

Glasses are considered ideal host materials for athermalisation as they are versatile and can be physically and compositionally tailored to meet various optical and thermal requirements. Generally, athermal glasses must exhibit negative refractive index change with temperature (dn/dT) to counter the effect of thermal expansion (α). Hence, phosphate-based glasses are ideal candidates for athermalisation as they exhibit negative dn/dT. Various compositions of ternary potassium aluminophosphate glasses, binary calcium phosphate and barium phosphate glasses and ternary barium borophosphate and sodium borophosphate glasses were examined in this research work to assess their potential for athermalisation. The optical and thermal properties of the phosphate glasses were studied, particularly α and dn/dT, and the compositional effects of the glasses on these properties were also investigated. Results showed that potassium aluminophosphate glasses were strongly influenced by the strengthening effect of Al^{3+} and the trend observed for the binary phosphate glasses depended on the availability of terminal oxygens to coordinate the cations, while the change in the properties of borophosphate glasses depended on the effect and coordination number of boron.

Glasses with negative dn/dT were achieved from the phosphate glasses studied. The results showed that α and dn/dT of the glasses were inter-dependent and semi-empirical equations relating these properties could be defined, enabling the estimation of dn/dT of any phosphate glasses within the same system. In addition, the results enabled the identification of the specific glass composition of a particular phosphate glass system with the minimum α value required in order for the glass to exhibit negative dn/dT. This research work represents the first comprehensive measurement of dn/dT of phosphate glasses.

Chapter 6 - This chapter reviews recent research progress in the synthesis, electrostatic layer-by-layer self-assembly and surface-relief-grating (SRG) formation behavior of azo polymers. The azo polymers were synthesized by some well-established methods such as post-polymerization azo-coupling reaction, functionalization of poly(acryloyl chloride), and polymerization through azo-coupling reactions. According to the polymer chain architecture, the azo polymers can be classified as epoxy-based, acrylate-based, urethane-based linear polymers and hyperbranched polymers. Some of the polymers were designed to contain ionizable groups. Multilayer films were prepared through the electrostatic layer-by-layer deposition of the azo polyelectrolytes and PDAC. SRGs were photofabricated by exposing both the spin-coated polymer films and multilayer films to an interference pattern of coherent Ar^+ laser beams. The SRG formation behavior was characterized by the inscription rates and the saturation levels of SRGs on the polymer films. The results show that the rates of SRG formation significantly depend on the types of the electron-withdrawing groups in the *para*-positions of the azobenzene units. The inscription rates decrease with the increase of the number of methyl substitutes on the *ortho*-positions of the azo groups. Meanwhile backbone structures also have some effect on the photoprocessibility through the ways such as altering the chromophore loading density and changing main-chain rigidity. The electrostatic multilayer films can form SRGs, but the inscription rates and the saturation levels are lower than those of the spin-coated films. Those results can be used to explain the SRG forming mechanism and to develop new SRG materials with better properties.

Chapter 7 - By using of Maxwell equations and the Bloch equations, we derive an analytical expression for stimulated emission in arbitrary 2-D photonic crystals. Using the analytical expression for stimulated emission we analyzed the transient emission, optical gain and gain threshold of two-level atom embedded in arbitrary structure photonic crystals. The transient emission can be expressed by the photon density of the states, the Rabi frequency and the relaxation coefficient. It is found that the sharp structures of the density of states near the band edge leads to a drastically enhancement of the emission and an abrupt change of the oscillation of the emission field when the ratio of Rabi frequency to decay rate is a proper value. Moreover, a transient gain is found to exist if the pumping pulse area is a certain value and the transition frequency is tuned to the band edge. It is also deduced that the optical gain in photonic crystal is proportional to the population inversion, and is inversely proportional to the group velocity and the relaxation coefficient of the phase, which indicates that the gain can be induced by the population inversion and be enhanced drastically due to the small group velocity or anomaly group velocity. Correspondingly, the lasing threshold expressed by the population inversion in arbitrary structure photonic crystals is obtained by an investigation of steady-state laser behavior. It shows that the lasing threshold expressed by population inversion is proportional to the second order of the group velocity, and proportional to the relaxation coefficient. The stimulated emission in the quasi-periodic photonic crystals is also investigated. It is found that there are some interesting phenomenons in the quasi-periodic crystals. The lasing peaks are more abundant and stronger than that in periodic crystals. More stimulated emission peaks appear when the size and the gain increase, and some lasing peaks are independent on incident direction. These lasing peaks maybe come from the localization of the disorder in the quasi-periodic photonic crystals.

In: Trends in Optical Materials Research
Editor: Owen G. Gardens, pp. 1-54
ISBN: 1-60021-290-5

Chapter 1

OPTICAL PROPERTIES OF METALLOPHTHALOCYANINES, ALQ$_3$ AND PERYLENE DERIVATIVES INVESTIGATED BY PHOTOLUMINESCENCE, ABSORPTION, REFLECTION AND PHOTOMODULATED SPECTROSCOPY

Michał Wojdyła[1], Mateusz Rębarz[1], Paweł Dalasiński[1], Wacław Bała[1], Zbigniew Łukasiak[1], Beata Derkowska[1] and Bouchta Sahraoui[2]

[1]Institute of Physics, N. Copernicus University,
Grudziądzka 5/7, 87-100 Toruń, Poland
[2]Laboratoire POMA, UMR CNRS 6136, Université d'Angers. 2,
Boulevard Lavoisier, 49045 Angers, France

ABSTRACT

Thin layers of metallophthalocyanines (CuPc, ZnPc, MgPc, CoPc), tris(8-hydroxyquinoline) aluminium (Alq3) and perylene derivatives: 3,4,9,10-perylene-tetracarboxylic-dianhydride (PTCDA) and 3,4,9,10-tetra(heptyloxycarbonyl) perylene (PeC7) deposited on glass, quartz and (001) silicon substrates were investigated by photoluminescence (PL), photoluminescence excitation (PLE), reflection (R), absorption (A) and sensitive modulated photoabsorption (PA)/photoreflection (PR) spectroscopy. These investigations were made in different temperatures. We have studied the optical parameters, the electronic structures and the mechanisms of energy transfer in these structures. Almost all samples exhibit strong luminescence in the temperature range from 13 K to 320 K and we also observed the thermal quenching of photoluminescence. We found that the strong temperature dependence of the PL spectra of thin films of Alq3 and perylene derivatives (PTCDA, PeC7) can be attributed to the reduced exciton-phonon interaction with decreasing temperature. Moreover, the PL spectra of thin films measured at room temperature were compared with results obtained for corresponding solutions.

The absorption spectra of all studied metallophthalocyanines (MPcs) show intense B (Soret) band in the UV region and Q band in the visible region. The absorption coefficient α and the extinction coefficient k were directly found from the transmission and reflection measurements and treats with the help of multioscillator Lorentz model modified by the Lorentz-Lorenz formula. We found that the position and intensity of the Q and B bands depend on the metallic substitution. Photomodulated spectra allowed us to enhance weak features, which are not clearly visible in usual absorption or reflectance spectroscopy. The energy of observed transitions and their broadening parameters were determined by fitting Aspnes formula to the experimental PA and PR data. The spectral distribution of complex refractive index and dielectric constants (real ε_1 and imaginary ε_2 part) was also found from reflection and transmission spectra.

The low temperature photoluminescence of metallophthalocyanines (ZnPc, MgPc) thin layers have been studied for the first time. The dependence of photoluminescence on temperature is connected with phonon interaction or/and the changes of the molecular symmetry of phthalocyanines.

INTRODUCTION

Organic materials with semiconducting properties such as molecular organic crystals (MOCs) or polymers are potential substitutes for regular components in novel electronic and optoelectronic devices. Metallophthalocyanines (MPcs), perylene derivatives or metal chelates are currently intensively investigated since they offer a wide range of potential applications in photovoltaic cells [1-4], organic light emitting diodes (OLEDs) [5-12], organic field effect transistors (OFET) [13-16] and gas sensors [17]. In many cases, organic semiconductor devices can replace the classical inorganic components. The advantage of organic materials over their inorganic equivalents is the growth on a large variety of non-reactive substrates [18-20]. Moreover, organic components can be even flexible [21-23]. This ability is caused by weak van der Waals-bonding between adjacent molecules, whereas the inorganic semiconductor devices are rigid due to the covalent bonding between neighbouring atoms.

Among modern organic devices, OLEDs are especially popular because they can be used in flat panel displays. In this context, the main advantages of organic materials are their excellent colour gamut and high fluorescence efficiency. In organic materials the light is produced by the fast decay of excited molecular states, and the colour of light depends on the difference in energy between those excited states and the molecular ground level. Thus, the modification of molecular structure by attaching some substituents, which changes electronic levels of molecule, allows easily obtaining various hues [24-26]. Other effective way is doping of based material by different guest molecules [27, 28]. In contrast to liquid crystals technology the OLEDs are able to produce light by themselves, do not require a backlight, can be very thin and need less energy. The promise of low-power consumption and excellent emissive quality with a wide viewing angle is unique among display technologies [29]. A major challenge in OLED manufacture is tuning the device to obtain an equal number of holes and electrons in the emissive layer. In order to make the holes and electrons arrive at the emission layer at the same time the OLEDs have to be constructed with suitable layers with accurately determined thickness. Thus, the crucial components in OLEDs are hole injection and transport layers (HIL, HTL), electron injection and transport layers (EIL, ETL) and emissive layer (EL). The functions of the individual organic films are distinct and can therefore be optimized independently. In order to develop organic devices, the physical phenomena taking place in organic materials need understanding. For instance, the efficiency of OLEDs is determined by charge balance, radiative decay of excitons, and light extraction, which depends on the refractive indices of the composed layers. In general, the nature of the excited electronic states is of great interest and it is directly related to optical processes such as light absorption and emission, photoconductivity, and electroluminescence.

Since very thin layers offer reduced resistance and high current densities for a low level of electrical bias voltage so organic materials are usually used as thin films in most cases. The research of electrical and optical properties in this form is very important. In spite of recent advances in thin films fabrication techniques such as spin-coating [30, 31], dip-coating [32,33] or Langmuir – Blodgett [26,34], the vacuum evaporation by resistive heating still seems to be most appropriate for depositing molecular materials [29].

The number of interesting organic compounds is very large [35] and our selection of specific materials is motivated by the application context that is each compound investigated by us is potentially used as OLED components. In this work we present the optical results of metallophthalocyanines: copper phthalocyanine (CuPc), magnesium phthalocyanine (MgPc), cobalt phthalocyanine (CoPc) and zinc phthalocyanine (ZnPc), perylene derivatives: 3,4,9,10-perylenetetracarboxylic dianhydride (PTCDA), 3,4,9,10-tetra-(heptyloxy-carbonyl)-perylene (PeC7), metal chelate: tris-(8-hydroxyquinolate)-aluminium (Alq3) which was doped by 4-di-cyanomethylene-2-methyl-6-(p-dimenthyaminostyryl)-4H-pyran (DCM). Aside from application advantages in organic optoelectronic devices, the mentioned compounds are excellent objects for verifying of quantum-mechanical models and principles of solid-state physics.

In this paper we investigate the different excited states, energy transfers between them and trapping processes in organic thin layers deposited onto the various substrates. We present the influence of the temperature on optical properties In order to compare the properties of polycrystalline and amorphous thin films with single molecule we also investigated these materials in solution form. Our compounds were dissolved in different solvents such as dimethyl sulfoxide (DMSO), ethanol, chloroform, and tetrahydrofuran (THF) (see Table 2). In contrast to solution, where single molecules are in weakly interacting surroundings, in the solid the close proximity of molecules leads to intermolecular interactions and formation of completely different types of optical excitations such as charge transfer (CT) excitons, excimers and other aggregate excitations [36-40]. This fact plays a major role in the determination of electronic and optical properties of the samples in solid phase. In spite of large variety of theoretical approaches [41-44] the fundamental phenomena in organic materials, in particular in thin layers, are relatively little understood [45,46]. Thus, the further studies in this area are indispensable because the new devices development depends on their microscopic foundations.

In our studies the following experimental techniques have been used: photoluminescence (PL), excitation (PLE), absorption and reflection (A, R), photomodulated absorption and reflection (PA, PR) spectroscopy. These methods allow investigating the electronic structure of materials and determining optical constants such as refractive index, extinction coefficient etc.

EXPERIMENTAL DETAILS

Fig. 1 shows the molecular structure of MPc, Alq_3, PTCDA and PeC7 used in our study. All these materials, with the exception of PeC7, were purchased from Aldrich Chemical Co. and used as received. The PeC7 was synthesized and chromatographically purified in the Institute of Dyes at Łódź University of Technology. The details of the synthesis are described in [47]. The glass, quartz and n-type (001) oriented silicon were used as substrates for organic thin films.

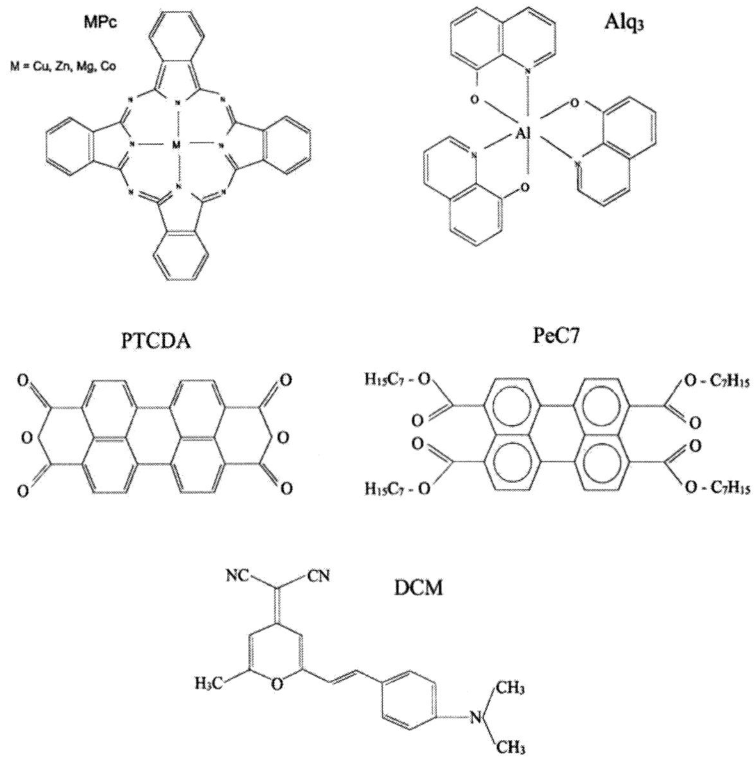

Figure 1. The molecular structure of investigated materials

The thin layers were prepared by thermal evaporation method. The purified powder was loaded into a quartz effusion cell with a nozzle of 3 mm in diameter on the top. Before deposition the substrates were cleaned in acetone, distilled water, isopropanol and distilled water again. Next, they were moved into the deposition chamber where the effusion cells were heated to proper temperature (see Table 1). The substrates were located 10 cm above the source and held at room temperatures during evaporation. The pressure in the vacuum chamber was about 10^{-5} Torr.

The solutions of the investigated materials were prepared by dissolving purified powder in various solvents via sonication. Different solvents for specific materials were used (see Table 2).

Table 1. Evaporation temperatures for investigated compounds.

Material	Evaporation Temperature [K]
MPc	640
PTCDA	600
PeC7	400
Alq3	430
DCM	400

Table 2. Different solvents used in dissolving of investigated materials

Material	Solvent
CuPc	Tetrahydrofuran (THF)
MgPc, CoPc, ZnPc, PTCDA	Dimethyl-sulfoxide (DMSO)
Alq3	Chloroform
PeC7	Ethanol

In our research different experimental techniques were applied so we used the apparatus consisted of various light sources and photodetectors. The Omnichrome He-Cd laser (λ = 325nm, 42 mW) was used for photoluminescence and photomodulation measurements as an excitation source. Transmission spectra were measured with halogen-tungsten (150 W) and deuterium (20W) lamps. For photomodulation spectroscopy and luminescence of MPc the phase sensitive detection technique was applied using lock-in amplifier (EG&G 7260) with reference signal (263 Hz). For the near infrared range the sensitive silicon photodiode (Hamamatsu S-8745) with preamplifier was used whereas in the UV spectral region a photocounting system with photomultiplier (Hamamatsu R-928) operated as a detector. For spectral selection the prism monochromator SPM-2 (Carl Zeiss Jena) was applied. The photoluminescence and absorption spectra were measured in the temperature range from 13K to 320K using the helium refrigerator (Cryogenic Inc). More detailed description of experimental set-up is elsewhere [48, 49].

RESULTS AND DISCUSSION

Mettalophthalocyanines

The first synthesis of metal free phthalocyanine (H_2Pc) was made in 1907 by Braun and Tcherniac who heated o-cyanobenzamide at high temperature [50]. The metal substituted copper phthalocyanine (CuPc) was originally manufactured by Diesbach and von der Weid in 1927 [51]. Phthalocyanines are poryphyrin deriviatives which characterize with high symmetry and extensively delocalized two-dimensional 18π-electron system [52], exhibiting a large number of unique properties. These materials are generally p-type semiconductors, which can be easily sublimed as high purity, chemically stable thin films without decompositions [53, 54]. The morphology of the phthalocyanine thin films is variable and strongly dependent on the sublimation conditions. The structural orientation has strong influence on the electrical as well as optical properties of MPcs layers [55]. MPcs films exist in several molecular forms which range from amorphous to highly crystalline. Until now, nine polymorph structures have been observed for CuPc [56]. There are two well-characterized polymorphs in the solid state: α and β-form, which differ from each other by tilt angle between the axis of the stack and the normal of the phthalocyanine ring as well as metal-metal and interstack distances. The preferred orientation is usually determined by sample preparation conditions such as the type of substrate, temperature of substrate, deposition rate, and quality of the substrate surface [57].

The interest of the phthalocyanines begins with discovery of their unique spectral properties. The single, high absorptive band located in the visible spectral range results in the purity and depth of colour. Beautiful blue or greenish hues, which characterize majority of phthalocyanine complexes enable using them in inks (especially ballpoint pens), dyes for plastics and metal surface, jeans and other clothing and even as coatings for decorative layers. Because of lack of toxicity Pcs dyes also found the application in food colouring, food coatings or contact lenses [58]. The tuning of the photophysical spectroscopic properties of phthaolcyanines over a very broad range can be easily realize by changing the peripheral substituents and/or the central metal atoms of the phthalocyanine macrocycle.

Recently, the thin films of metallophthalocyanine have been extensively studied due to their potential application in modern optoelectronics and microelectronic devices. Nowadays many researchers indicate metallophthalocyanines as a material for production organic field effect transistors [54, 59, 60]. However field-effect electron and hole mobility of phthalocyanine-based transistor are still low in comparison with inorganic technology. The high linear optical absorption coefficient over a wide range of the solar irradiation spectrum and sufficient charge transport properties of MPcs results in their photovoltaic properties, which could be exploited in production of cheap, efficient and flexible solar cells, photosensors or photoconducting agents in photocopying machines [61-63]. The efficiency of the phthalocyanine-based organic solar cells can be improved by incorporation of various metal nanoclusters at the ITO-metallophthalocyanine interface [64-65]. Nowadays MPcs are materials which play very important role in the field of charge transport. Recent investigations have proved that using the CuPc as a hole transport layer or hole injection buffer layer we can considerably improve efficiency and guarantee stability of OLED [66,67]. Due to efficient emission certain phthalocyanines have been also proposed as an emissive layer of infrared electroluminescent diode [68, 69] or dopants in polymer based OLEDs [70]. Promising applications of phthalocyanines are based on their electrochromism properties. The changes in an absorption spectrum due to the presence of an electric field (and/or faradaic reactions) can be potentially exploited for building multicolour electrochromic, low power and flat display devices [58,71].

Metallophthalocyanine layers also pay attention of scientists due to their nonlinear properties [72].The two-dimensional π-electron conjugated system of phthalocyanines give rise to the large third order nonlinearity optical (NLO) susceptibility $\chi^{<3>}$ owing to the modified chemical and molecular structures [73]. Such nonlinear materials are very important because of their potential application in optical communications, optical storage, optical computing, harmonic generation, optical switching or optical limiting which can be used in protection of eyes or optical sensors for high-intensity laser beams [74-76].

Phthalocyanines are also characterized by very interesting thermo-physical properties. Low melting point, low decomposition temperature and high chemical stability makes these materials excellent candidates for production of high density optical computer read/write discs [77]. It was also observed that the semiconducting properties of phthalocyanines are modulated by the adsorption and desorption of gases. This suggests their exploitation in the chemical gas sensors [17]. Phthalocyanines and other metallo macrocycle molecules are also very interesting in respect of their possible application in the field of molecular electronics [78]. Phthalocyanine-based molecular electronic devices may play significant role in generation and processing of information. Some metallophthalocyanines, for example ZnPc, have been found to be highly promising from the biomedical point of view. This compound belongs to a second generation of photoactive agents, mainly used in medicine as a sensitizer in photodynamic laser cancer therapy [79-81].

The absorption properties of metallophthalocyanines thin films have been recently extensively studied but there is no consensus about number observed transitions and their physical origination [52]. We present the sensitive modulated spectroscopy for characterization of the low dimension CuPc and ZnPc structures. Due to the derivative nature of this method the broad features and shoulders of absorption or reflection spectrum can be replacing with very sharp lines with characteristic shape. That is why very small features, which are hardly identified in the usual stationary absorption spectra, are strongly enhanced and can be analyzed in detail.

Although photoluminescence emission observed from phthalocyanine complexes in solution have been many times reported [81-84], a little attention has been paid to the photoluminescence spectroscopy of MPcs thin layers. Generally there is a lack of data concerning temperature dependence of photoluminescence spectra of MPcs thin films in the literature. Thus, our studies are important completion of existing knowledge about optical properties of MPcs.

The basic features of electronic spectra for many differently substituted phthalocyanines are quite similar. The appearance of the main absorption bands, conventionally labelled as Q, B, L, N and C, is a typical feature of phthalocyanine absorption spectra in both free and metallic derivatives forms. Fig. 2-5 show comparison of absorption spectra of metallophthalocyanines (ZnPc, MgPc, CuPc and CoPc) thin films deposited on quartz substrates with absorption spectra of corresponding solutions. The MgPc, ZnPc and CoPc were dissolved in DMSO. CuPc (without peripheral substitutions) is particularly insoluble in most solvents, making its absorption spectra difficult to measure. From this reason unreliable spectra have been observed for CuPc dissolved in DMSO. However absorption spectrum of CuPc dissolved in THF exhibits a sharp peak which can be associated with well-known Q absorption band usually observed for phthalocyanines (see Fig. 4). It should be pointed out that this peak appears on the broad background, which can be associated with strong scattering of incident light by undissolved particles. The absorption spectra of ZnPc and MgPc solutions (Fig. 2 and 3) are typical of many phthalocyanine complexes. Generally for "unperturbed" metallophthalocyanine, six degenerate excited states (all $\pi \rightarrow \pi^*$ in nature) are accessible between 1.55eV and 6.2eV, named in order of increasing energy Q, B_1, B_2, N, L, and C. Gouterman's group [85, 86] used a four orbital model, based on the top two occupied molecular orbitals (a_{2u} and a_{1u}) and the lowest degenerated unoccupied orbital (e_g) to calculate theoretically excited states in metallophthalocyanines. According to this model the first two allowed $\pi \rightarrow \pi^*$ bands (Q and B) arise from transitions out of a_{1u} and a_{2u} into the same e_g orbital, respectively. The single, sharp absorption peak in the visible region (near 1.8eV) of all MPc solutions spectra is associated with Q band. Two absorption peaks to the blue of the Q band ZnPc, MgPc and CoPc solutions are assigned as vibrational components [87] and marked Q_{vib}.

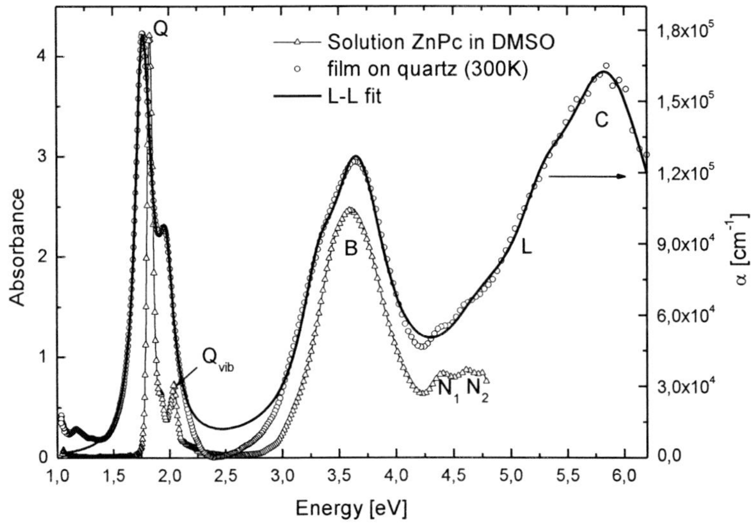

Figure 2. The absorption spectrum of ZnPc solution (triangle points) and thin film (circle points). The solid line represents Lorentz-Lorenz fitting assuming seven-oscillator model

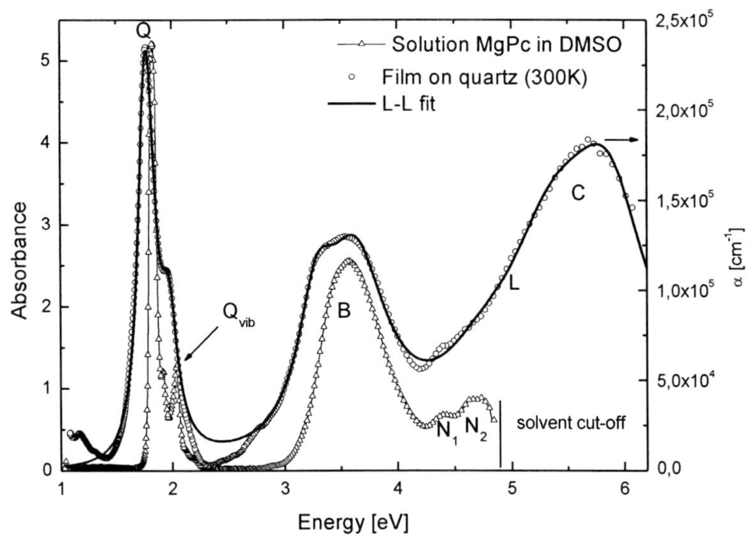

Figure 3. The absorption spectrum of MgPc solution (triangle points) and thin film (circle points). The solid line represents Lorentz-Lorenz fitting assuming seven-oscillator model

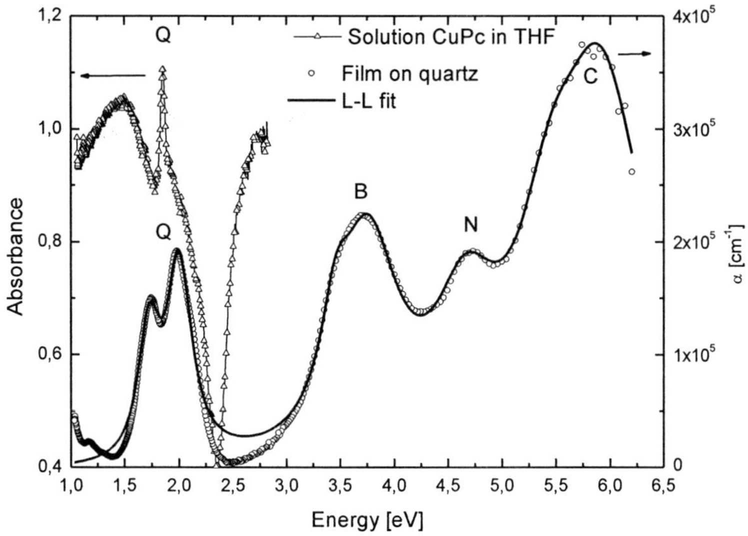

Figure 4. The absorption spectrum of CuPc solution (triangle points) and thin film (circle points). The solid line represents Lorentz-Lorenz fitting assuming seven-oscillator model

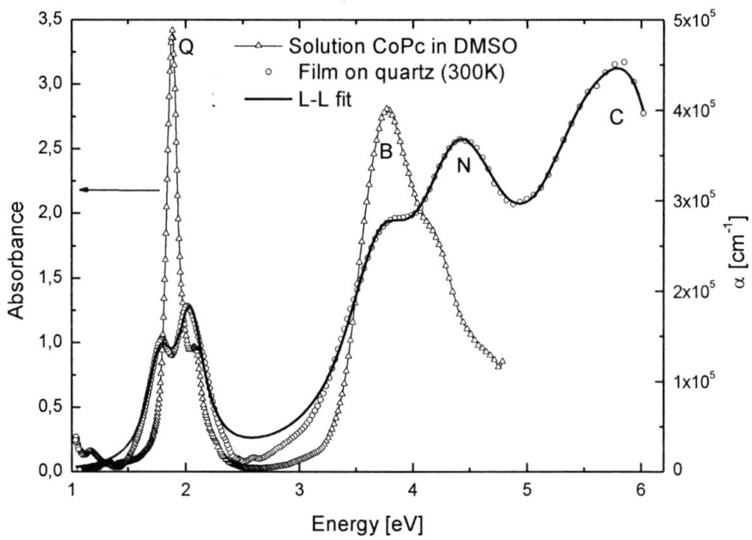

Figure 5. The absorption spectrum of CoPc solution (triangle points) and thin film (circle points). The solid line represents Lorentz-Lorenz fitting assuming six-oscillator model

In contradiction to Gouterman's calculations the spectral deconvolution based upon magnetic circular dichroism (MCD) and absorption data has indicated that there are at least two electronic transitions overlapped in the B region of monomer spectrum [88, 89]. Strongly broadened peak near 3.8eV in CoPc and 3.6eV in others MPc solutions spectra is due to both a_{2u} to e_g (B_1) and b_{2u} to e_g (B_2) transitions called B (or Soret) band. Due to overlapping the width of the B band is significantly greater than the Q band. Henriksson et al. in a study [90] used a self-consistent field molecular-orbital (SCF-MO) method to calculate state energies, oscillator strength and state symmetry for CuPc. These results also support the multiple bands in the B and higher energy regions of CuPc. It was also suggested that states in the B band region arise from several transitions. Moreover the valence electron on the Cu atom were included using the Peel method so also metal-related transitions were predicted. In separate study [91] Edwards and Gouterman supported that transitions involving d electrons of central metal atom are also possible. Thus bands between 4.5eV and 5.7eV can be due to $d \rightarrow \pi^*$ transitions in CuPc. In the time-dependent density functional theory (TDDFT) calculations reported by Kiet et al. [92] for ZnPc it was confirmed that several orbital transitions contribute to the observed B, N and L bands. All the transitions are $\pi \rightarrow \pi^*$ or $n \rightarrow \pi^*$ types. The TDDFT predicts up to six allowed transitions between 3eV and 4eV ($3b_{2u} \rightarrow 7e_g$, $4a_{2u} \rightarrow 7e_g(B_2)$, $2b_{1u} \rightarrow 7e_g(B_1)$, $3a_{2u} \rightarrow 7e_g$, $18e_u \rightarrow 7e_g(B_3)$, $1a_{1u} \rightarrow 8e_g$). Only one allowed transition at 4.34eV has been predicted between 4eV and 5eV ($1a_{1u} \rightarrow 7e_g$ N-band) however we can observe clear doublet structure for solution in this region. In higher energy region of the ZnPc, the computed spectrum predicts that the L band is due to three $\pi \rightarrow \pi^*$ transitions ($6e_g \rightarrow 3b_{1u}(L_1)$, $6e_g \rightarrow 3b_{1u}(L_2)$ and $5e_g \rightarrow 3b_{1u}(L_3)$). For phthalocyanines with the central metal atom, which is devoid of d electrons (MgPc) or has a d^{10} configuration (ZnPc), the mixing between metal and ligand orbitals is negligible and the electronic transitions take place between ligand-centred orbitals so the ligand to metal charge transfer as well as d-d transitions should not complicate the spectra [58, 88]. The positions of observed bands of investigated metallophthalocyanines in solutions are presented in Table 3. In cases of MgPc and ZnPc the peak positions are quite similar especially in low energy region. This is because the 3d subshell is filled and deep enough to form rather pure molecular orbitals in these molecules.

Phthalocyanines containing certain transitions metal (CoPc, CuPc) have more complex electronic structure because the metal 3d-like orbital lies between the Pc HOMO and LUMO. As a result of that the spectra of these compounds can contain extra features arising from charge transfer transitions [93]. Absorption spectra of MPcs thin films deposited on quartz substrate enabled observation of additional peaks because the absence of solvent allowed measuring band up to the 6.2eV spectral region. Solution spectra are limited to the 4.96eV because of cut-off DMSO solvent. Molecular crystals are composed of single molecules held together by van der Walls forces. Because of weak nature of this bonding the properties of individual molecule are partially retained in the solid state [94, 95]. Nevertheless the energy of interaction between the molecules in the aggregate imposes certain differences. From Fig. 2 – 5 we can see that the UV-ViS spectra of thin films are characterized by broadened bands, much broader than the solution data. This effect can be explained by aggregation, molecular distortion and exciton (or Davydov) coupling between adjacent rings.

Table 3. Peak positions of MPcs solutions absorption spectra

Solutions	Q[eV]	Q_{vib}[eV]	B[eV]	N_1[ev]	N_2[eV]
ZnPc	1.83	1.92, 2.04	3.6	4.61	4.74
MgPc	1.83	1.92, 2.04	3.57	4.39	4.7
CoPc	1.88	2.08	3.77	~4.2	~4.6
CuPc	1.85	-	-	-	-

In solid state of Pcs the stronger molecular interaction occurs so molecules create aggregates. Dimerization generally results in broadening, shifting and splitting of bands relative to the monomer spectra. The spectral shift can be either to the blue or to the red regions, depending on the mutual orientation of the monomer transition moment. The Davydov coupling broadens and splits each of the electronic transitions observed in solution according to the number of translationally inequivalent molecules in the unit cell [95]. This splitting is in addition to the level splitting produced by interaction energy between two adjacent molecules [94]. For β-polymorph form it is known to be two but for α-form up to four molecules in the unit cell, so we can expect that the excited states of metallophthalocyanines are splitting into two or four energy levels. In the Q-band region two absorption peaks of deposited MPc films spectra are clearly visible and can be attributed to the two Davydov components of the first π-π^* ($a_{1u} \rightarrow e_g$) transition on the phthalocyanine macrocycle [53,58,96,97]. The splitting of Q-band can be also associated with molecular distortion which results in the lift of the degeneracy of two LUMO orbitals [98]. Such deformation upon crystallization can also increase bands splitting. The observation of this splitting in the higher energy region is much more difficult because of strongly broadening and overlapping of bands. Absorption spectra of MPc films were fitted (see Fig. 2-5) by Lorentz-Lorenz mixing model using well-known formulas [99,100]:

$$\frac{\varepsilon - 1}{\varepsilon + 2} = A + B\nu^2 + \frac{1}{\pi}\sum_j J_j \left[\frac{1}{\nu_{0j} - \nu - i\Gamma_j} + \frac{1}{\nu_{0j} + \nu + i\Gamma_j} \right] \tag{1}$$

$$\alpha = 4\pi\nu \,\mathrm{Im}\sqrt{\varepsilon} \tag{2}$$

where ε is a dielectric constant, J_j is an intensity factor, ν_{0j} and Γ_j are resonant wavenumber and linewidth of j-th transition, respectively. The A and B are parameters, which take into account influence of absorption bands above 6 eV. During fitting procedure it was assumed that all investigated MPcs absorption spectra (except the CoPc one) are composed of seven oscillator components (j=7 in equation 1). In case of CoPc film the most reliable fit was obtained assuming only six transitions involved in absorption spectrum because B band appears as shoulder rather than separated single peak. The calculated fitting parameters for absorption spectra measured at 300K are shown in Table 4.

As we can observed in Fig. 2 – 5 and deduced from calculated fitting parameters (see Tab. 4), the metal ion plays a crucial role in determining the shape and positions of particular bands both for MPcs thin films and solutions. It is caused by different degree of interaction between metal ion and the phthalocyanine π system which can depend on the number of electron in the outer shell of the central metal [101,102]. In case of thin films a problem is more complicated because a surface morphology, which depends on the metal substitution, can also have meaningful influence on observed absorption spectra [102]. Senthilarasu et al. [103] showed out that the absorption spectra of ZnPc thin films can be also dependent on thickness of the layer. The increase of extinction coefficient with increasing thickness has been attributed to an increase of particle size or changes a root mean square (RMS) strain and the lattice constant.

Table 4. The parameters obtained from fitting the Lorentz-Lorenz mixed model for the MPcs absorption spectra

	CuPc-film			ZnPc-film			MgPc-film			CoPc-film		
band	J	E [eV]	Γ[eV]	J	E [eV]	Γ[eV]	J	E [eV]	Γ [eV]	J	E [eV]	Γ [eV]
Q_1	546	1.756	0.120	770	1.800	0.078	960	1.813	0.074	512	1.803	0.137
Q_2	1322	2.042	0.124	420	1.976	0.082	500	1.977	0.087	1097	2.081	0.132
B_1	282	3.486	0.167	246	3.328	0.196	408	3.302	0.206	-	-	-
B_2	1700	3.802	0.319	1029	3.684	0.294	1120	3.649	0.329	1220	3.765	0.356
N	925	4.686	0.360	299	5.290	0.344	247	4.713	0.439	3123	4.512	0.469
L	920	5.494	0.379	251	4.742	0.405	1250	5.391	0.573	2985	5.584	0.597
C	3141	5.994	0.452	1579	5.880	0.556	918	5.893	0.412	2502	5.992	0.382

The variation of different MPcs absorption spectra can also depend on the phthalocyanine cavity size and molecule symmetry which determine state energies and oscillator strength values. Cu and Co metals have ion size similar to that of phthalocyanine cavity size so they are accommodated in the cavity without any ring contraction or expansion and represent phthalocyanine ring in their equilibrium state [87]. As a result their structure is planar and posses approximately D_{4h} symmetry. ZnPc is an example of structure where the metal ion is larger than equilibrium cavity size of the ring and cannot be accommodated without ring expansion nevertheless zinc ion is still located in-plane [104,105]. In the other side there are certain gas-phase electron diffraction study [106] and DFT calculations [107] of ZnPc pointed to a C_{4v} structure. In case of MgPc crystal, X-ray investigations, which were carried out at different temperatures, showed that this molecule is not planar and forms pyramidal structure. However the displacement of the magnesium from the plane depends on the temperature. Moreover intermolecular interactions lead to the formation of the dimers statistically disordered in the crystal [108].

Fig. 6 presents the comparison between absorption and emission spectra of MgPc dissolved in DMSO measured at room temperature. We can see the mirror symmetry between these two spectra. As we can notice that MgPc solution exhibits luminescence from red to near-IR region. We can also see that a vibronic structure is clearly visible on both absorption and emission spectra. The molecular luminescence occurs from the lowest excited singlet, corresponding to the Q band, to different vibrational levels of ground electronic state [82]. Maximum of luminescence (Q_{00} transition) is centred at 1.825eV whereas the transition to the higher vibrational level of ground state is positioned at 1.668eV. In case of MgPc it should be pointed out that the emission from high-lying vibrational levels of the lowest excited electronic singlet state was also observed [109]. Similar PL spectrum was recorded for ZnPc solution with bands positioned at 1.817eV and 1.682eV (see Fig. 7).

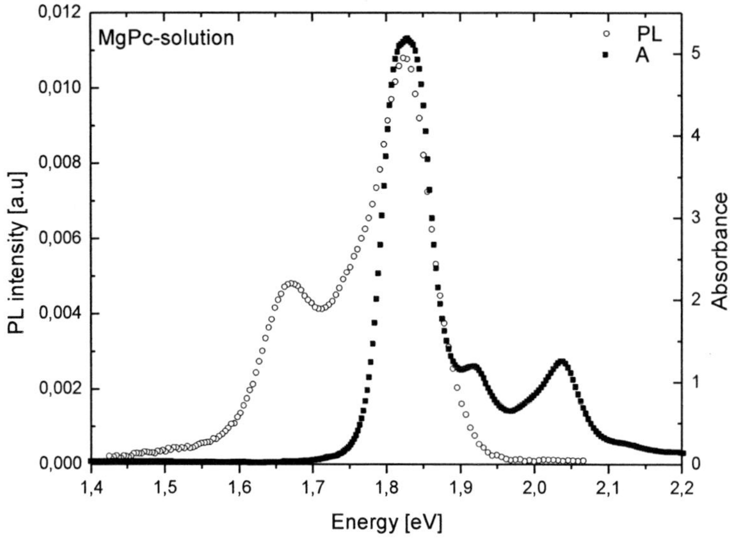

Figure 6. Photoluminescence (circle) and absorption (square) spectra of MgPc solution

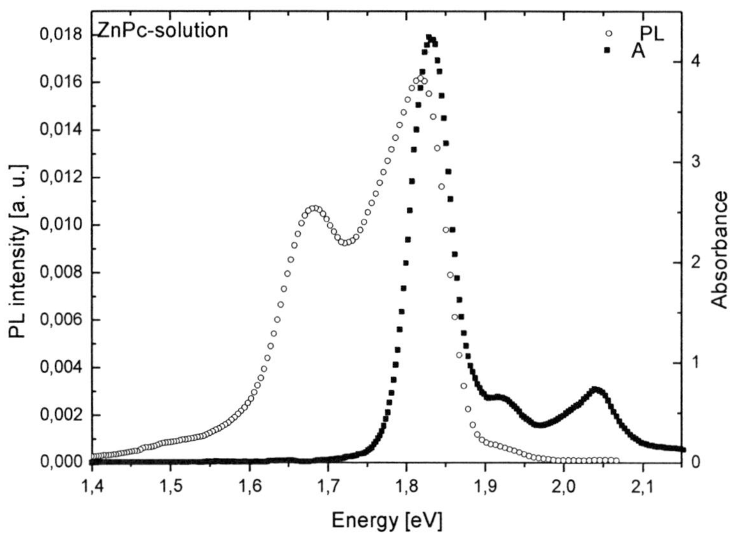

Figure 7. Photoluminescence (circle) and absorption (square) spectra of ZnPc solution

We have found that ZnPc PL solution spectra are strongly dependent on the molar ratio. PL spectra measured for concentration about 10^{-3}M and 10^{-5}M are presented in Fig. 8. It is clearly seen that spectrum at higher concentration show a red shift. It can be explained by strong tendency of ZnPc molecules in DMSO solvent to aggregation.

The comparison between absorption and emission spectra of MgPc thin film is presented in Fig. 9. In the same figure we also presented the emission from MgPc solution. It can be easily noticed that PL spectrum undergoes a dramatic change going to the crystal form.

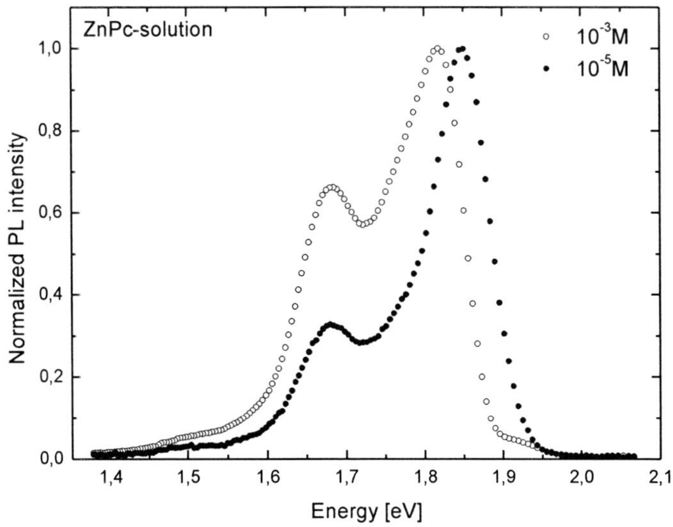

Figure 8. Normalized photoluminescence spectra of ZnPc dissolved in DMSO at two different concentrations 10^{-3}M (circle) and 10^{-5}M (square)

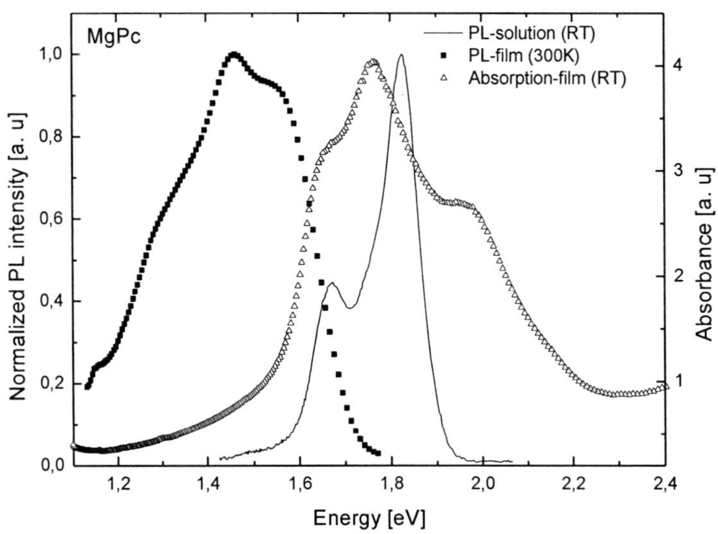

Figure 9. Absorption (triangle) and photoluminescence (square) spectra of MgPc thin film. Solid lines present the photoluminescence of MgPc dissolved in DMSO

First of all, the PL spectrum of film is red shifted and significantly more broadened in comparison with corresponding solution. These effects can be associated with very strong aggregation of molecules in solid where excimer interaction occurs due to the short interplanar distance. The observed fluorescence spectrum of the film displays three main bands. The emission at about 1.53eV and 1.45eV can be attributed to the fluorescence from the two Davydov components of first singlet exciton band (Q). The strongly broadened band at about 1.3eV can be assigned to an excimer emission. Similar broad, structureless spectrum was observed for metal-free phthalocyanine (H_2Pc) solid [110]. The Stokes shift between the lowest absorption (1.66eV) and the highest PL (1.57eV) maxima of film is much larger than for dissolved MgPc. Such large Stokes shift can be attributed to contribution of external phonon modes connected with lattice vibrations [111]. The photoluminescence spectra of MgPc film measured in the temperature range from 13K to 320K are presented in Fig. 10. We assumed the PL spectrum of MgPc is composed of some electronic transitions from crystal exciton Q band to different vibrational levels of ground electronic state. Since the bands overlap, the curve fitting technique seems to be the only way to properly resolve individual bands. We chose the simple Gaussian lineshape to fit the observed PL bands (see Fig. 11 and 12 as an example). In case of MgPc film we assumed that there are five components at energies (broadenings) at 13K: 1.53(0.11) eV, 1.45(0.10) eV, 1.36(0.11) eV, 1.26(0.10) eV, 1.15(0.10) eV which have contribution to the PL envelope. Particular transitions were denoted by consecutive Arabic numerals (see Fig. 11).

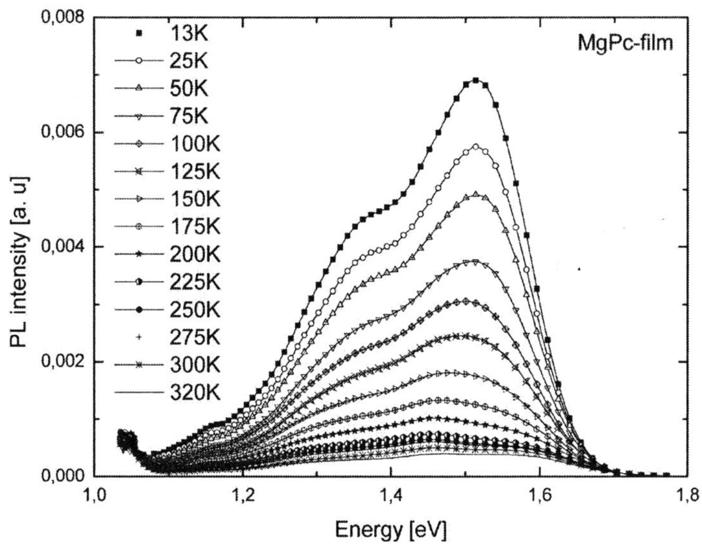

Figure 10. Photoluminescence spectra of MgPc thin film as a function of temperature from 13K to 320K

We found the PL intensities of all bands decrease with increasing temperature. This thermal quenching for particular transitions is shown in Fig. 13. The strong temperature dependence of the PL spectra of MgPc film can be attributed to the weaken of the exciton-phonon interaction with decreasing temperature. Comparing the PL spectra measured at 13K (Fig. 12) and 320K (Fig. 11) we found that the contribution of the peak no. 4 to whole spectrum at higher temperatures is significantly greater than at 13K despite the fact that integral intensity associated with this peak decreases with raising temperature. It means that at higher temperatures the emission from the lower lying Davydov band dominates the spectrum. It can be explained by Boltzman redistribution between excimer or defect state and lower lying Davydov band which leads to increase the occupation of this band resulting in enhancement of emission from it.

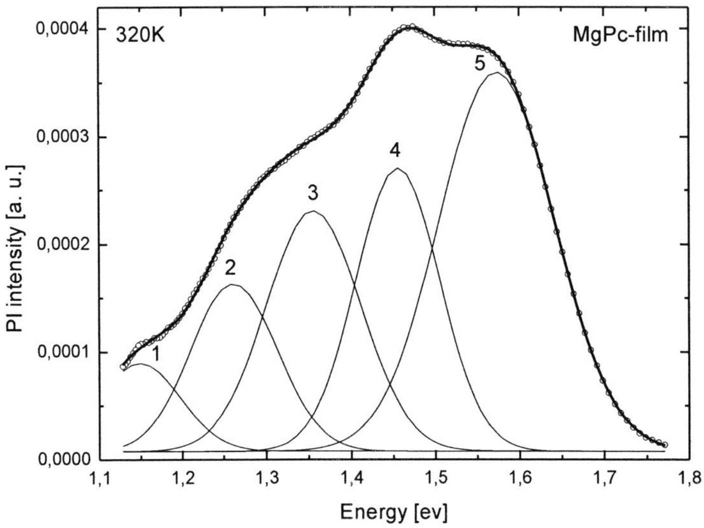

Figure 11. The gaussian fitting of the MgPc thin film photoluminescence spectrum measured at 320K

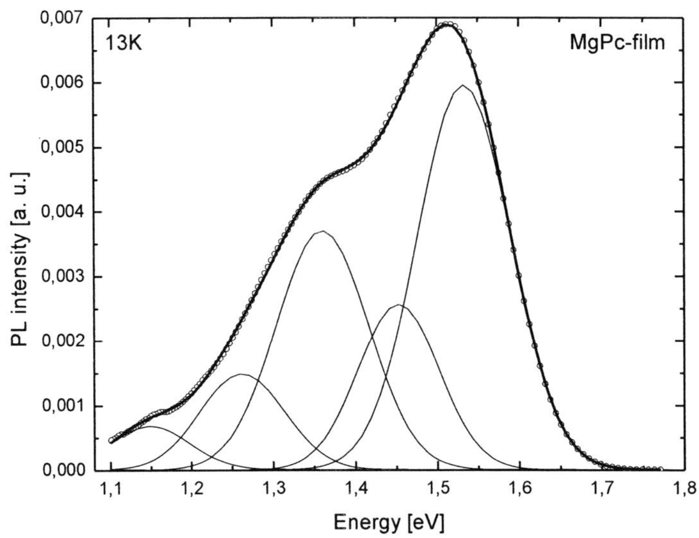

Figure 12. The gaussian fitting of the MgPc film photoluminescence spectrum measured at 13K

The calculated peaks positions and broadening parameters remain almost constant with changing temperature for all peaks except the peak 5. The changes of these band parameters versus temperature are presented in Fig. 14. This peak broadens with increasing temperature due to increased exciton-phonon interactions. The blue shift of this band with increasing temperature can originate from Boltzman redistribution of occupied levels [112].

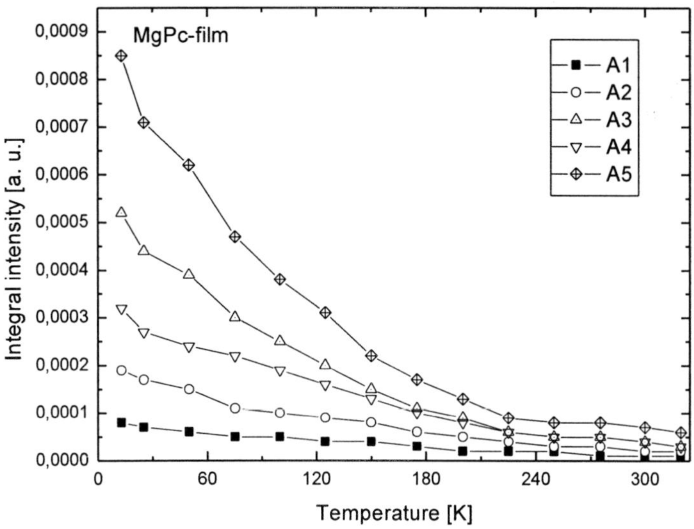

Figure 13. The thermal quenching of separated PL bands of MgPc thin film

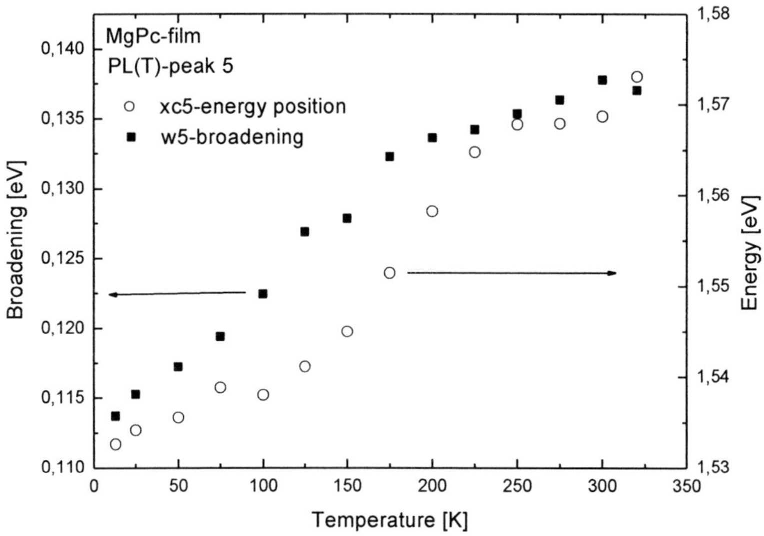

Figure 14. The shift of the PL peak (5) and its broadening with changing of temperature

The temperature dependences of the photoluminescence spectra of the zinc phthalocyanine (ZnPc) thin layer grown on quartz substrate are presented in Fig. 15. We can notice that the investigated thin film shows PL emission from the violet to near-IR region in the wide temperature range (13K-320K).

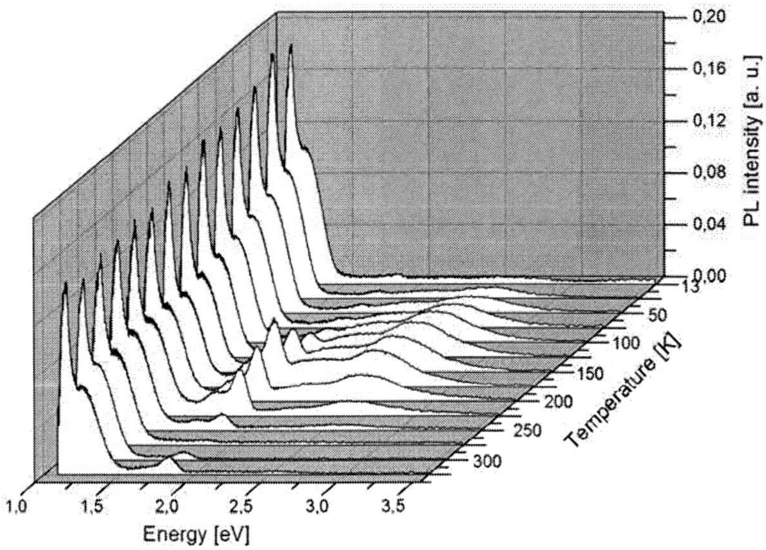

Figure 15. Photoluminescence spectra of the ZnPc film as a function of temperature from 13K to 320K

The PL spectra of ZnPc film measured at 200K are dominated by three main bands. Some of these bands are characterized by vibronic structure. We identified six maxima at: 1.14, 1.24, 1.72, 1.853, 2.16, 2.55 eV with a shoulder at 2.96 eV. The particular transitions were denoted by consecutive Arabic numerals (see Fig. 16). It is clearly visible in Fig. 15 that some bands in the red and UV region appear only at certain temperature range. Below and above this range these bands disappear.

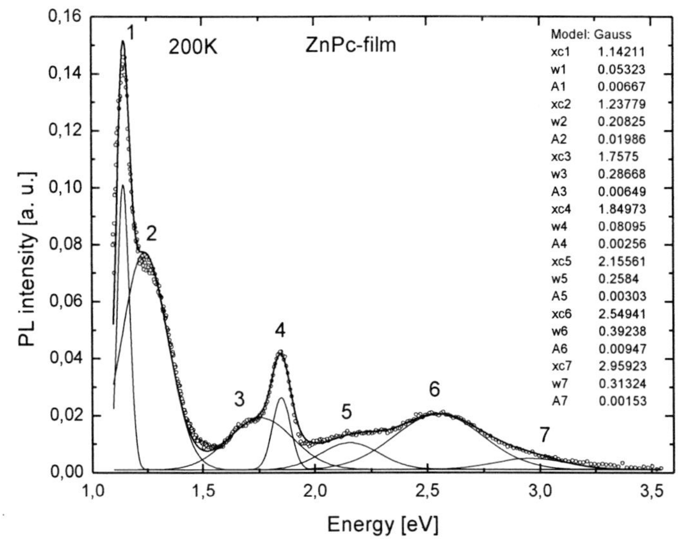

Figure 16. The gaussian fitting of the photoluminescence spectrum of the ZnPc thin film measured at 200K

The emission in the near-IR region (transition 1) occurs from the first singlet exciton band to the ground electronic state (S_0). The broad band at 1.24 eV can be attributed to the excimer fluorescence. The emission in the violet-orange region (transitions no. 5, 6, 7) can be assigned to the fluorescence from the higher lying singlet (S_2-Soret) state to the different vibrational levels of the ground electronic state (S_0). The red emission (peaks no. 3 and no. 4) resembles quite well the solution PL spectrum so can be attributed to the fluorescence from a singlet excited state (S_1) of monomeric ZnPc.

In order to estimate the energy peak positions, broadening parameters and intensity of particular transitions the PL data were fitted by superposition of seven Gaussian components. The broadening parameters of each transition have almost constant values in the whole measured range of the temperature. The example fitting and the resulting parameters of ZnPc spectrum measured at 200K are shown in Fig. 16. It can be seen that the assumed model fits to the experimental data quite well. The plot of the integral intensity of each transition vs. temperature is presented in Fig. 17. The integral intensity of peak no. 1 decreases with raising temperature probably due to existence certain trapping states (Fig. 42) or exciton-phonon interaction. It is clearly visible the intensity of the transition in the near-IR region (peaks no. 2) strongly decreases with increasing temperature up to 175K. Above this temperature the changes are insignificant. We have noticed, that the intensity of other features of photoluminescence spectrum (3, 4, 5, 6, 7) increases until about 175K. Above this temperature the intensities of these transitions attenuate. These results can suggest that the energy migrates from the excimer to the monomer, when the energy levels are appropriate.

The energy peaks positions of transitions no. 1, 2, 7 remain almost invariable with changes of the temperature, whereas the others bands shift with temperature especially about 175K. These changes for transitions no. 3, 4, 5 and 6 are shown in Fig. 18. As we can see the energy peak position of transition no. 6 at 13K has value about 2.553eV and shifts to the red with the increase of temperature achieving minimum (2.517eV) at 175K. Above 175K the energy peak position shifts to higher energies up to 2.585 eV at 320K. The energy peak position of others transitions are invariable up to 175K. Above it they exhibit rapid change of energy. This behaviour can be correlated with change of molecular symmetry resulting from metal ion position in molecule ring and its distortion of planarity versus temperature. The most probable structure of ZnPc at room temperature is D_{4h} planar symmetry. At low temperature the zinc atom adopts a position out of the plane of the ring and about 140-230K significant doming occurs, giving the molecule C_{4v} symmetry. These changes in molecular symmetry lift the degeneracy of the excited state changing both energy and probability of particular transitions.

In case of CuPc and CoPc the emission was undetectable both in solutions and deposited layers. The absence of emission band can be correlated with increased spin orbit interaction and intersystem crossing.

Modulated photoreflectance and photoabsorption spectroscopy are very powerful, contactless and non-destructive methods for characterization of optical transitions in semiconductors. Photoabsorption, as a very sensitivity method, gives much useful information about electronic structure of investigated materials. In particular, it is possible to determine the energy and broadening parameter of the observed features that appear at energies corresponding to the band gap critical points. The main advantage of this method is that even weak features, which are not clearly visible in usual, stationary absorption spectra, can be strongly enhanced in PA even at room temperature [113]. Thus, we are able to determine the accurately energy and broadening parameter of excitonic transitions from the detailed line shape fit to the experimental data. The absorption spectrum of the CuPc layer grown on the glass substrate measured in the Q-band region is presented in Fig. 19. The observed absorption band is dominated by two distinct peaks about 1.75 eV and 1.99 eV which have been attributed to the two Davydov components of the first π to π^* transition.

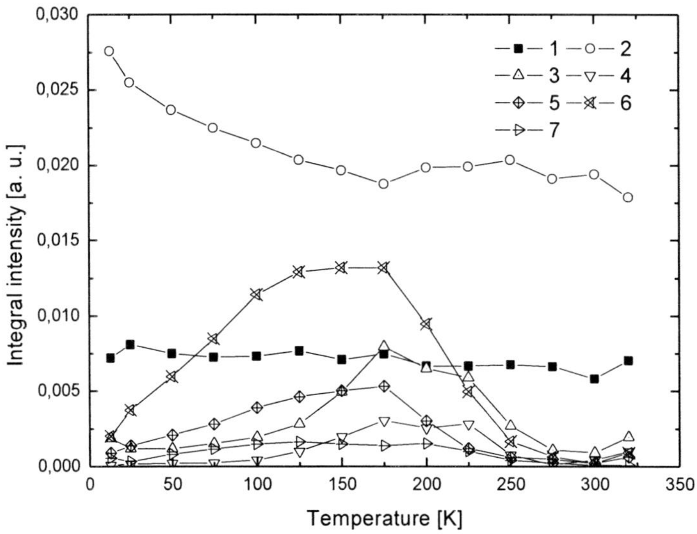

Figure 17. The changes of integral intensities of particular transitions as a function of temperature

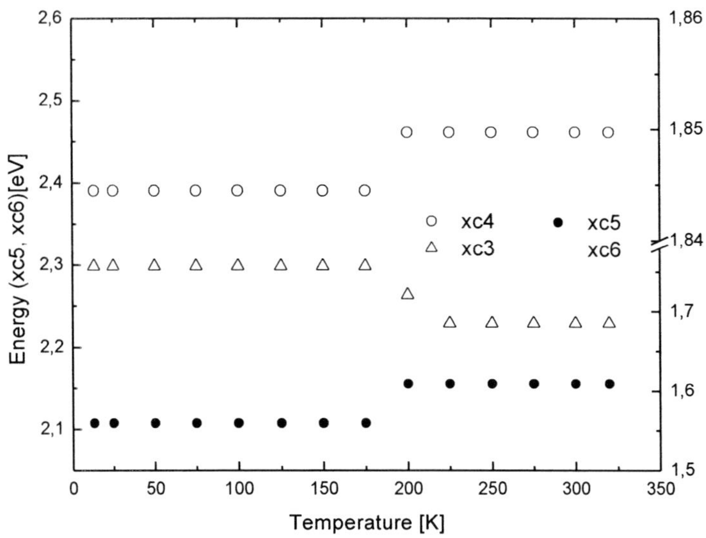

Figure 18. The temperature dependence of peak positions of the ZnPc thin film photoluminescence spectrum

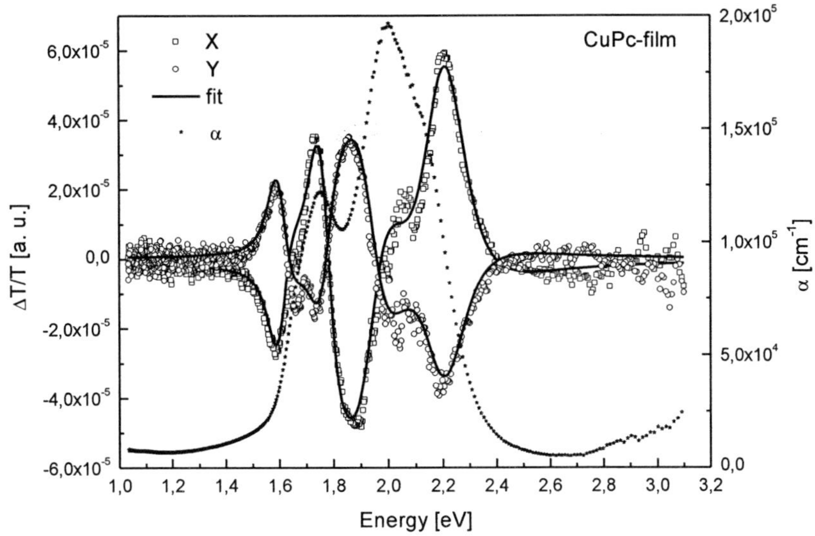

Figure 19. Photoabsorption in-phase (X) and out-phase (Y) spectra of CuPc thin film fitted by the Aspnes formula (solid line) compared with the absorption spectrum (stars)

There are also, in the absorption spectrum, weakly visible shoulders at energy about 1.6 eV and 2.1 eV. The shoulder at energy 2.1 eV can be associated to the vibrational interval whereas the lower energetically shoulder at energy ~1.6eV has been assigned to the charge transfer excitation [114]. Fig. 19 also presents the experimental phase-resolved PA spectra of CuPc layer obtained for the in-phase-X (square points) and out-phase-Y (circle points) channels of the lock-in amplifier and corresponding fits (solid lines). These spectra were fitted using third derivative line shape proposed by Aspnes and given by formula [115-117]:

$$\frac{\Delta T}{T} = \sum_{j}^{n} \mathrm{Re}\left[A_j e^{i\varphi_j} (E - E_{gj} + i\Gamma_j)^{-m_j}\right] \tag{3}$$

where E_g is energy band gap, Γ is broadening parameter, A and φ are the amplitude and phase factor, respectively. Factor j denotes the number of spectral features. Parameter m depends on the type of the critical point and the order of the derivative. During the fitting procedure m was assumed equal 2 like for excitonic transitions [118]. Before measurement the phase ϕ_{ref} was chosen equal -35^0 in order to obtain comparable signal amplitudes for the X and Y components.

It can be noted that all transitions, even weakly visible in usual absorption spectrum, have been strongly enhanced in case of modulated photoabsorption. The obtained photoabsorption in-phase (X) and out-phase (Y) spectra were fitted assuming superposition of four components (j=4 in equation 3) using the same model parameter values except the transition amplitude A. The obtained fitting parameters are collected in Table 5.

Decomposition of fitted function (equation 3) on particular transitions for Y channel is presented in Fig. 20. Each of the components of the photoabsorption signal has its own phase delay ϕ_j and time constant τ_j. As a result, each of the components has its specific vector representation in X-Y plane. The measured values of in-phase (X) and out-phase (Y) signals are generally projections of a vector sum of all existing components on the particular axis's (see Fig. 21).

Table 5. The parameters obtained from fitting of the Aspnes model to photoabsorption in-phase and out-phase spectra

Transition No.	Amplitude]	Phase [rad]	Energy [eV]	Broadening [eV]
1	-8.902E-8(Y) 9.291E-8(X)	-5.394	1.607	0.064
2	1.405E-7(Y) -2.942E-7(X)	0.779	1.766	0.079
3	5.862E-7(Y) -5.869E-7(X)	-7.936	1.939	0.128
4	6.264E-7(Y) -1.118E-6(X)	-6.528	2.199	0.147

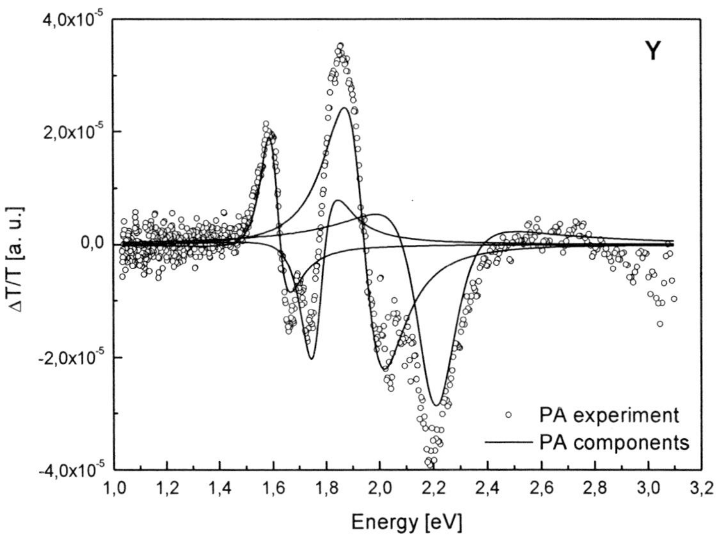

Figure 20. Decomposition of out-phase (Y) spectrum into particular transition

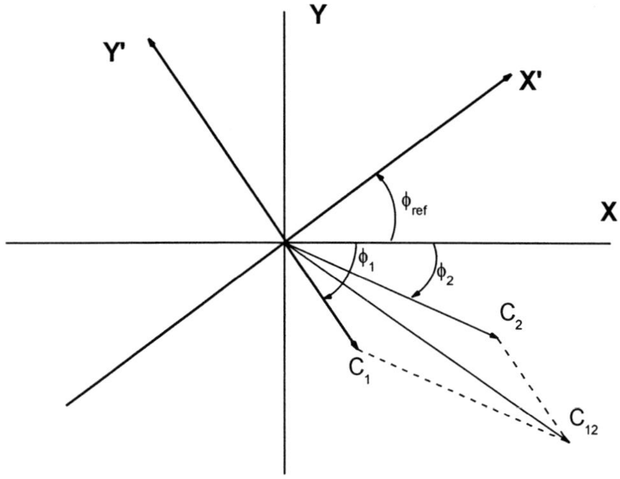

Figure 21. Vector representation of the two components with different phase delay

The accuracy of the fitting can be verified by comparison the phase diagram constructed from experimentally obtained points with theoretically model, which was fitted to X and Y spectra. The experimental phase diagram posses the complicated, irregular form (see Fig. 22) but the shape has been well reconstructed by fitted function. The phase shifts and time constants associated with particular transitions can be found from the ratios of imaginary (Y_j) to real (X_j) transition amplitudes A, according to equation [119]:

$$\tan(\phi_j) = \frac{Y_j}{X_j} = -2\pi f \tau_j \tag{4}$$

where X_j and Y_j are in-phase and out-phase contributions for single component of spectrum, respectively. Calculated phase shifts and corresponding time constants are collected in Table 6.

Table 6. The determined values of phase delay angles and corresponding time constants

Transition No.	Phase shift [^{0}] ($\phi_j=\phi_{ref}-35^0$)	Time constant [ms] (τ_j)
1	-79	3.1
2	-61	1.1
3	-80	3.4
4	-64	1.2

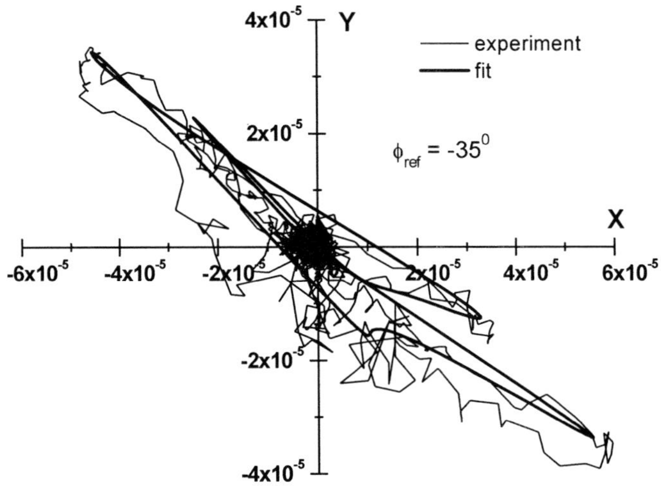

Figure 22. The reconstruction (thick curve) of the experimental phase diagram (thin curve) obtained from the Aspnes function fitted to X and Y spectra

There is also a possibility of suppressing particular components of the PA spectrum by selecting ϕ_{ref}, which satisfies an equation (ϕ_{ref} - ϕ_j)=90^0. We can find the phase delay (ϕ_j) and characteristic time constant (τ_j) using equation 4 from the reference phase, which suppressed a certain component. In order to find suitable ϕ_{ref} corresponding to the total suppressing of appropriate component of the PA spectrum we can measure many spectra at different angles ϕ_{ref}, which are preset in the lock-in amplifier. PA measurements are time-consuming so such approach is rather ineffective. More efficient way to find these angles is the computer generation of the series of spectra at various angle ϕ_{ref} using equations, which transform co-ordinate system by rotation [49, 120]. In the new, rotated X'-Y' co-ordinate system the projections of particular in-phase (X) and out-phase (Y) components are given by equations:

$$X' = X\cos(\phi_{ref}) + Y\sin(\phi_{ref}) \tag{5}$$

$$Y' = -X\sin(\phi_{ref}) + Y\cos(\phi_{ref}) \tag{6}$$

Fig. 23 presents the generated spectra for the angles ϕ_{ref}, which caused the total suppression of amplitude connected with particular transition. Vertical lines denote the resonance energy positions. Using this method we obtained the same values of phase delay angles and corresponding time constants (Table 6). Calculated times constants corresponding with bands which were associated with the first π-π^* intramolecular transition (3), and the charge transfer transition (1) are comparable and we can notice that they are about three times larger than for the others (2 and 4). However their conclusive interpretation remains difficult respect to variety process giving contribution to the observed time constants [120]. In case of ZnPc, photoreflectance spectroscopy confirmed the existence three transitions in the Q-band region of studied material. Modulated photoreflectance spectrum of ZnPc layer is presented in Fig. 24. This spectrum is noised because of the presence of the parasitic photoluminescence signal which is difficult to eliminate in modulated measurements. The obtained PR spectrum can be decomposed (dashed line) on three transitions in the Q band region. Bands parameters were determined by fitting Aspnes formula (equation 3) and are collected in Table 7. It can be noticed that transition about 2.26 eV, weakly visible as a shoulder on absorption spectrum, is strongly enhanced in case of modulated reflectance spectroscopy. This observation confirms derivative nature of modulated spectroscopy and its extraordinary sensitivities.

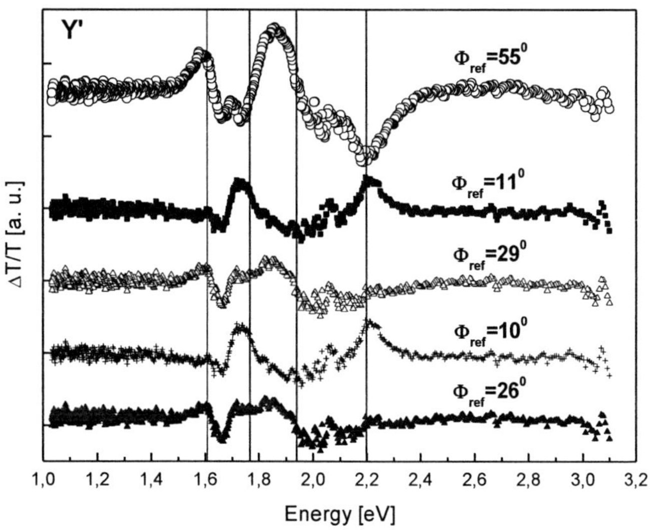

Figure 23. Computer generated spectra for different angles ϕ_{ref} which correspond to the total suppression of particular transitions

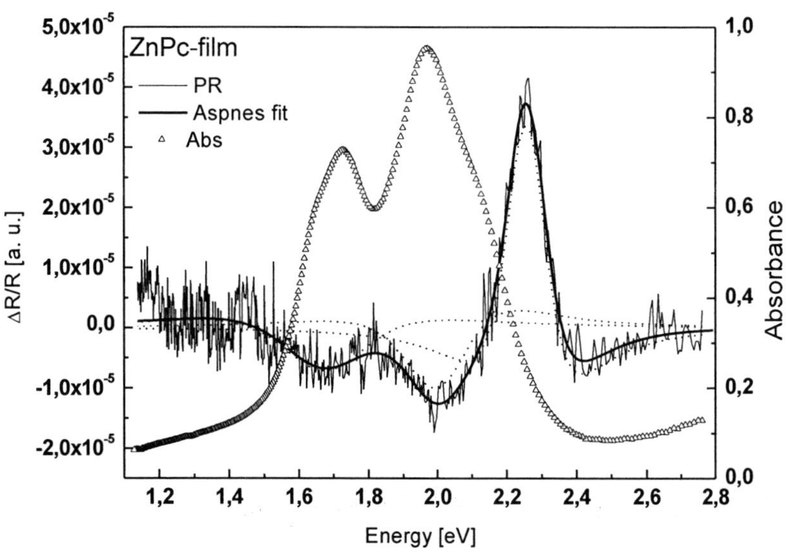

Figure 24. The photoreflectance (PR) experimental spectrum (noised line) fitted by the Aspnes formula (solid line) distributed on particular transitions (dashed lines). The absorption spectrum (triangle points) in the Q band range is shown for comparison

Table 7. Values obtained from fitting of third derivative Aspnes model for photoreflectance (PR) spectrum of ZnPc thin layer

Transitions No.	Amplitude	Phase [rad]	Energy [eV]	Broadening [eV]
1	$-3.22\cdot10^{-7}$	3.929	1.674	0.290
2	$-2.23\cdot10^{-7}$	4.304	2.015	0.217
3	$-2.19\cdot10^{-7}$	0.923	2.260	0.133

The optical constants of metallophthalocyanine layers: refractive indexes, extinction coefficients and layer thicknesses have been calculated based on transmission and reflectance measurements. All these spectra show no usable interference extremes because the investigated films were too thin so the calculation of optical constants using for example envelope methods [121] was impossible. However calculations could be performed by minimanization of error function, which can be written as [122].

$$err = \sum_{i=1}^{N}((T_{\exp}(\lambda_j) - T_{calc}(n(\lambda_j),k(\lambda_j),d,\lambda_j))^2 + (R_{\exp}(\lambda) - R_{calc}(n(\lambda_j),k(\lambda_j),d,\lambda_j))^2) \tag{7}$$

For the air-film-substrate system the transmission and reflection are given by equations [123]:

$$T_{calc} = \frac{n_2}{n_0}\frac{[(1+g_1)^2 + h_1^2][(1+g_2)^2 + h_2^2)}{e^{2\alpha_1} + (g_1^2 + h_1^2)(g_2^2 + h_2^2)(e^{-2\alpha_1} + C\cos 2r_1 + D\sin 2r_1)} \tag{8}$$

$$R_{calc} = \frac{(g_1^2 + h_1^2)e^{2\alpha_1} + (g_2^2 + h_2^2)(e^{-2\alpha_1} + A\cos 2r_1 + B\sin 2r_1)}{e^{2\alpha_1} + (g_1^2 + h_1^2)(g_2^2 + h_2^2)e^{-2\alpha_1} + C\cos 2r_1 + D\sin 2r_1)} \tag{9}$$

$$\left.\begin{aligned}
g_1 &= \frac{n_0^2 - n_1^2 - k_1^2}{(n_0 + n_1)^2 + k_1^2} \qquad & h_1 &= \frac{2n_0k_1}{(n_0 + n_1)^2 + k_1^2} \\
\alpha_1 &= \frac{2\pi k_1 d}{\lambda} & & \\
g_2 &= \frac{n_1^2 - n_2^2 + k_1^2}{(n_1 + n_2)^2 + k_1^2} & h_2 &= \frac{-2n_2k_1}{(n_1 + n_2)^2 + k_1^2} \\
r_1 &= \frac{2\pi n_1 d}{\lambda} & &
\end{aligned}\right\} \tag{10}$$

$$A = 2(g_1g_2 + h_1h_2) \qquad C = 2(g_1g_2 - h_1h_2)$$

$$B = 2(g_1h_2 - g_2h_1) \qquad D = 2(g_1h_2 + g_2h_1)$$

where d is the layer thickness, λ-wavelength of the incident radiation, n_0, n_1, n_2 are the refractive indices of air, layer and quartz respectively, k_1 is the extinction coefficient of film.

The experimental and calculated reflection and transmission spectra for investigated metallophthalocyanine layers are presented in Fig. 25-28. As it is seen from these figures the fitting is quite satisfactory.

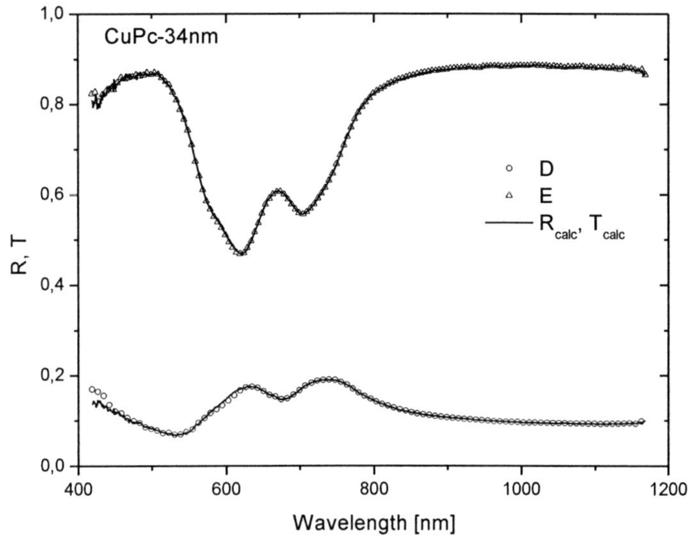

Figure 25. Transmission (triangle) and reflection (circle) spectra of CuPc thin film

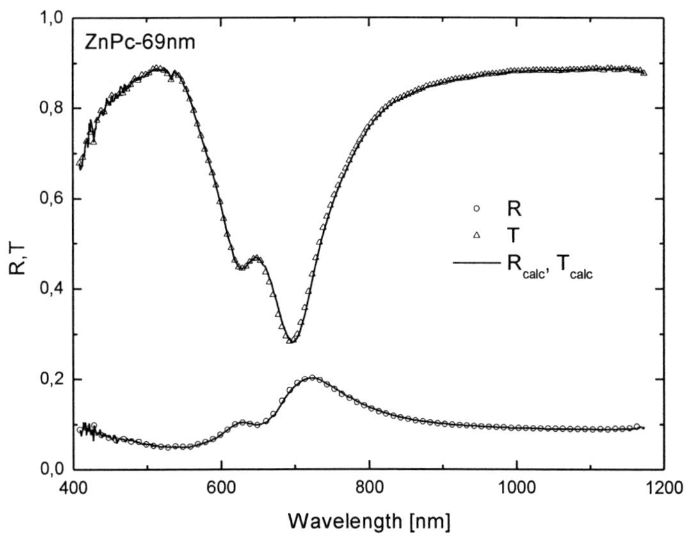

Figure 26. Transmission (triangle) and reflection (circle) spectra of ZnPc thin film

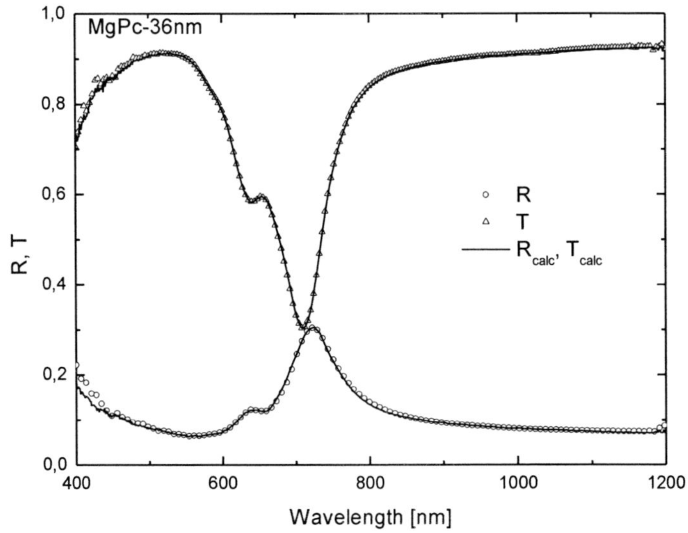

Figure 27. Transmission (triangle) and reflection (circle) spectra of MgPc thin film

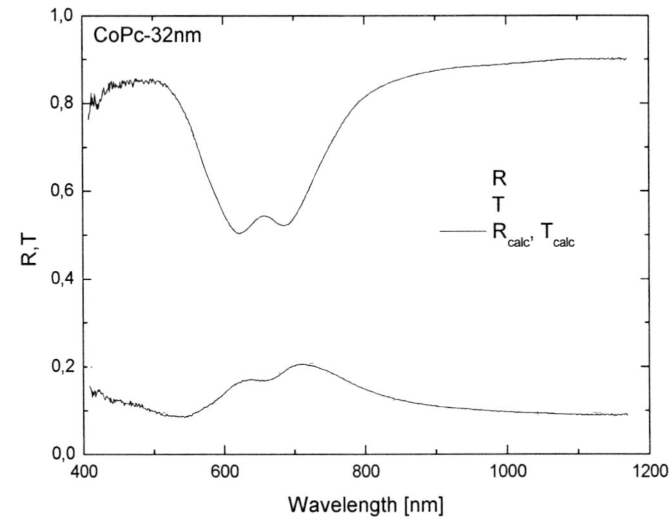

Figure 28. Transmission (triangle) and reflection (circle) spectra of CoPc thin film

Calculated refractive indexes and extinction coefficients for metallophthalocyanines are presented in Fig. 29 and 30, respectively. As we can see their behaviour is in satisfying correlation with the absorption coefficient. Moreover they show an anomalous dispersion in the absorption region and normal dispersion in the transparent region. The values of refractive indexes for CuPc, ZnPc and CoPc are close each other especially in the transparent region (λ>900nm). The MgPc refractive index adopts the biggest values which significantly differ from the others because it is caused by the very strong and relative narrow Q absorption band of the investigated film.

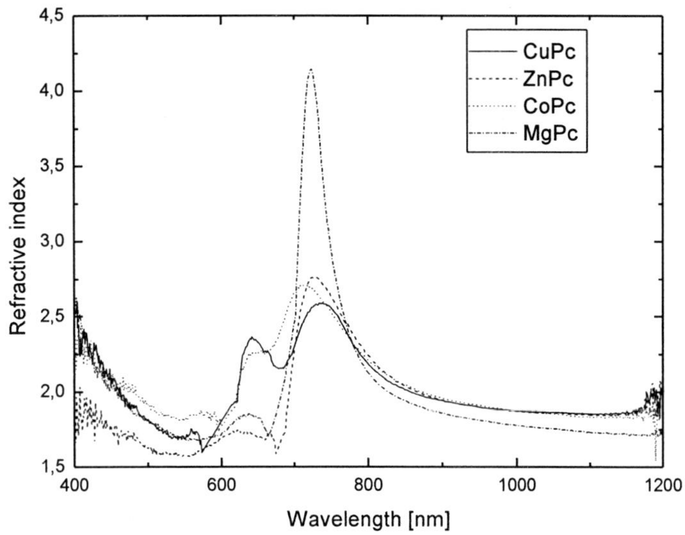

Figure 29. Calculated refractive indices of MPcs films

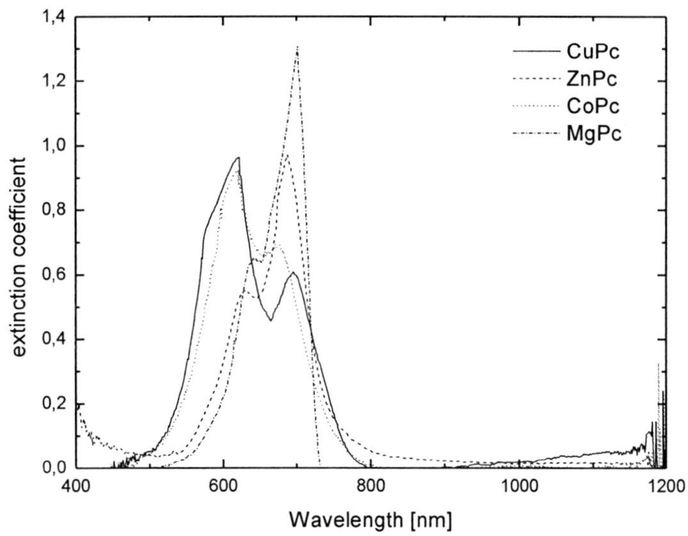

Figure 30. Calculated extinction coeficients of MPcs films

The real (ε_1) and imaginary (ε_2) part of dielectric constant can be extracted using determined values of complex refractive index and equations:

$$\varepsilon_1 = n^2 - k^2 \quad (11)$$

$$\varepsilon_2 = 2nk \quad (12)$$

Spectral distributions of dielectric constants are shown in Fig. 31-34.

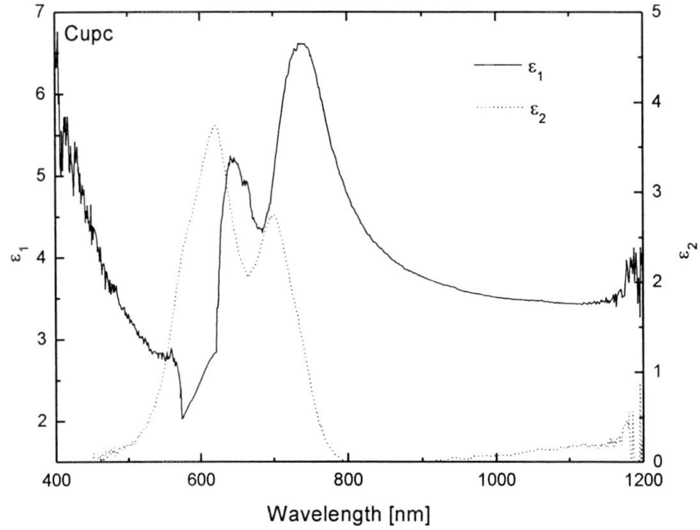

Figure 31. The spectral distribution of real (ε1) and imaginary (ε2) part of dielectric constant of CuPc film

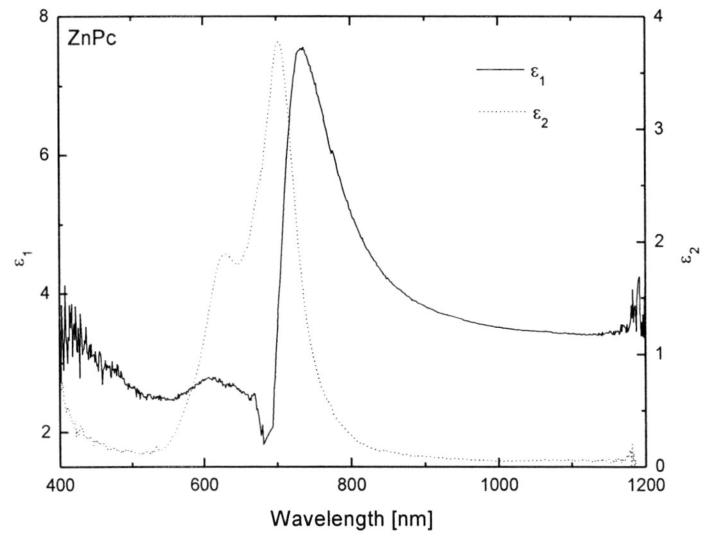

Figure 32. The spectral distribution of real (ε1) and imaginary (ε2) part of dielectric constant of ZnPc film

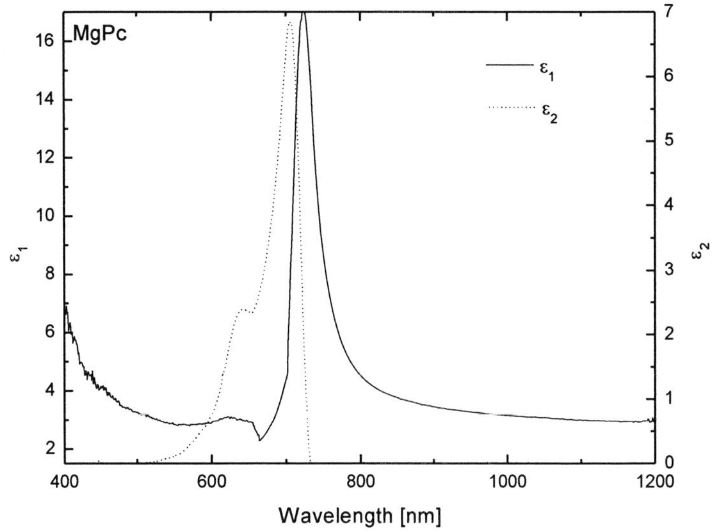

Figure 33. The spectral distribution of real (ε_1) and imaginary (ε_2) part of dielectric constant of MgPc film

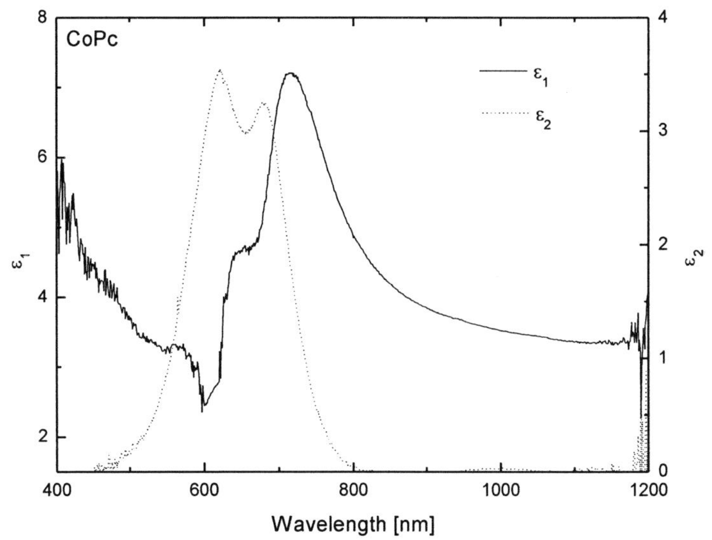

Figure 34. The spectral distribution of real (ε_1) and imaginary (ε_2) part of dielectric constant of CoPc film

Alq_3

The tris(8-hydroxyqinoline) aluminium (Alq_3) is one of the most important organic materials for electronic and optical applications. Alq_3 belongs to a class of metal chelates in which the central metal ion, Al^{3+}, is surrounded by three ligands (8-hydroxyquinoline anions). Much progress of interest of this material has been made since the discovery of an Alq_3-based

OLED structure by Tang and co-workers in 1987 [2]. In these structures Alq_3 can play two main roles: it is used as the electron-transport layer, giving rise to the transport of charge carriers, and the emitting layer. In case of emitting layers we often use Alq_3 doped with the fluorescent dyes (for instance DCM) where by direct excitation or via energy transfer from the host, the guest molecules emit their characteristic luminescence. One of the most important features of that material is its propensity to form good thin films upon vacuum deposition. Because of its remarkably thermal stability it can sublimed without decomposition at a temperature of about 600K [124]. The thin layers consist of a solid solution of two geometric isomers (meridianal and facial) that hinder the recrystallization of films and explain their long term stability in a glassy form [124]. The form of the layer (crystalline or amorphous) depends on the deposition conditions. The intermolecular interactions in organic materials are governed by relatively weak van der Waals forces and hydrogen bonds. From this reason, the overlap of the molecular orbitals is small. Thus, the molecules retain their identity and the properties of a solid-state film are strongly influenced by the arrangement and concentration of functional groups within the contained molecules [125]. Others common features resulting from the small intermolecular coupling are the small dielectric constant and a high exciton binding energy (several hundred meV) [125]. Delocalized π-electrons determine the semiconducting properties. The HOMO (π orbitals) and LUMO (π* orbitals) states have a strong dipole moment and are localized on the ligands. In case of Alq_3 the HOMOs are localized on the phenoxide ring (O) of the quinoline, whereas the LUMOs are principally on the pyridyl side (N) [125].

The absorption spectrum of Alq_3 thin layer is characterized by two strong bands: A - 3.17 eV ≈ 391 nm and B - 4.67 eV ≈ 266 nm (Fig. 35). There is a large discrepancy of positions of these bands in the literature (A: *380 nm* [126], *390 nm* [127], *385 nm* [128], *390 nm* [129], *394 nm* [130]; B: *264 nm* [127], *260 nm* [128], *261 nm* [129], *276 nm* [130]). It can be explained by differences of the layer structures. The conditions of the film deposition (for example the temperature of the substrate) influence the form of the layers. Some of them are crystalline and some are amorphous. The *A* and B transitions occur between lower state and two higher states: $1L_a$ and $1L_b$ (Platt notation), respectively. On the other hand we can say that the band at *3.17 eV* is connected with a transition from a ground state (S_0) to an excited state (S_1) [131]. This band was also assigned to the CT (Charge Transfer) from the quinolate ring containing oxygen to the pyridyl ring [126]. Additionally there are two others weak bands at C - 3.88 eV ≈ 320 nm (*320 nm* [129], *321 nm* [126]) and D - 3.72 eV ≈ 333 nm (*334 nm* [129], *336 nm* [126]) in the spectrum. There is a good agreement in literature as regard the origin of *A* and *B* transitions, which are localized inside the ligands. However there is a problem to say exactly what an origin of two smaller bands is. They were assigned to an electronic transition and its vibronic progressions. It turns out that the lowest electronic transitions are $\pi - \pi^*$ ones and they take place in the quinolate rings.

In case of the absorption spectrum of Alq_3:DCM thin layer we observe a weak additional band at about 500 nm (Fig. 36). The absorption threshold is shifted to longer wavelengths what implies absorption in DCM molecules. The position of other bands is not changed in comparison with pure Alq_3 absorption but the intensity of these bands is smaller. The comparison of absorption spectra performed for Alq_3 layers and solutions shows the absorption properties are determined by molecular structure. That fact results in the conclusion that the transitions take place in single particles.

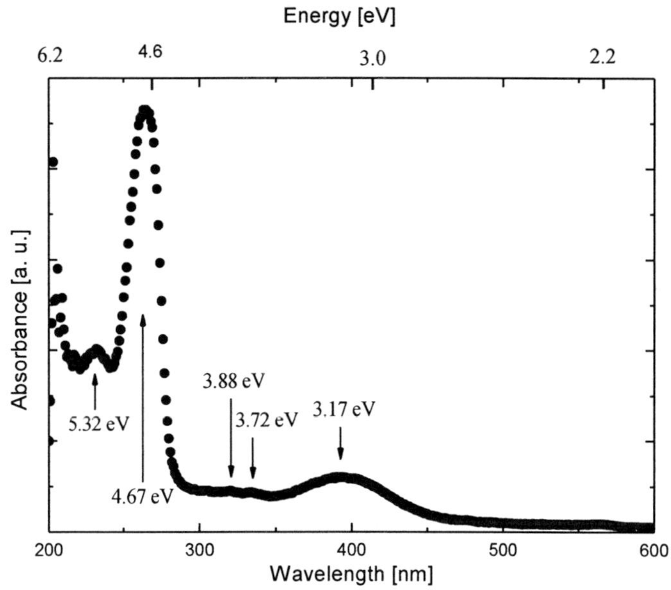

Figure 35. The absorption spectrum of Alq_3 thin layer

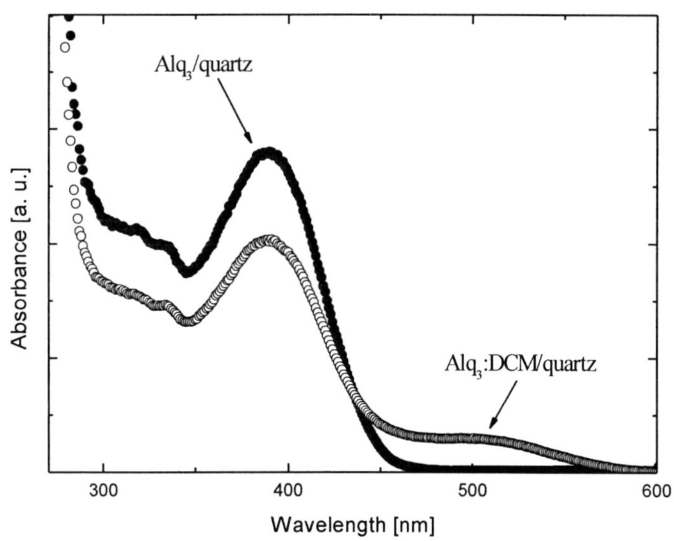

Figure 36. The comparison between the absorption spectra of Alq_3 and Alq_3:DCM thin layers

PL and PLE spectra for Alq_3 thin layer are presented in Fig. 37. The typical mirror symmetry is observed. The Alq_3 exhibits very good photoluminescence properties. After excitation, by means of blue or UV radiation, the material emits the strong green light. A relatively wide PL spectrum is characterized by asymmetric, smooth shape with a maximum at about 530 nm and FWHM about 120 nm. Similarly to the absorption, the position of the PL maximum depends on the structure of a layer (amorphous or crystalline) and can be placed between 500 nm and 540 nm [132]. Since the results of some structures similar to Alq_3 (i.e.

Gaq_3 or Inq_3) exhibit almost identical spectra, scientists suggested that excited states responsible for luminescence are localized in ligands. The HOMO states are localized on the phenoxide side while LUMO on pyridyl side [128].

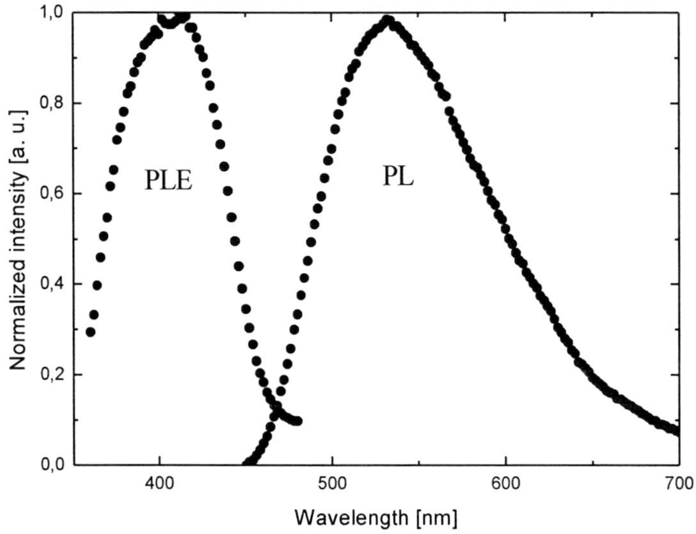

Figure 37. PL and PLE spectra of Alq_3

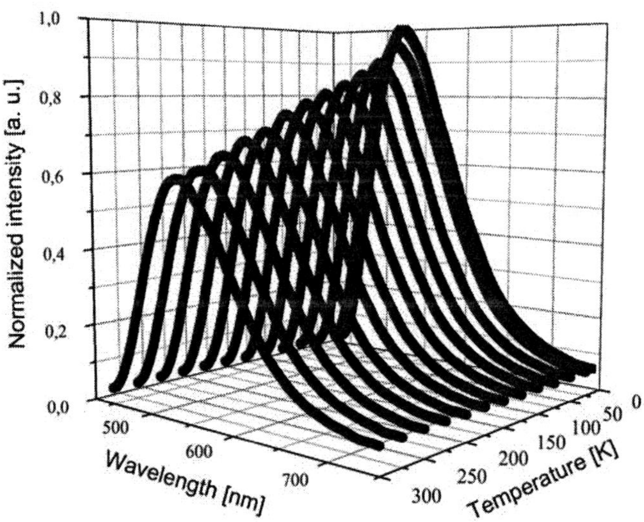

Figure 38. The temperature dependence of PL of Alq_3 thin layer

Fig. 38 shows the series of the PL spectra measured at the temperature range from 13K to 300K. These spectra were measured at wavelength range from 450 nm to 750 nm. At low temperatures bigger intensity of PL line is clearly observed. It is also distinctly visible that the PL intensity decreases with increasing temperature. The observed behaviour can be explained by means of a configuration coordinate model which is schematically shown in Fig. 42. The curves S_0, S_1 and T are adiabatic potentials of the initial and final states of molecule. The transition AB corresponds to the excitation of the electron – hole pair. The transition BC over barrier E_1 corresponds to the capture of carrier by defect. When the temperature increases the electrons can easily skip to the T state and return with phonon emission to ground state S_0. This behaviour of carriers results in increasing of nonradiative recombination and decreasing of a luminescence. The transition CD is the thermal excitation of a carrier over the crossover of S_0 and T curves, after which the nonradiative recombination takes place (transition DA). E_2 is energy of this thermal excitation. Such two steps quenching mechanism can be described by the following formula [133,134].

$$I(T) = \frac{I_0}{1 + A\exp\left(-\frac{E_1}{kT}\right) + B\exp\left(-\frac{E_2}{kT}\right)} \tag{13}$$

where A and B are constants, E_1 and E_2 are thermal activation energies. Calculated energies have following values: $E_1 = 3.28\ meV$ and $E_2 = 47\ meV$. The plot of the PL intensity versus inverse temperature along with fitting is shown in Fig. 39.

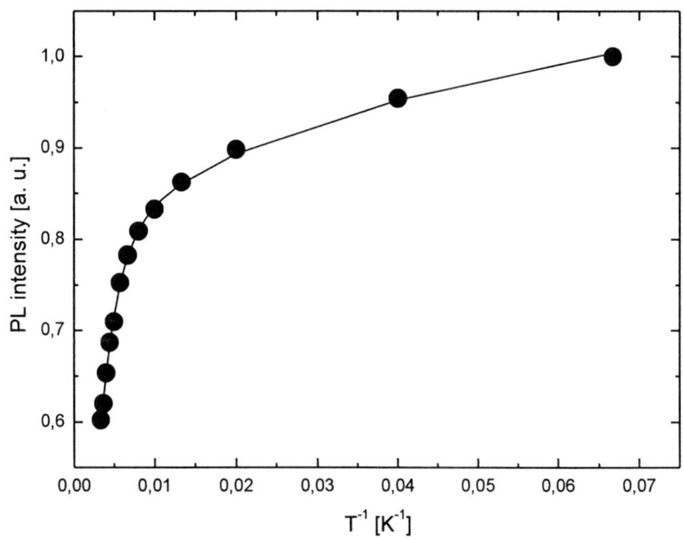

Figure 39. The thermal quenching of PL of Alq_3 thin layer

The strong temperature dependence of the PL spectra is attributed to the reduced exciton-phonon interaction with decreasing temperature. In planar molecules (Alq_3) the strong coupling exists between excitons and intermolecular phonon modes. Phonon-enhanced internal conversion provides a nonradiative transition path for excitons, leading to quenching of the integral PL at higher temperatures. The PL temperature dependence also shows a small blueshift of the whole spectrum (Fig. 40) and diminishing of the FWHM (Fig. 41) with decreasing temperature. As regard the shift of the spectrum there can be found two processes responsible for such behaviours: thermal expansion and renormalization of band energies connected with mentioned electron – phonon interaction [135]. Both processes compete each other and result in observed small change. The wide PL spectrum (about 528 nm at RT) observed in case of amorphous thin film of Alq_3 result from its structure which is characterized by a considerable disordered of molecules in a layer.

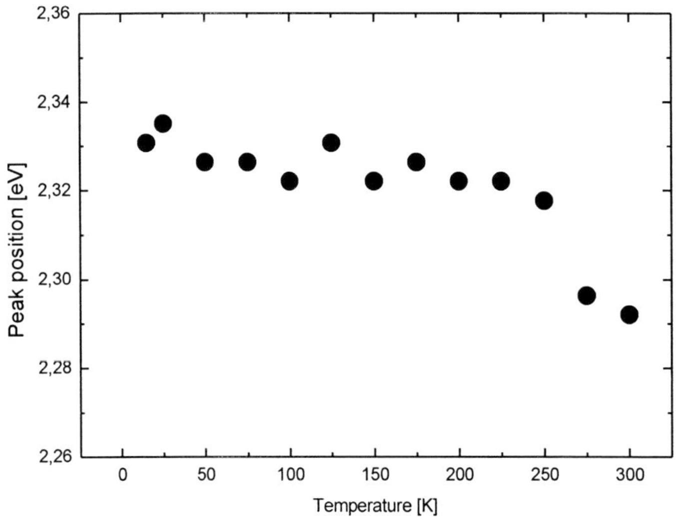

Figure 40. The PL peak position of Alq_3 thin layer as a function of temperature

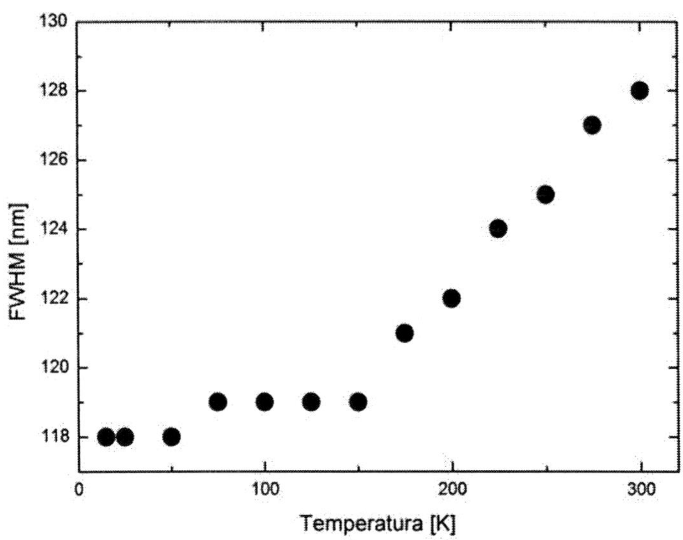

Figure 41. The full width half maximum of PL of Alq_3 thin layer as a function of temperature

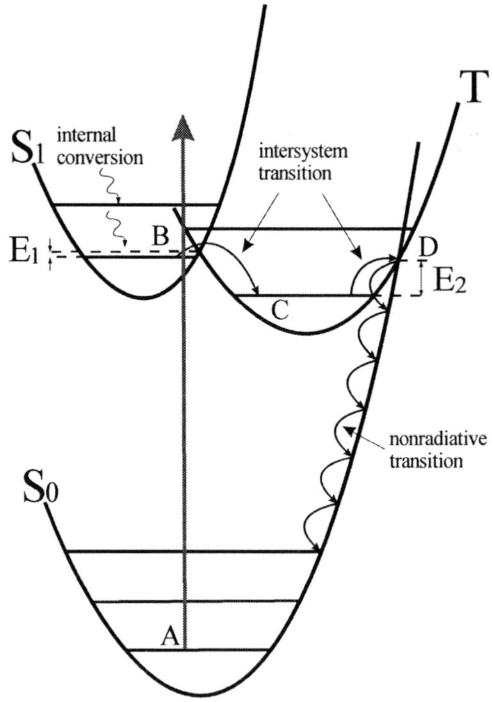

Figure 42. The explanation of the thermal quenching effect in Alq_3

As it can be seen in Fig. 43 the existence of dopants in Alq_3 amorphous thin films changes its photoluminescence properties. The DCM molecules inserted to the Alq_3 layer cause a shift of a whole spectrum towards red wavelengths. Such behaviour can be explained by means of nonradiative Förster energy transfer between Alq_3 and DCM molecules. In this case Alq_3 is a donor and DCM is an acceptor. To obtain strong and efficient energy transfer there should be meet some conditions. One of the most important is that the absorption spectrum of the acceptor and the emission one of the donor must overlap each other. That is the levels in DCM are placed inside optical band gap energy in Alq_3. Although in Alq_3:DCM system both molecules are optical excited the luminescence originates mainly from DCM molecules (see Fig. 43). It is caused by fast relaxation from Alq_3 levels to DCM excited state which is much faster than emission to the ground level [136]. However, it is known that DCM molecules exhibit strong tendency to formation of aggregates [137] which emit longer wavelengths radiation than monomer one. That is why Alq_3:DCM PL spectrum is positioned between spectra of pure components. Such complexes can be also formed in solution. Fig. 44 presents the dependence of the PL peak position on DCM concentration in Alq_3:DCM solution (CH_3Cl). The growth of dopants concentration increases the red shift of the PL spectrum (Fig. 44).

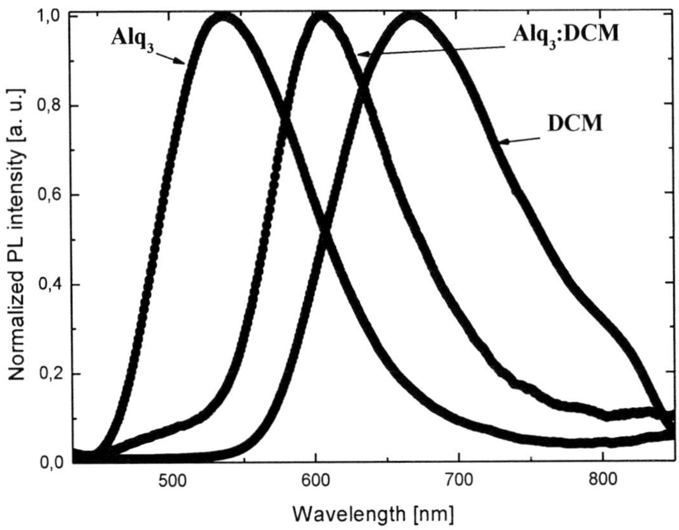

Figure 43. Normalized PL spectra of Alq_3, Alq_3:DCM and DCM thin layers

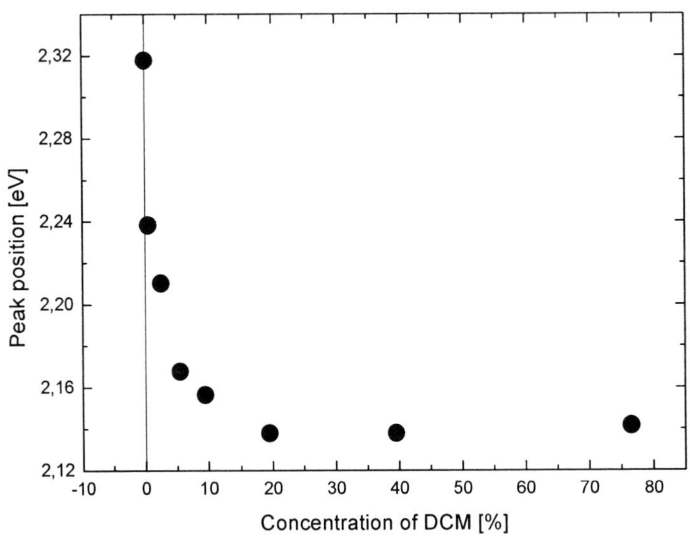

Figure 44. The peak position of PL spectra of Alq_3:DCM solution ($CHCl_3$) for different concentration of DCM molecules

The transmission and reflection spectra of Alq_3 layer are shown in Fig. 45 and Fig. 46 respectively. It is clearly visible that both spectra exhibit distinct interference fringes. These fringes can be used to calculate optical constants and layer thickness using procedure described by Minkov [121] in case of reflection and Swanepoel [138] in case of transmission. The refractive index of the investigated layer was computed using both methods. Fig. 47 shows the comparison of calculated refractive index obtained from reflection and transmission spectra separately. A good agreement of the refractive index determined from both methods is clearly visible. The calculated thickness of the film from both spectra was about 1700 nm.

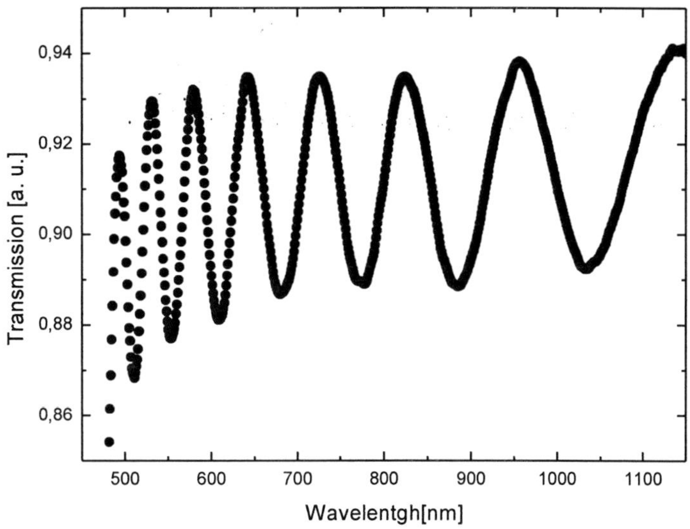

Figure 45. The transmission spectrum of Alq_3 thin layer

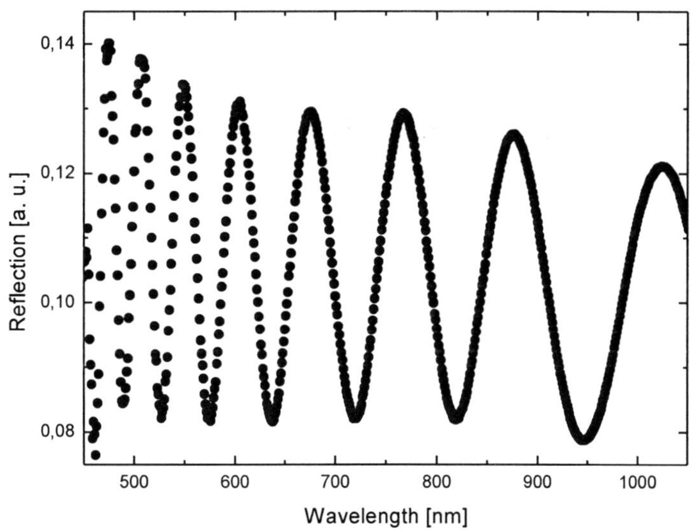

Figure 46. The reflection spectrum of Alq_3 thin layer

In an absorption threshold region we can distinguish two parts (Fig. 48). The first is weak absorption region (1 $cm^{-1} < \alpha <$ 104 cm^{-1}), the second one is high absorption region ($\alpha \geq \sim$104 cm^{-1}) [139]. In case of the first region described by the so-called Urbach relation the absorption depends exponentially on the photon energy. As regard the second one the absorption can be described by means of the well-known Tauc law [140]. Using such description it is possible to determine the optical energy gap (Fig. 49) which in case of Alq_3 thin layer is about 2.65 eV.

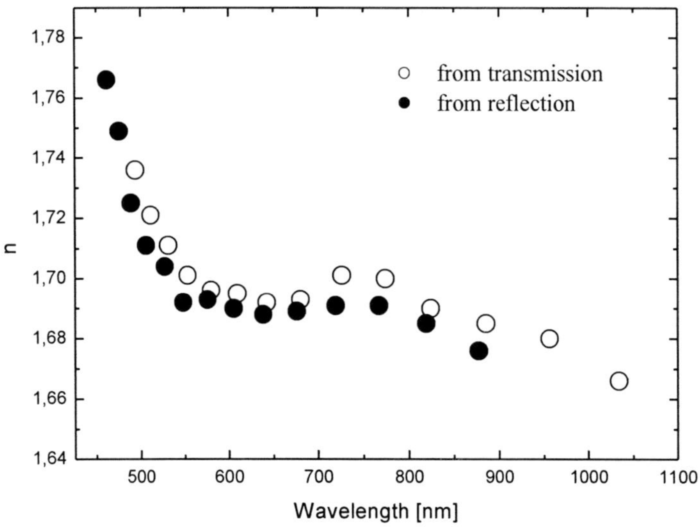

Figure 47. The refractive index of Alq_3 thin layer calculated from transmission and reflection spectra using Swanepoel and Minkov method

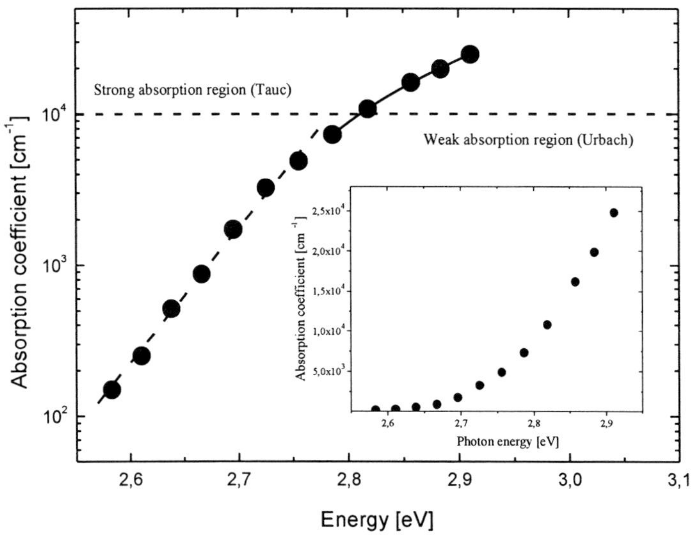

Figure 48. The absorption coefficient (logarithmic scale) of Alq_3 thin layer determined from Connell and Lewis relation (linear scale - inside)

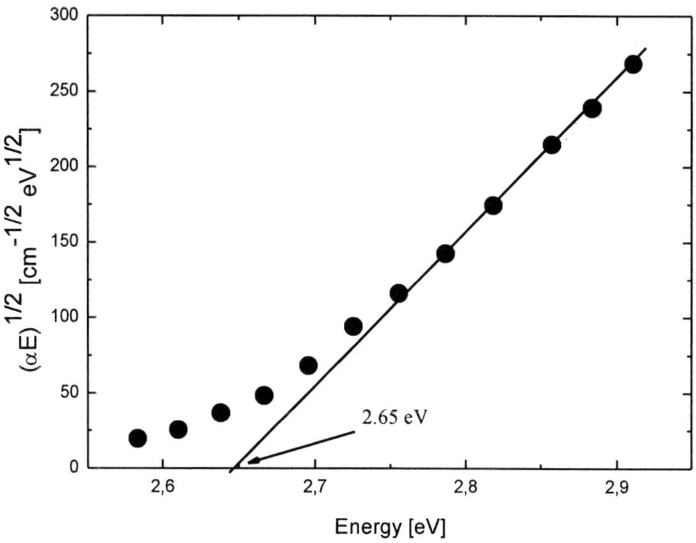

Figure 49. Determination of optical band gap energy for Alq_3 thin layer according the Tauc formula.

Perylene Derivatives

One of the most intensively studied perylene derivative is PTCDA (see Fig.1). It forms well ordered polycrystalline films on a variety of substrates [18, 19]. This compound is characterized by the significant intermolecular interaction of the π – electron system, resulting in the interesting optical and electronic properties. This interaction is caused by unusually small interplanar molecular stacking distance which is 0.321 nm [141]. It is shorter than that in pure perylene, resulting in a larger π - orbital overlap of neighbouring molecules [44].

The absorption of PTCDA thin layer (thickness = 2000Å) deposited on the quartz plate is shown in Fig. 50. In the low-energy region, this spectrum has a broad feature centred at 2.58 eV and a narrow peak at 2.26 eV. The position of the first one agrees well with the calculated excited state at 2.559 eV connected with the HOMO to LUMO electronic transition. The high-energy excited states, which are visible as absorption peaks above 3 eV, arise from other HOMO (-) to LUMO (+) transitions [111]. The peak at 2.26 eV can be attributed to a charge transfer exciton of the extended PTCDA structure [142,143].

The formation of such aggregate states can be confirmed by the concentration dependence of absorption in solution. The results for PTCDA dissolved in DMSO at different concentration from 0.2 to 1.5 μM are presented in Fig. 51. The spectrum at the highest concentration has well separated peaks with a shoulder on the high energy side. The positions of these maxima are 2.11, 2.39, 2.56 and 2.75 eV. It is evident that peak positions are constant for all solutions. Moreover, the shapes of high-energy peaks (2.39, 2.56 and 2.75 eV) are independent on concentration of solution. In addition, the spacing between these peaks is nearly equal what suggests that they can be attributed to transitions from ground singlet state (S_0) to different vibronic levels of excited singlet state (S_1). However, there are considerable differences in attenuation of individual bands. The lowest-energy peak diminishes rapidly, with large rate relative to three high-energy bands, when the PTCDA concentration decreases.

Furthermore, this peak disappears at sufficiently small concentration. The distinctly different behaviour of the lowest-energy band as compared to the high-energy peaks indicates that this transition is connected with an aggregate state. Such excitations can appear when the concentration of solution is high enough to form physical dimer. This process is enabled because of arising of the π-π orbital overlap of closely stacked PTCDA molecules. Similar states were observed in other organic crystals such as perylene [144].

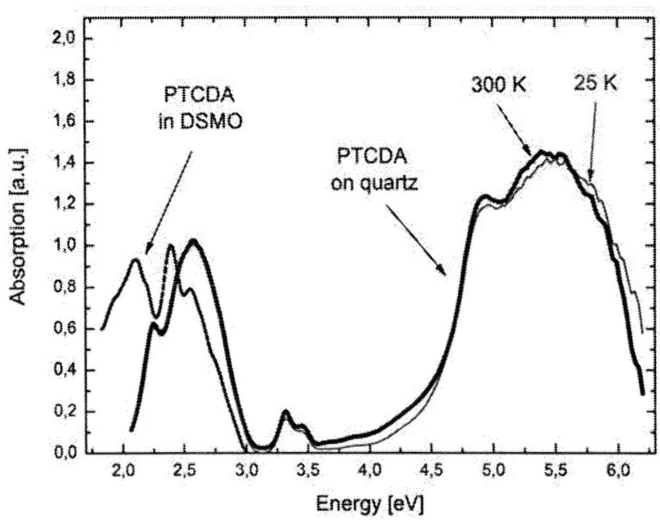

Figure 50. The absorption of PTCDA thin film (thickness = 2000Å) on quartz at two temperatures: 25K (grey) and 300K (black). The dashed line represents the absorption of PTCDA dissolved in DSMO (concentration about 1.5 μM)

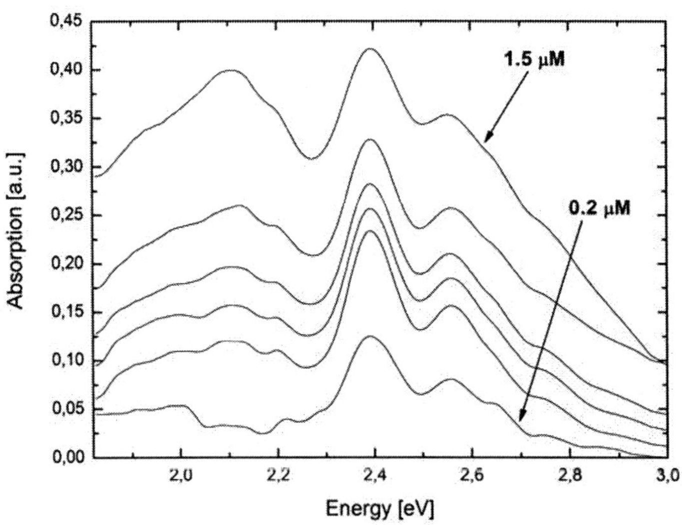

Figure 51. The absorption spectra of PTCDA dissolved in DSMO at different concentration k=1.5, 1.1, 0.9, 0.7, 0,5 and 0.2 μM

The formation of aggregate states such as CT excitons in thin film is very likely especially as the distance between adjacent molecules in solid state is much shorter than in the solution. The thin layer absorption peak connected with CT state as well as the other bands are strongly red-shifted relative to the solution. It is clearly visible in Fig. 50, where the absorption spectrum of dissolved PTCDA is superimposed on the thin film spectrum, for comparison. This shift is originated from the change in the interaction energy of single molecule with all of the surrounding molecules in crystal when the transition to the excited state takes place [95]. The value of this energy is negative; therefore the absorption observed in crystalline layer is shifted to the lower energy in comparison to the monomer spectrum. Moreover, the absorption spectra for polycrystalline films are considerably broadened and are devoid of clear vibronic structure (see Fig. 50). This fact is caused by the increased coupling of phonon modes with electronic transitions in thin layer. In addition, the low-energy absorption of PTCDA thin film was found to be independent on temperature from 13 to 325 K. Moreover, it is shown in Fig. 50 that the influence of the temperature on the intensities, broadenings and positions of high-energy absorption bands is slight. Aside from CT excitons, the completely different type of excitations such as self – trapped excitons (STE) can appear in thin film. Such states have lower energy than the corresponding free excitons so in case of PTCDA they are included in the low-energy tail extends to below 2.0 eV (see Fig. 50). Bulovic et. al. [142] analyzed this tail using the Urbach absorption rule [145] and showed that the exciton-phonon coupling constant in PTCDA is larger than the critical value necessary to form a self-trapped exciton. Thus, due to strong exciton-phonon coupling, the self-trapping of CT excitons is very likely in thin films.

The presence CT and self – trapped CT excitons affects strongly the luminescence of the PTCDA thin layer. The PL spectra of thin film (thickness = 2000 Å) at 13K and the room temperature along with the absorption spectrum of the same layer are presented in Fig. 52. The absorption and photoluminescence do not reveal the mirror symmetry (see Fig. 52), whereas such attribute was observed in dissolved PTCDA [142]. In addition, the Stokes shift between the lowest absorption and the highest PL band is much larger than for the solution. This behaviour can be connected with additional external phonon modes in solid phase [111].

The PL spectrum at room temperature is characterized by the main peak at 1.75 eV and two shoulders positioned at 1.55 and 1.97 eV. We noticed that shape of these bands does not depend on the excitation energy in the range from 2.25 to 3.0 eV so we concluded that the relaxation from S_1 state to CT and self – trapped (CT^{ST}) states takes place before emission occurs. Thus, observed bands can be assigned to transitions from CT^{ST} state to different vibronic levels of molecular ground state S_0 [142]. The latest study of the time resolved photoluminescence in PTCDA [146] showed that there exists the additional PL channel at higher temperatures, which can be associated with excimer state.

The photoluminescence properties of PTCDA were found to be strongly dependent on temperature from 13 to 300K. A set of PL spectra of the PTCDA thin layer as a function of temperature is presented in Fig. 53. The PL spectrum at the lowest temperature (13K) has the major maximum centred at 1.70 eV and two shoulders positioned at 1.54 and 1.86 eV. The main bands characterizing PL spectra have different positions at various temperatures. The overall intensity decreases with rising temperature and it is strongly reduced at high temperatures due to a large nonradiative deexcitation rate. When the temperature increases the probability of the non-radiative transitions such as the internal conversion (S_1 – S_0), intersystem crossing into the triplet state (S_1 – T) or some trapping states rises so these processes can be effectively compete with radiative transitions. It is also clearly seen in Fig. 53 that the different PL channels broaden when the temperature increases. This broadening is caused by the enhancement of exciton – phonon interactions [142].

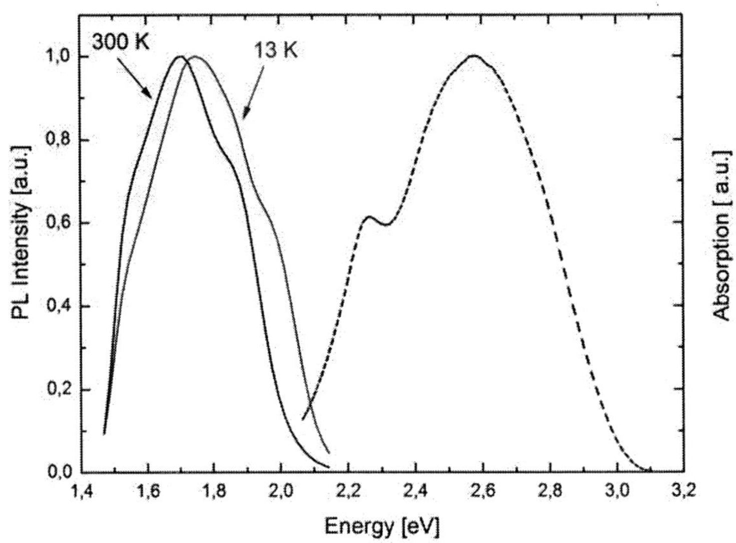

Figure 52. The PL spectra of PTCDA thin film on quartz at two different temperatures 13K (grey), 300K (black) and absorption spectrum (dashed line) of the same layer

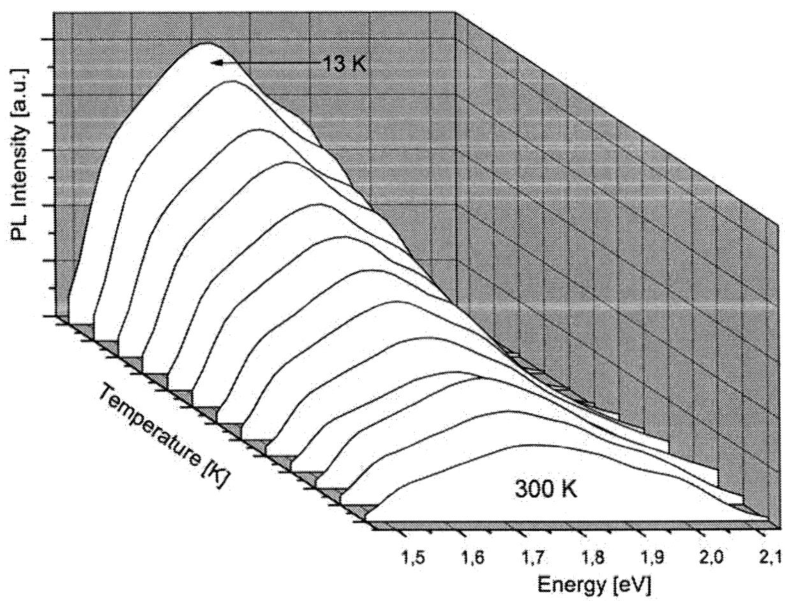

Figure 53. The PL spectra of PTCDA thin film on the quartz substrate measured at different temperatures: 13, 25,50,75,100,125,150,175,200,225,250,275 and 300K

Among various perylene derivatives the 3,4,9,10-tetra-(heptyloxy-carbonyl)-perylene (PeC7) (see Fig.1) is distinguished by excellent fluorescent properties. This compound emits light in a spectral region advantageous for the human eye (yellow emission) with a very high efficiency [147]. For this reason, PeC7 thin films can be used as emissive layers in OLEDs. PeC7 easily forms thin films on different substrates by thermal evaporation as well as Langmuir – Blodgett techniques [26]. The other field of the application for this material is liquid crystal technology. PeC7 has a good dichroic and may form columnar hexagonal liquid crystalline phase in certain temperature range. These advantages allow using this perylene derivative in "guest-host" liquid crystal devices (LCD) [147]. Besides, PeC7 has very interesting electronic properties such as a strong tendency to creation of self-aggregates, which are revealed especially in the fluorescence spectrum [26].

The absorption spectrum of PeC7 dissolved in ethanol (with concentration about 10^{-4} M) is presented in Fig. 54. It is characterized by clearly resolved structure with two intense peaks at 2.65 and 2.82 eV. Moreover, two weaker bands at 2.99 and 3.16 eV are also visible. The equal distance between consecutive peaks indicates that they are connected with monomer transitions from the ground state (S_0) to different vibronic levels of excited state (S_1). The photoluminescence spectrum of the same solution exhibits the classical Stokes shift and the mirror symmetry in relation to absorption. Two intense bands centred at 2.55 and 2.41 eV with shoulder at 2.24 eV are evident from Fig. 54. Both the emission and absorption spectra of PeC7 change dramatically in thin polycrystalline films in comparison to that in solution. The absorption spectrum of PeC7 thin layer (thickness = 2500 Å) is characterized by three clearly resolved bands at 2.52, 2.73 and 2.87 eV. The whole spectrum is considerably broadened in relation to the solution spectrum. This feature usually is connected with additional interactions between individual molecule and surroundings which is stronger in solid state than in solution. The same interaction causes the shift of molecular excited state towards lower energy and the splitting of that in molecular crystals.

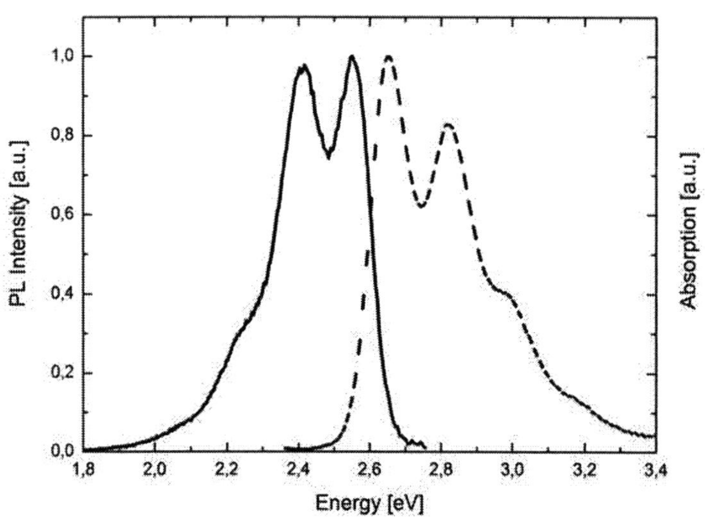

Figure 54. The photoluminescence and absorption of PeC7 dissolved in ethanol

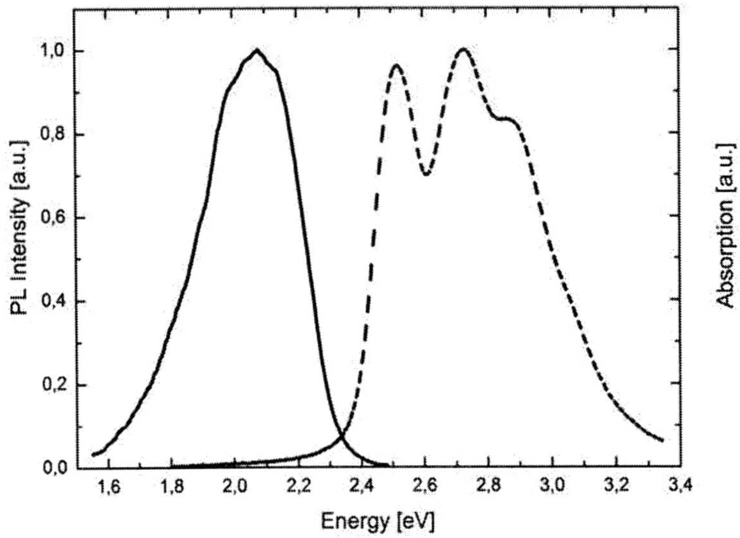

Figure 55. The photoluminescence and absorption of PeC7 thin film on quartz

This effect, called the Davydov splitting [95], is evident in absorption spectrum for PeC7 thin film (see Fig. 55) where two main bands (2.52 and 2.73 eV) have arisen as a result of the shift and the splitting of monomer absorption band (2.65 eV). The PL spectrum of PeC7 thin film with maximum at 2.08 eV is broad and unstructured in comparison to corresponding spectrum in solution (see Fig. 54 and 55). Moreover it is significantly red-shifted in relation to absorption spectrum. The magnitude of this shift is much bigger than in solution, so we concluded that emission of PeC7 thin film have to originate from some aggregate excitations laying lower than monomer ones. The formation of the different type of aggregates is very likely in various organic molecular crystals [144,145]. The broad and structureless emission is characteristic for excimer [94] which arises from connection of two molecules when one of them is excited. The ground state of such dimer is dissociative so excimer fluorescence is very broad. The probability of excimer formation is high if the distance between neighbouring molecules is small enough. This condition is well realized in most of organic molecular crystals such as α-perylene [144], so we suggest that emission in PeC7 polycrystalline layer occurs from excimer state.

The absorption and emission properties of PeC7 thin films were found to be dependent on the temperature. In particular, the photoluminescence spectra are strongly affected by temperature changes in the range from 20 to 300K. The typical set of PL spectra for PeC7 thin layer at different temperatures is shown in Fig. 56.

It is evident that shape of these spectra is dependent on the temperature. The positions of individual bands as well as their intensities vary with changing temperature. The integral PL intensity increases insignificantly in the temperature range of 13 - 60 K. Next it is quickly thermally quenched, accompanied by a large broadening of the whole spectrum. The reduction of intensity is mainly caused by the increase of probability of nonradiative transitions at higher temperatures. These transition paths, leading to the luminescence quenching, are connected with a strong exciton – phonon interaction at higher temperatures. The thermal quenching of luminescence is clearly seen in Fig. 57 where the integrated photoluminescence intensity versus temperature for typical PeC7 layer is presented. It is clearly seen that the integral intensity at room temperature is reduced about 40% in comparison to maximum value at 60K. We noticed that the PL spectrum at 20 K, aside from normal excimer band, has clearly visible peak centred at 2.14 eV. This additional low-temperature band can be connected with the existence of different type of excimer. Such excitation can be treated as a variant of the normal excimer state in which degree of molecular superposition is less than in normal excimer [94].

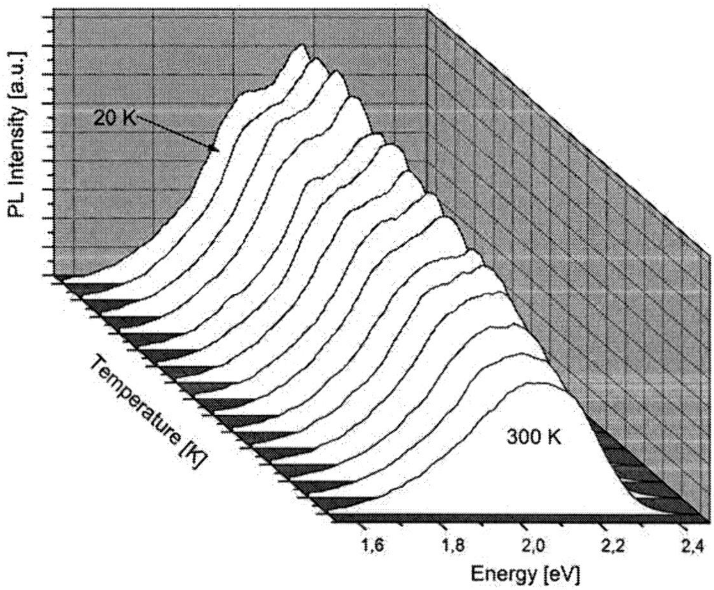

Figure 56. The set of PL spectra of PeC7 thin film on quartz versus temperature with step 20K

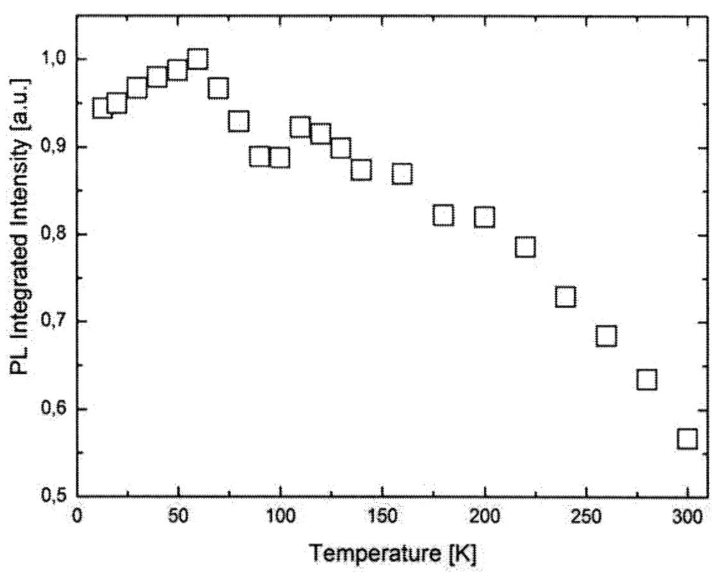

Figure 57. The thermal quenching of the integrated photoluminescence intensity

This state lies higher in the energy scale so the emission originates from it is blue-shifted relative to the transitions from normal excimer. The absorption of PeC7 thin layer is also affected by temperature though the changes are not as significant as in case of emission. From comparison of absorption spectra measured at different temperatures (Fig. 58) we noticed that the major bands shift to higher energy about 0.03 eV when temperature increases from 13 to 300K. Furthermore, it is visible the relative high of individual peaks little vary with changing temperature.

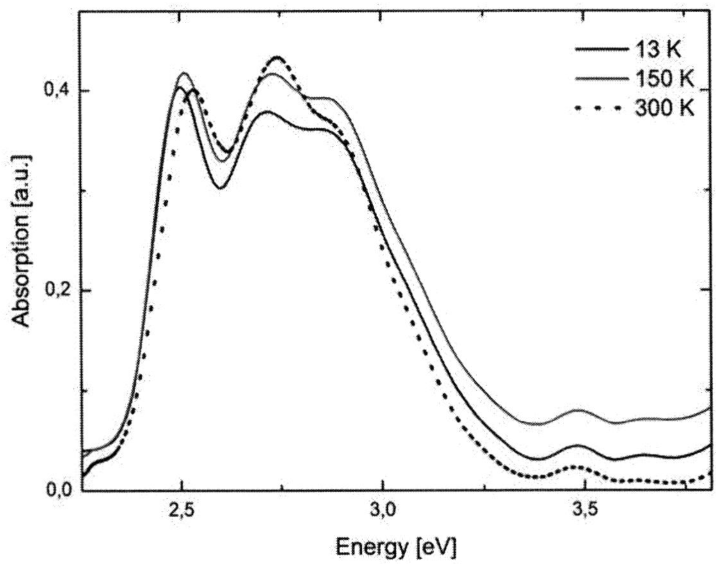

Figure 58. Absorption spectra of PeC7 thin layer on quartz at three selected temperatures: 13K (black line), 150 K (grey line) and 300 K (dashed line)

CONCLUSION

The systematic study of electronic structure of metallophthalocyanines (CuPc, ZnPc, MgPc, CoPc), perylene derivatives (PTCDA, PeC7) and Alq_3 was performed in solution as well as thin films. We found that going to the crystal the absorption and photoluminescence spectra were dramatically changed in case of MPcs and perylene derivatives. Thin layers spectra were significantly broadened and shifted relative to monomer. Moreover the splitting (Davydov splitting) of some bands was observed. These effects were connected with aggregation of molecules in solid state. The tendency to form such complexes was supported by the measurements of high concentrated solutions. Our studies of molecular thin films confirmed the existence of various types of optical excitations such as charge transfer excitons, excimers, and self-trapped excitons. Photoluminescence spectra of all investigated films were found to be dependent on the temperature. In almost all studied samples the integral PL intensity decreases with rising temperature. This thermal quenching was explained by the enhancement of nonradiative processes connected with existence some trapping states and phonon-exciton coupling. The temperature dependences revealed mechanisms of energy transfer between different excited states. Moreover in case of ZnPc film complexity of PL spectra in certain temperature range was explained by change of molecular symmetry resulting from metal ion position in molecule ring and its distortion of planarity.

We also found that DCM dopants in Alq_3 enable readily tune the emission in a wide spectral range due to effective Förster energy transfer between host and guest molecules.

Optical constants of MPcs and Alq_3 films, such as refractive index, extinction coefficient and layer thickness were calculated straightforward from transmission and reflection spectra. We found that the different metal ions in phthalocyanine ring influence the position and intensity of the main absorption bands and optical constants.

Sensitive modulated spectroscopy, which enabled to emphasize weakly visible transitions, was also used to determine the electronic structure of metallophthalocyanine. Modulated photoabsorption and photoreflectance measurements gave us abundant spectra reflecting the complex nature of MPcs excited states. The combination of stationary and modulated spectroscopy might be promising in understanding of the fundamental phenomena in organic materials.

According to our knowledge, modulated photoabsorption and photoreflectance spectroscopy were applied for the first time to study MPcs. Furthermore the low temperature dependence of MPcs thin films PL spectra has never been presented before.

REFERENCES

[1] Simon, J.; Andre, J.J. *Molecular Semiconductors*; Springer: Berlin, 1985.

[2] Tang, C. W. *Appl. Phys. Lett.* 1986, 48, 183.

[3] Gebeyehu, D.; Maennig, B.; Drechsel, J.; Leo, K.; Pfeiffer M. *Sol. Energ. Mat. Sol.* C. 2003, 79, 81.

[4] Drechsel, J.; Maennig, B.; Kozlowski, F.; Gebeyehu, D.; Werner, A.; Koch, M.; Leo, K.; Pfeiffer, M. *Thin Solid Films* 2004, 451, 515.

[5] Shen, Z.; Burrows, P. E.; Bulovic, V.; Forrest, S. R.; Thompson, M. E. *Science* 1997, 276, 2009.

[6] Sheat, J. R. *Science* 1997, 277, 191.

[7] Burroughes, J.H.; Bradley, D.D.C.; Brown, A.R.; Marks, R.N.; McKay, K.; Friend, R.H.; Burn, P.L.; Holmes, A.B. *Nature* 1990, 347, 539.

[8] Strukelj, M.; Jordan, R.; H. Dodabalapu, A. *J. Am. Chem. Soc.* 1996, 118, 1213.

[9] Kido, J.; Kimura, M.; Nagaim, K. *Science* 1995, 267, 1332.

[10] Tang, C. W.; Slyke, S. A. V.; Chen, C. H. *J. Appl. Phys.* 1989, 65, 3610.

[11] Tang, C. W.; Slyke, S. A. V. *Appl. Phys. Lett.* 1987, 51, 913.

[12] Pfeiffer, M.; Leo, K.; Zhou, X.; Huang, J.; Hofmann, M.; Werner, A.; Blochwitz-Nimoth, J. *Org. Electron.* 2003, 4, 89.

[13] Katz, E. K.; Lovinger, A. J.; Laquindraum, J. G. *Chem. Mater.* 1998, 10, 457.

[14] Dimitrakopoulos, C. D.; Furman, B. K.; Graham, T.; Hegde, S.; Purushothaman, S. *Synthetic Met.* 1998, 92, 47.

[15] Horowitz, G.; Kouki, F.; Spearman, P.; Fichou, D.; Nogues, C.; Pan, X.; Garnier, F. *Adv. Mater.* 1996, 8, 242.

[16] Haddon, R. C.; Perel, A. S.; Morris, R. C.; Palstra, T. T. M.; Hebard, A. F.; Fleming, R. M. *Appl. Phys. Lett.* 1995, 67, 121.

[17] Liu, C. J.; Peng, C. H.; Ju, Y. H.; Hsieh, J. C. Sensor Actuat. *B-Chem.* 1998, 52, 264.

[18] Forrest, S.R. *Chem. Rev.* 1997, 97, 1793.

[19] Lovinger, A.J.; Forrest, S.R.; Kaplan, M.L.; Schmidt, P.H.; Venkatesan, T. *J. Appl. Phys.* 1984, 55, 476.

[20] Hirose, Y.; Forrest, S.R.; Kahn, A. *Phys. Rev.* B 1995, 52, 14040.

[21] Forrest, S.R. *Nature* 2004, 428, 911.

[22] Wong, F.L.; Fung, M.K.; Tong, S.W.; Lee, C.S.; Lee, S.T. *Thin Solid Films* 2004, 466, 225.

[23] Xie, Z.Y.; Wong, F.L.; Fung Li, Y.Q.; Wong, F.L.; Hung, L.S. *Mater. Sci. Eng.* 2004, 106, 219.

[24] Burrows, P.E.; Bulovic, V.; Gu, G.; Kozlov, V.; Forrest, S.R.; Thompson, M.E. *Thin Solid Films* 1998, 331, 101.

[25] Wewer, M.; Stienkemeier, F. *J. Chem. Phys.* 2004, 120, 1239.

[26] Hertmanowski, R.; Biadasz, A.; Martyński, T.; Bauman, D. *J. Mol. Struct.* 2003, 646, 25.

[27] Shoustikov, A.; You, Y.; Burrows, P.E.; Thompson, M. E.; Forrest, S.R. *Synthetic Met.* 1997, 91, 211.

[28] Bulovic, V.; Deshpande, R.; Thompson, M.E.; Forrest, S.R. *Chem. Phys. Lett.* 1999, 308, 318.

[29] Hung, L.S.; Chen, C.H. *Mater. Sci. Eng.* 2002, 39, 143.

[30] Chang, C. C.; Pai, C. L.; Chen, W. C.; Jenekhe, S. A. *Thin Solid Films* 2005, 479, 254.

[31] Bała, W.; Kreja, L.; Łukasiak, Z.; Czarnowski, P.; Dalasiński, P. *Proc. SPIE Int. Soc. Opt. Eng.* 2003, 5136, 295.

[32] Brinker, C. J.; Hurd, A. J. *J. Phys. III France* 1994, 4, 1231.

[33] Rębarz, M.; Dalasiński, P.; Bała, W.; Łukasiak, Z.; Wojdyła, M.; Kreja, L. *Opt. Appl.* 2005, 35 (3).

[34] Antunes, P.A.; Constantino, C.J.L.; Aroca, R.F.; Duff, J. *Langmuir* 2001, 17, 2958.

[35] Inokuchi, H. *Org. Electron.* 2006, 7, 62.

[36] Weiss, D.; Kietzmann, R.; Mahrt, J.; Tufts, B.; Storck, W.; Willig, F. *J. Phys. Chem.* 1992, 96, 5320.

[37] Matsui, H. *Pure Appl. Chem.* 1995, 67, 429.

[38] Herz, L. M.; Silva, C.; Friend, R. H.; Phillips, R. T. *Phys. Rev.* B 2001, 64, 195203.

[39] Proehl, H.; Dienel, T.; Nitsche, R.; Fritz, T. *Phys. Rev. Lett.* 2004, 93, 97403.

[40] Scholz, R.; Kobitski, A.Yu.; Vragovic, I.; Wagner, H.P.; Zahn, D.R.T. *Org. Electron.* 2004, 5, 99.

[41] Haskal, E.I.; Shen, Z.; Burrows, P.E.; Forrest, S. R. *Phys. Rev.* B 1995, 51, 4449.

[42] Shen, Z.; Forrest, S.R.; *Phys. Rev. B* 1997, 55, 10578.

[43] Hennessy, M.H.; Soos, Z.G.; Pascal Jr., R.A.; Girlando, A. *Chem. Phys.* 1999, 245, 199.

[44] Vragovic, I.; Scholz, R. *Phys. Rev. B* 68 2003, 155202.

[45] Knupfer, M.; Schwieger, T.; Fink, J. *Phys. Rev. B* 2002, 66, 35208.

[46] Kobitski, A. Yu.; Scholz, R.; Vragovic, I.; Wagner, H. P.; Zahn, D. R. T. *Phys. Rev. B* 2002, 66, 15320.

[47] Stolarski, R.; Fiksiński, K. *Dyes Pigments* 1994, 24, 295.

[48] Dalasiński, P.; Łukasiak, Z.; Rębarz, M.; Wojdyła, M.; Bratkowski, A.; Bała, W. *Opto-Electron. Rev.* 2004, 12, 429.

[49] Wojdyła, M.; Derkowska, B.; Bała, W. *Opt. Appl.* 2005, 35 (3).

[50] Braun, A.; Tcherniac, J. Ber. Deut. *Chem. Ges.* 1907, 40, 2709.

[51] Gould, R. D. Coordin. *Chem. Rev.* 1996, 156, 237.

[52] Djurisic, A. B.; Kwong, C. Y.; Lau, T. W.; Guo, W. L.; Li, E. H.; Liu, Z. T.; Kwok, H. S.; Lam, L. S. M.; Chan, W. K. *Opt. Commun.* 2002, 205, 155.

[53] El-Nahass, M.; M. Bahabri, F. S.; Al-Harabi, R. Egypt. *J. Sol.* 2001, 24, 11.

[54] Xiao, K., Liu, Y., Yu, G., Zhu, D. *Appl. Phys.* A 2003, 77, 367.

[55] Bała, W.; Grodzicki, A.; Piszczek, P.; Wojdyła, M.; Bratkowski, A.; Szybowicz, M.; Runka, T.; Drozdowski, M. *J. Mol. Struct.* 2006, 782, 177.

[56] Debe, M.; Poirier, R.; Kam, K. *Thin Solid Films* 1991, 197, 335.

[57] Szybowicz, M.; Runka, T.; Drozdowski, M.; Bała, W.; Grodzicki, A.; Piszczek, P. ; Bratkowski, A. *J. Mol. Struct.* 2004, 704, 107.

[58] Leznoff, C. C.; Lever, A. B. P. *Phthalocyanines properties and applications*; VCH: New York, 1989; Vol. 1-3.

[59] Yasuda, T.; Tsutsui, T. *Chem. Phys. Lett.* 2005, 402, 395.

[60] Okuda, T.; Shintoh, S.; Terada, N. *J. Appl. Phys.* 2004, 96 (6), 3586.

[61] Ray, K.; Tracey, S.; M. Hodgson, S. N. B. J. *Phys. D: Appl. Phys.* 2003, 36, 1409.

[62] Hu, W.; Matsumaura, M. J. *Phys. D: Appl. Phys.* 2004, 37, 1434.

[63] Kwong, C. Y.; Djurisic, A. B.; Chui, P. C.; Lam, L. S. M.; Chan, W. K. *Appl. Phys. A* 2003, 77, 555.

[64] Stenzel, O.; Stendal, A.; Voigtsberger, K.; von Borczyskowski, C. *Sol. Energ. Mat. Sol. C.* 1995, 37, 337.

[65] Westphalen, M.; Kreibig, U.; Rostalski, J.; Luth, H. *Sol. Energ. Mat. Sol. C.* 2000, 61, 97.

[66] Kim, S. K.; Chung, T. G.; Chung, D. H.; Lee, H. S.; Song, M. J.; Park, J. W.; Lee, J. U.; Kim, T. W. *Opt. Mater.* 2002, 21, 159.

[67] Tadayyon, S. M.; Grandin, H. M.; Griffiths, K.; Norton, P. R.; Aziz, H.; Popovic, Z. D. *Org. Electron.* 2004, 5, 157.

[68] Fujii, N.; Ohmori, Y.; Yoshino, K. *Electron Devices IEEE Transactions on.* 1997, 44 (8), 1204 .

[69] Fujii, A.; Yoshida, M.; Ohmori Y.; Yoshino, K. *Jpn. J. Appl. Phys.* 1996, 35, 37.

[70] Flora, W. H.; Hall, H. K.; Armstron, N. R. *J. Phys. Chem. B* 2003, 107 (5), 1142 .

[71] Tominaga, T.; Hayashi, K.; Toshima, N. *J. Porphyr. Phthalocya.* 1997, 1 (3), 239.

[72] Tian, M.; Yanagi, S.; Sasaki, K.; Wada, T.; Sasabe, H. *J. Opt. Soc. Am. B* 1998, 15, 846.

[73] Nalwa, H.S.; Kobayashi, S. *J. Porphyr. Phthalocya.* 1998, 2, 21.

[74] Kajzar, F.; Swalan, J.D. *Organic thin films for waveguiding nonlinear optics*; CRC, 1996.

[75] Kuzyk, M.G.; Dirk, C.W. *Characterization techniques and tabulations for organic nonlinear optical materials;* Marcel Dekker, Inc., 1998.

[76] Moloney, J.V. *Nonlinear optical materials*; Springer: New York, 1998.

[77] Gu, D.; Chen, Q.; Tang, X.; Gun, F.; Shen, S.; Liu, K.; Xu, H. *Opt. Commun.* 1995, 121, 125.

[78] Matsuda, M.; Hanasaki, N.; Ikeda, S.; Tajima, H.; Naito, T.; Inabe, T. *J. Phys.* IV France 2004, 114, 541.

[79] Kaestner, L.; Cesson, M.; Kassab, K.; Christensen, T.; Edminson, P.D.; Cook, M.J.; Chambrier, I.; Jori, G. *Photoch. Photobio. Sci.* 2003, 6, 660.

[80] Kobayashi, N. *Curr. Opin. Solid St. M.* 1999, 4, 345.

[81] Hajdur, C.; Wagnieres, G.; Ihringer, F.; Monnier, Ph.; van den Bergh, H. *J. Photoch. Photobio. B.* 1997, 38, 196.

[82] Kaneko, Y.; Nishimura, Y.; Takane, N.; Arai, T.; Sakuragi, H.; Kabayashi, N.; Matsunaga, D.; Pac, C.;Takumaru, K. *J. Photoch. Photobio. A.* 1997, 106, 177.

[83] Wiktorowicz, K.; Waszkowiak, A.; Cofta, J.; Frąckowiak, D. *Curr. Top. Biophys.* 2003, 27, 37.

[84] Wróbel, D.; Boguta, A. *J. Photoch. Photobio. A.* 2002, 150, 67.

[85] McHugh, A. J.; Gouterman, M.; Weiss, C. *Theor. Chim. Acta* 1972, 24, 346.

[86] Gouterman, M.; Wagniere, G. H.; Snyder, L. C. *J. Mol. Spectrosc.* 1963, 11, 2.

[87] Tackley, D. R.; Dent, G.; Smith, W. E. *Phys. Chem. Chem. Phys.* 2001, 3, 1419.

[88] VanCott, T.; C Rose, J. L.; Misener, G. C.; Williamson, B. E.; Schrimpf, A. E.; Boyle, M. E.; Schatz, N. *J. Phys. Chem.* 1989, 93, 2999.

[89] Mack, J.; Stillman, M. J. *J. Phys. Chem.* 1995, 99, 7935.

[90] Henriksson, A.; Roos, B.; Sundbom, M. *Theor. Chim. Acta* 1972, 27, 303.

[91] Edwards, L.; Gouterman, M. *J. Mol. Spectrosc.* 1970, 33, 292.

[92] Kiet, A., N.; Pachter, R., *J. Chem. Phys.* 2001, 114 (24) 10757.

[93] Liao, M. S.; Scheiner, S. *J. Chem. Phys.* 2001, 114, (22) 9780.

[94] Pope, M.; Swenberg, C. E. *Electronic Processes in Molecular Crystals*; Clarendon Press: Oxford, 1982.

[95] Davydov, A.S. *Theory of molecular excitons*; McGraw-Hill Book Company: New York, 1962.

[96] Nuesch, F.; Carrara, M.; Schaer, M.; Romero, D. B.; Zuppiroli, L. *Chem. Phys. Lett.* 2001, 347, 311.

[97] Stendal, A.; Beckers, U.; Wilbrandt, S.; Stenzel, O.; von Borczyskowski, C. *J. Phys. B: At. Mol. Opt. Phys.* 1996, 29, 2589.

[98] Mi, J.; Guo, L.; Liu, Y.; Liu, W.; You, G.; Qian, S. *Phys. Lett.* A 2003, 310, 486.

[99] Stenzel, O.; Wilbrant, S.; Stendal, A.; Beckers, U.; Voigtsberger, K.; von Borczyskowski, C. *J. Phys. D*: 1995, 28, 2154.

[100] Wojdyła, M.; Derkowska, B.; Rębarz, M.; Bratkowski, A.; Bała, W. *J. Opt. A: Pure Appl. Opt.* 2005, 7, 463.

[101] Kumara, G. A.; Thomasa, J.; Unnikrishnanb, N. V.; Nampooria, V. P. N.; Vallabhana, C. P. G. *Mater. Res. Bull.* 2001, 36, 1.

[102] Seoudi, R.; El-Bahy, G. S.; El Sayed, Z. A. *Opt. Mater.* 2006, in Press.

[103] Senthilarasu, S.; Sathyamoorthy, R.; Lalitha, S.; Subbarayan, A.; Natarajan, K. *Sol. Energ. Mat. Sol. C.* 2004, 82, 179.

[104] Ruan, Ch.; Mastryukov, V.; Fink, M. *J. Chem. Phys.* 1999, 111, 3035.

[105] Scheidt, W. R.; Dow, W. *J. Am. Chem. Soc.* 1977, 99, 1101.

[106] Milhill, A.; Buell, W.; Fink, M. *J. Chem. Phys.* 1993, 111, 6416.

[107] Tackley, D. R.; Dent, G.; Smith, W. E. *Phys. Chem. Chem. Phys.* 2000, 2, 3949.

[108] Janczak, J.; Kubiak, R. *Polyhedron* 2001, 20, 2901.

[109] Rieckhoff, K. E.; Menzel, E. R.; Voigt, E. M. *Phys. Rev. Lett.* 1972, 28 (5) 261.

[110] Menzel, E. R.; Jordan, K. *J. Chem. Phys.* 1978, 32, 223.
[111] Kobitski, Y. (2002). *Time-resolved Photoluminescence and Theoretical Study of Excitons in PTCDA* [pdf]. http://deposit.ddb.de/cgi-bin/dokserv?idn=968799213
[112] Bulovic, V.; Burrows, P.E.; Forest, S.R.; Cronin, J.A. Thompson, M.E. *Chem. Phys.* 1996, 210, 1.
[113] Misiewicz, J.; Sitarek, P.; Sęk, G. *Introduction to the photoreflectance spectroscopy of semiconductor structures;* OWPW: Wrocław, 1999.
[114] Knupfer, M.; Schwieger, T.; Piersert, H.; Fink, *J. Phys. Rev.* B 2004, 69, 165210.
[115] Derkowska, B.; Wojdyła, M.; Płóciennik, P.; Sahraoui, B.; Bała, W. *Opto-electron. Rev.* 2004, 12, 405.
[116] Wojdyła, M.; Derkowska, B.; Bała, W.; Bratkowski, A.; Korcala, A. *Opt. Mater.* 2005, in Press
[117] Haraguchi, M.; Nakagawa, Y.; Fukui, M. Superlattice. *Microst.* 1993, 13 (3), 341.
[118] Misiewicz, J.; Sęk, G.; Kudrawiec, R.; Sitarek, P. *Thin Solid Films* 2004, 450, 14.
[119] Kuzmenko. R.; Ganzha, A.; Schreiber, J.; Hildebrandt, S. *Phys. Solid State* 1997, 39 (12), 1900.
[120] Hildebrandt, S.; Murtagh, M.; Kuzmenko, R.; Kircher, W.; Schreiber, *J. Phys. Status Solidi* 1995, 152, 147.
[121] Minkov, D. A. *J. Phys. D: Appl. Phys.* 1989, 22, 1157.
[122] Stenzel, O.; Petrich, R. *J. Phys. D: Appl. Phys.* 1995, 28, 978.
[123] Senthilarasu, S.; Velumani, S.; Sathyamoorthy, R.; Subbarayan, A. ; Ascencio, J.A.; Canizal, G.; Sebastian, P.J.; Chavez, J.A.; Perez, R. *Appl. Phys.* A 2003, 77, 383.
[124] Chen, C.H.; Shi, J. Coordin. Chem. Rev. 1998, 171, 161.
[125] Riechel, S. (2002). Organic semiconductor lasers with two-dimensional distributed feedback [pdf].
a. http://edoc.ub.uni-muenchen.de/archive/00000172/01/Riechel_Stefan.pdf [126] Ravi Kishore, V. V. N.; Periasamy, N.; Bhattecharjee, B.; Das, R.; Paulose, P. L.; Narasimhan, K. L. *Chem. Phys. Lett.* 2003, 367, 657.
[126] Aziz, A.; Narasimhan, K. L. *Synthetic Metals*, 2001, 122, 53.
[127] Halls, M. D.; Schlegel, H. B. *Chem. Mater.* 2001, 13, 2632.
[128] Ravi Kishore, V. V. N.; Aziz, A.; Narasimhan, K. L.; Periasamy, N.; Meenakshi, P. S.; Wategaonkar, S. *Synthetic Met.* 2002, 126, 199.
[129] Stampor, W.; Kalinowski, J.; Marconi, G.; Marco, P. Di.; Fattori, V.; Giro, G. *Chem. Phys. Lett.* 1998, 283, 373.
[130] Akai, I.; Harai, K.; Kouno, K.; Karasawa, T.; *J. Lumin.* 2004, 108, 11.
[131] Brinkmann, M.; Gadret, G.; Muccini, M.; Taliani, C.; Masciocchi, N.; Sironi, A. *J. Am. Chem. Soc.* 2000, 122, 5147.
[132] Lozykowski, H. J.; Shastri, V. K. *J. Appl. Phys.* 1991, 69, 3235.
[133] Bała, W.; Łukasiak, Z.; Rębarz, M.; Dalasinski, P.; Bratkowski, A.; Bauman, D.; Hertmanowski, R. *Opto-Electron. Rev.*, 2004, 12, 445.
[134] Guha, S.; Cai, Q.; Chandrasekhar, M.; Chandrasekhar, H. R.; Kim, H.; Alvarenga, A. D.; Vogelgesang, R.; Ramdas, A. K.; Melloch, M. R. *Phys. Rev. B* 1998, 58, 7222.
[135] Kozlov, V. G.; Forrest, S. R. *Curr. Opin. Solid St. M.* 1999, 4, 203.
[136] Bulovic, V.; Deshpande, R.; Thompson, M. E.; Forrest, S. R. *Chem. Phys. Lett.* 1999, 308, 317.
[137] Swanepoel, R. *J. Phys. E Sci. Instrum.* 1983, 16, 1214.

[138] Marquez, E.; Gonzalez-Leal, J. M.; Jimenez-Garay, R.; Lukic, S. R.; Petrovic, D. M. *J. Phys. D Appl. Phys*. 1997, 30, 690.

[139] Ramirez-Malo, J. B.; Marquez, E.; Villarez, P.; Jimenez-Garay, R. *Phys. Solid State* 1992, 133, 499.

[140] So F.F.; Forrest, S.R. *Mol. Cryst. Liq. Crys. B* 1992, 2, 205.

[141] Bulovic, V.; Burrows, P.E.; Forest, S.R.; Cronin, J.A.; Thompson, M.E. *Chem. Phys.* 1996, 210, 1.

[142] Hoffmann, M.; Schmidt, K.; Fritz, T.; Hasche, T.; Agranovich, V.M.; Leo, K. *Chem. Phys*. 2000, 258, 73.

[143] Tanaka, J. B. *Chem. Soc. Jpn*. 1963, 36, 1237.

[144] Song, K.S.; Williams, R.T. *Self - trapped excitons*; Springer: Berlin, 1993.

[145] Kobitski, A. Yu.; Scholz, R.; Zahn, D. R. T.; Wagner, H. P. *Phys. Rev*. B 2003, 68, 155201.

[146] Hertmanowski, R.; Chudziński, Ł.; Martyński, T.; Stempniewicz, P.; Wolarz, E.; Bauman, D.; *Liq. Cryst*, 2004, 31, 791.

[147] Hochstrasser, R.M. *Rev. Mod. Phys*. 1962, 34(3), 531.

In: Trends in Optical Materials Research
Editor: Owen G. Gardens, pp. 55-90
ISBN: 1-60021-290-5

Chapter 2

PHOTOLUMINESCENCE AND ELECTROLUMINESCENCE STUDIES ON THE MORPHOLOGY OF ORGANIC MOLECULES IN THIN SOLID FILMS

Taiju Tsuboi*
Faculty of Engineering, Kyoto Sangyo University,
Kamigamo, Kita-ku, Kyoto 603-8555, Japan

ABSTRACT

Photoluminescence, excitation and absorption spectra are investigated for thin films lightly and heavily doped with TPD, Ir(ppy)$_3$, Alq$_3$, PtOEP and mCP and for these neat films at various temperatures between 10 and 290 K. The photoluminescence spectra are compared with the electroluminescence spectra of organic light emitting diodes with emitting layers of Ir(ppy)$_3$ and PtOEP. Aggregates such as dimer and trimer are observed in the neat films of Ir(ppy)$_3$ and Alq$_3$. On the other hand, dimer and trimer are formed in TPD, PtOEP- and mCP-doped films and these neat films, together with monomer. In the case of PtOEP, large size aggregates which consist of many PtOEP molecules are also formed even in lightly doped films. Such a difference between the group of Ir(ppy)$_3$ and Alq$_3$ and the group of TPD, PtOEP and mCP is caused by difference of competition and balance among solid state solvation effect, dipole-dipole interactions between dopant molecules, and intermolecular covalent bonding. It is suggested that molecular geometrical structure gives rise to the difference.

1. INTRODUCTION

Organic light emitting diode (OLED) is used for creating flexible, full-color and thin flat panel display. Comparing with the currently popular liquid-crystal display, the most peculiar advantage of the OLED display is the possibility of roll-up display on the curved surface. The

OLED display is considered as a near future display with high performance including high brightness, fast response, light weight, and low voltage operation.

The display is an application of electroluminescence (EL) generated from OLEDs. The OLEDs have the multi-thin-layer structure that consists of metal cathode, electron injection layer, transport layer, hole blocking layer, emitting layer, hole transport layer, hole injection layer, and transparent indium-tin-oxide (ITO) anode. The materials of these layers are organic semiconductors with low carrier mobility of 10^{-3}-10^{-7} cm^2/Vs. Applied voltage to OLEDs induces the recombination of the injected electrons and holes in the emitting layer, giving rise to EL from the emitting materials.

Thin layers in OLED devices are fabricated by thermal evaporation in vacuum, spin-coating, or ink-jet printing. The optical properties of OLEDs have been explained using the properties of isolated organic molecules so far although each molecule is expected to have interaction with not only neighboring molecules in the confined layer but also molecules of neighboring layers. It has not been confirmed whether each molecule is present independently from neighboring molecules without formation of aggregate (e.g. dimer grain and island-like structure) in amorphous film and whether the constituent molecules form a lattice that repeats periodically in space. The electrical and optical properties of OLEDs depend on not only the location and distribution of each molecule but also the structure and geometry of aggregated molecules in thin film because they change the structure and alignment of the electronic energy levels of the constituent molecules. Therefore it is necessary to clarify the morphology of molecules in thin amorphous films, i.e. to clarify whether each molecule is distributed randomly without strong interaction between the neighboring molecules or several molecules form aggregates. No detailed spectroscopic investigation, however, has been made for the morphology of organic thin films except film of tris(8-hydroxyquinoline) aluminum (Alq_3) doped in N,N'-bis(3-methylphenyl)-N,N'-bis(phenyl)-benzidine (TPD) and of [2-methyl-6-[2-(2,3,6,7 -tetrahydro-1H, 5H-benzo [ij] quinolizin-9-yl) ethenyl]-4H-pyran-4-ylidene] propane-dinitrile (DCM2) doped in Alq_3 [1] and neat film of Alq_3 [2].

Baldo et al have suggested that the polar Alq_3 and DCM2 molecules form ordered domains by the dipole-dipole interaction among the neighboring molecules in less polar molecular host [1]. Using a model of locally ordered domains, they explained the peak shifts of the Alq_3 and DCM2 dopant emission bands which were observed with increasing the concentration of dopant. On the other hand, Brinkmann et al have suggested that there is a large distribution of molecular packings by π-π overlaps between facing ligands in amorphous Alq_3 films [2].

Thin film of Alq_3 shows large Stokes shift (about 130 nm) and nearly Gaussian-shape emission band without any vibronic structure even at low temperature like 10 K as shown in Fig. 1. Such a spectrum is quite different from those of the other organic molecules that show much smaller Stokes shift and asymmetric line shape with sharp drop at shorter wavelength side and well-resolved vibronic structure at the longer wavelength side. Therefore it is not clear whether the model of domain or packing formation for Alq_3 is applicable to the other organic molecules used for OLED devices.

In this Chapter we study the morphology of various organic thin films from the photoluminescence (PL), EL and absorption spectra at not only room temperature but also low temperature like 10 K.

* E-mail: tsuboi@cc.kyoto-su.ac.jp Fax: +81-75-705-1914

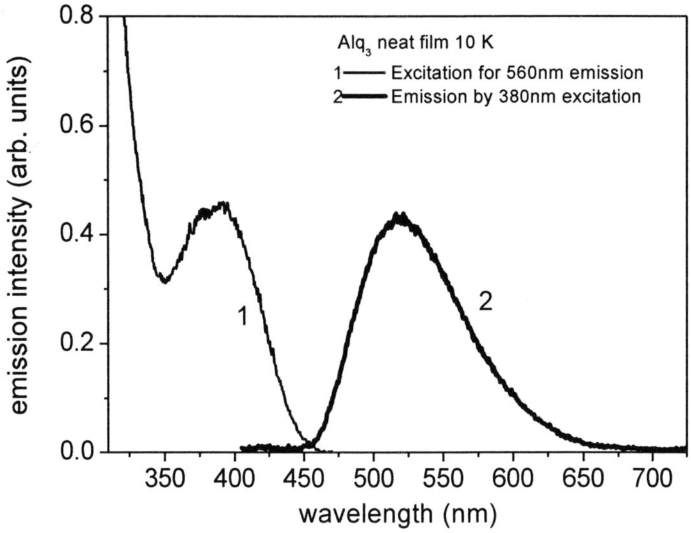

Fig. 1. Photoluminescence and excitation spectra of a 10-nm thick Alq_3 neat film at 10 K. Excitation was made at 380 nm and excitation spectrum was obtained for 560 nm emission.

2. General Characteristics of Emission Spectra of Organic Molecules

Most organic molecules in solution show the absorption and PL spectra with vibronic structure and vibrational progressions. The emission and absorption spectra are nearly mirror images of each other. These characteristics are explained using the adiabatic potential-energy parabola curves with vibrational quantum states for the ground and excited states and using the electronic transitions between the two states under the Franck-Condon principle [3, 4].

The vibronic structure of the absorption band is due to the electronic transitions from the vibrational quantum number n=0 in the ground state to the vibrational quantum numbers m =0, 1, 2, 3, , , in the excited state. The intense line is usually observed for the 0-0 transition (i.e. zero-phonon transition). The vibronic structure of the emission band is due to the electronic transitions from the vibrational quantum number m=0 in the excited state to the vibrational quantum numbers n=0, 1, 2, 3, , , , in the ground state. The intense emission line is usually the 0-0 line and the 0-1, 0-2, 0-3 lines locate at the lower energy side of the 0-0 line.

Such an emission line shape is also observed in organic solid films as seen in Fig. 2 where the absorption and fluorescence spectra due to 4,4'-N,N'-dicarbazole-biphenyl (CBP) in film are presented. The vibronic structure of 0-0, 0-1 and 0-2 lines is clearly observed in the two spectra together with usual structure of 0-0 line higher than 0-1 line. Mirror image is found between the absorption and emission spectra, which are plotted against the photon energy, except a region of low-energy tail of the absorption spectrum. The long low-energy tail arises from the weak absorption bands due to triplet state. Observation of the mirror image indicates that the absorption bands of CBP observed at 325-370 nm arise from the same S_1 singlet state. The mirror symmetry also indicates that only one center (i.e. CBP monomer) is responsible for the spectra and additional several centers such as dimer and aggregate are not.

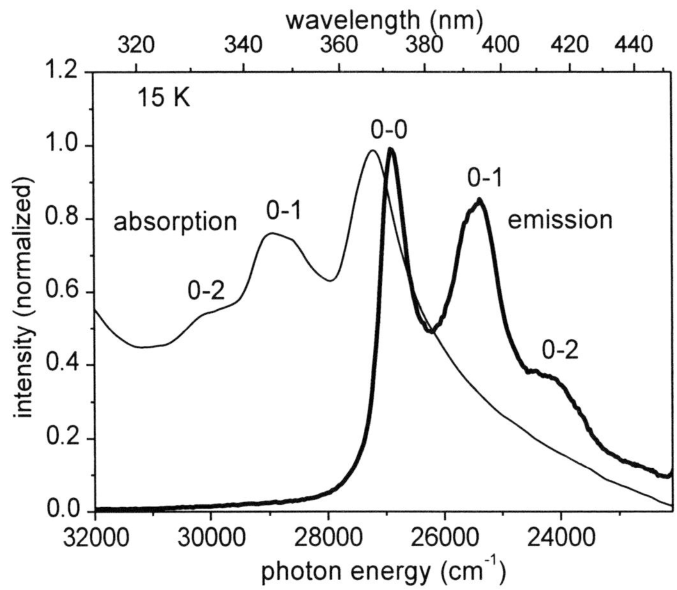

Fig. 2. Absorption and photoluminescence (PL) spectra of CBP at 15 K. The PL spectrum was obtained by excitation at 300 nm. The vibrational lines are indicated as 0-0, 0-1 and 0-2 (see text).

When we measure the PL and EL spectra of various organic thin films, however, we realized that such an idealized emission spectrum is not observed for all organic materials. In the following sections we show several examples.

3. Emission Spectra of TPD

Figure 3 shows the PL and excitation spectra of 100 % TPD film (i.e. neat film) of a thickness of 91 nm. The film was spin-coated on the glass plate. The PL spectrum at 10K shows the usual vibronic structure where the peak intensity decreases as going from the 0-0 line to 0-1 line and to 0-2 line. The 0-0 line peak locates at 404 nm and never changes the peak position for variation of temperature, while the 0-1 line shifts to short wavelength with increasing temperature from 10 K to 290 K, i.e. the peak is at 426 and 417 nm at 10 and 290 K, respectively. The peak-height ratio of the 0-1 line to the 0-0 line increases considerably with increasing temperature from 10 K to 290 K. At 290 K, the 0-1 line height is nearly the same as the 0-0 line height. The ratio of the 0-1 line to the 0-2 line, however, changes little. The absorption cross-section spectrum at 290 K is shown in curve 9 [5]. From the absorption spectrum, it is suggested that such a rapid decrease of the 0-0 line is caused by the self-absorption as discussed later.

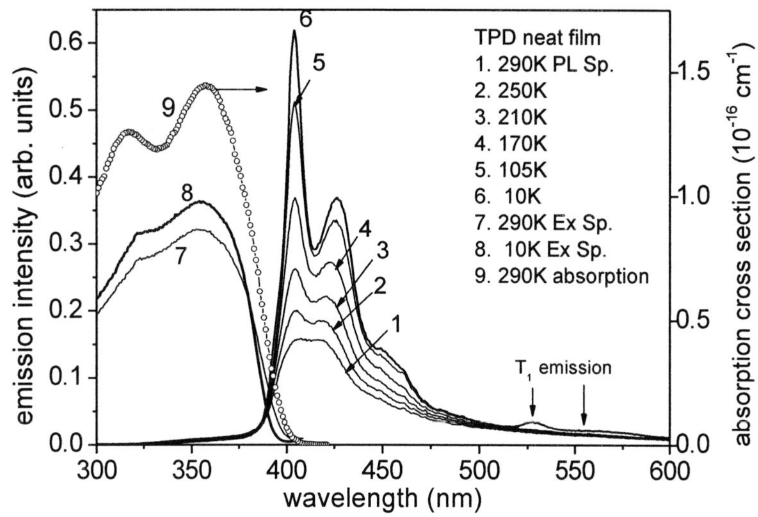

Fig. 3. Photoluminescence spectra of a neat TPD film at various temperatures (curves 1-6), which was excited at 230 nm, together with the excitation spectra for 430 nm emission at 290 and 10 K (curves 7 and 8, respectively) and the absorption cross section spectrum (curve 9, right scale). Vertical arrows indicate the emission from the triplet state T_1.

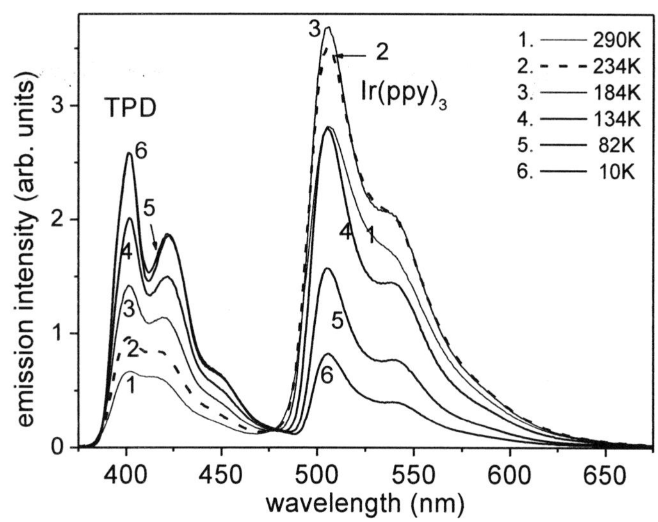

Fig. 4. Photoluminescence spectra of a 60-nm thick film of TPD doped with 5wt% $Ir(ppy)_3$ at various temperatures, which was excited at 360 nm.

Emission bands appear at 520-600 nm at low temperatures with decreasing temperature from 150 K. The intensity increases with decreasing temperature as shown in Fig.3. These weak bands are attributed to the phosphorescence due to the transition from the triplet state T_1 to the ground state.

Figure 4 shows the PL spectra of a spin-coated film of 5wt% *fac* tris(2-phenylpyridine) iridium [$Ir(ppy)_3$] doped in TPD at various temperatures between 10 and 290 K. The emission band due to $Ir(ppy)_3$ is observed at 500-650 nm [6] besides the TPD band. Almost the same TPD emission spectra as Fig.3 are obtained, except (1) the intensity ratio of the 0-0 and 0-1 line at 290 K, i.e. the 0-0 line intensity is higher than the 0-1 line and (2) the peak position of the 0-0 line is at 402 nm that shifts to the shorter wavelength from 404 nm of the neat TPD film. Unlike the case of TPD, the intensity ratio between the 0-0 and 0-1 lines does not change in the Ir(ppy)$_3$ emission for the variation of temperature.

In Figs. 3 and 4 are shown the PL spectra of TPD which is doped as host. We compare them with the spectra of TPD doped as guest. Figure 5 shows the PL spectra of a spin-coated film of 5wt% TPD doped in polystyrene (PS). The emission intensity increases with increasing temperature from 10 K to 290 K. The intensity ratio of 0-1 line to the 0-0 line change with increasing temperature from 10 K, but, unlike the case of 100% TPD film, the 0-0 line is still higher at 290 K than the 0-1 line.

The peak of the 0-0 line in Fig.5 is at 397 nm. It is noted that this peak wavelength was largely blue-shifted from the 404 nm 0-0 line of the 100% TPD film. The same is true for the 0-1 line, i.e. the 0-1 line peak is at 417 nm at 10 K for the dopant TPD, while at 426 nm at 10 K for the 100% TPD. There is peak shift but no change in the whole spectral line shape between the two films.

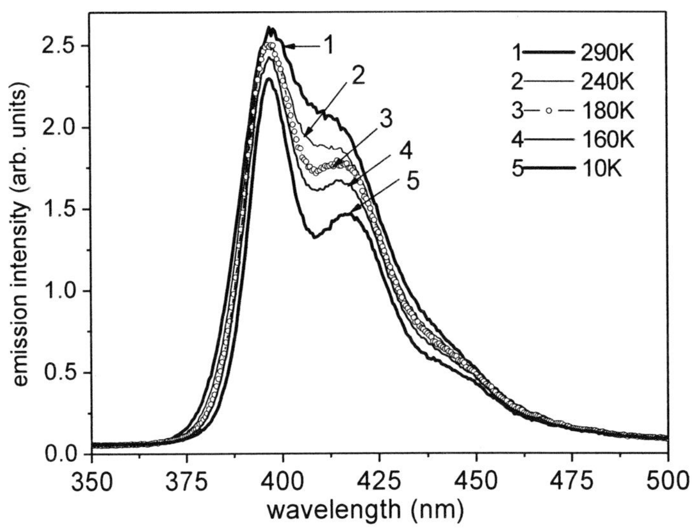

Fig. 5. Photoluminescence spectra of TPD in a 90-nm thick film of polystyrene (PS) doped with 5wt% TPD at various temperatures, which was excited at 360 nm.

Figure 6 summarizes the emission spectra of the three films of 5wt% TPD doped in PS (called film TPD-A), TPD doped with 5wt% Ir(ppy)$_3$ (film TPD-B) and 100% TPD at 10 K. It is clearly observed by excitation at 360 nm that the intense peak wavelength is changed among the three films, i.e. 397, 402 and 404 nm for films TPD-A, TPD-B and neat film, respectively, and that the whole spectrum shows a red-shift with increasing the TPD concentration. It is found that both TPD-B and neat films have a shoulder at 397 nm which is the same position as the 0-0 line observed in the film TPD-A. The 397 nm band is attributed to monomer. Discussion is given in Sec.17 on the assignment of the emission bands of TPD-B and neat film.

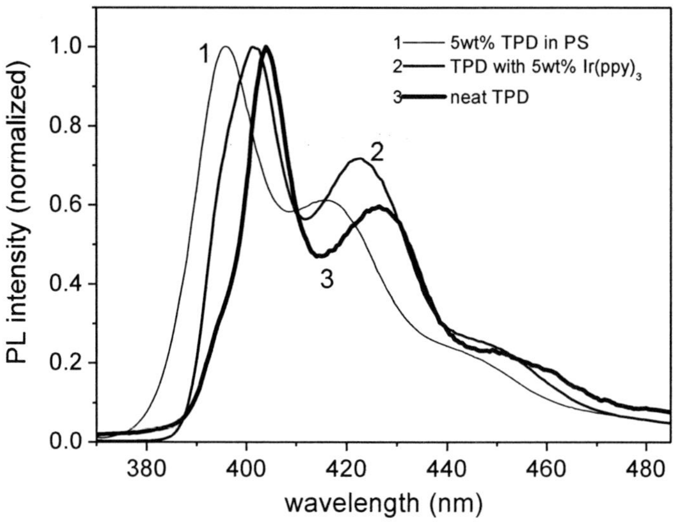

Fig. 6. Photoluminescence (PL) spectra of the spin-coated films of 5wt% TPD doped in polystyrene (PS), TPD doped with 5wt% $Ir(ppy)_3$ and 100% TPD at 10 K, which were excited at 360 nm.

4. Self-Absorption or Appearance of Dimer ?

As shown in Fig.3, the 0-0 line has the almost same peak height as 0-1 line at 290 K. This is quite different from the cases of low temperatures. The absorption band tail of TPD extends to longer wavelength (up to about 430 nm) with increasing temperature from 10 K to room temperature as seen in curves 8 and 9 of Fig.3. Therefore self-absorption is conceivable as one of the reasons for this unusual line shape. To check this possibility, we calculate the PL line shape corrected with the self-absorption.

The PL intensity A(λ) (λ : emission wavelength) emitted from the molecules in film is changed to the PL intensity B(λ) after the absorption k(λ) in the film of thickness d. The observed spectrum is B(λ), which is expressed by

$$B(\lambda) = A(\lambda)[1-\exp(-k(\lambda)d)]/k(\lambda)d.$$

From this equation we can derive the A(λ) line shape before self-absorption.

Figure 7 shows the room temperature absorption and PL spectra (curves 1 and 2, respectively) of TPD neat film spin-coated on glass substrate. Curve 3 shows the PL spectrum A(λ), i.e. line shape corrected with absorption. The main 0-0 line peak of the derived PL spectrum is at 402 nm, which is shifted from the observed peak wavelength at about 407 nm. This line shape is similar to the PL spectrum observed at 220 K. Thus quite reasonable PL spectrum is derived after the correction.

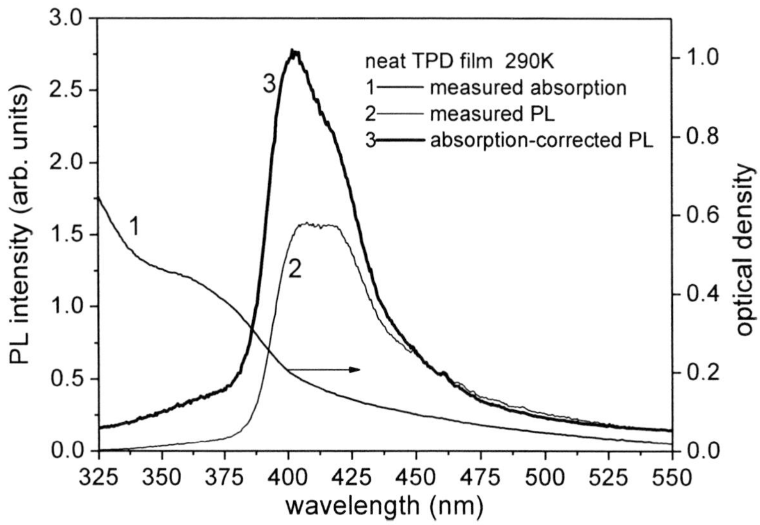

Fig. 7. Absorption spectrum of neat TPD film at 290 K (curve 1), and the observed and expected emission spectra of TPD which are obtained from the same film at 290 K (curves 2 and 3, respectively) Curve 3 was derived after correction of self-absorption. The absorption spectrum contains the absorption by glass substrate.

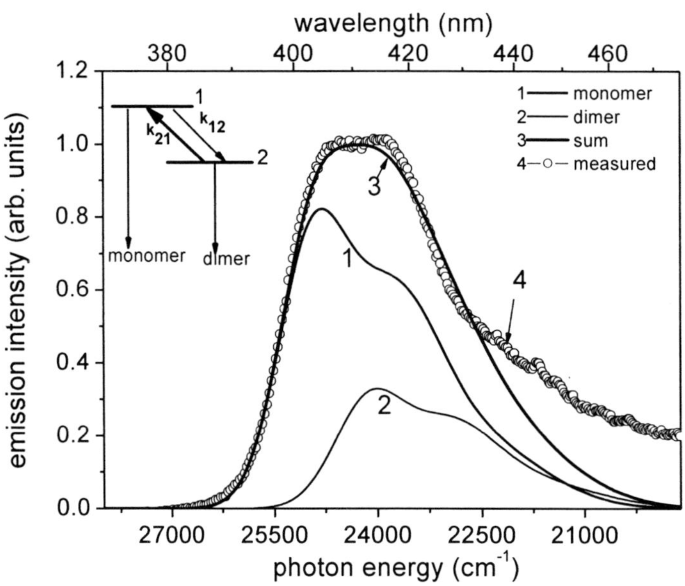

Fig. 8. The photoluminescence (PL) spectra of monomer and dimer of TPD (curves 1 and 2, respectively) and the sum of the two spectra (curve3), compared with the PL spectrum of TPD neat film observed at 290 K (curve 4) (see text). Energy transfer between the emitting states of the monomer and dimmer is shown in inset. The parameter k_{12} is the transfer rates from the monomer to dimmer, and k_{21} is the reversed rate.

Another reason might be possible for the unusual line shape of TPD observed at 290 K in neat film. It is formation of dimer and aggregate of TPD molecules. We assumed that monomer and dimer coexist in the neat film, and their PL band line shapes are the same but the band of dimer is red-shifted by 800 cm^{-1} from the monomer band whose 0-0 line is at 24950 cm^{-1} (corresponding to 400.8 nm). The line shape due to the monomer was estimated from the PL spectrum at 290 K of Fig.5. A good fit is obtained between the calculation and observation as shown in Fig.8 where we choose a value of 0.4:1 as the intensity ratio of the dimer band to the monomer band.

This dimer model also gives a good fitting of the PL spectrum at 10 K by changing the intensity ratio to 4:1. Unlike the case of 290 K, it is estimated that the dimer-band intensity is much larger than the monomer-band intensity at 10 K. This is reasonable because the excited molecules are relaxed to the lower emitting state of dimer at low temperature, while the endothermic energy transfer occurs from the dimer level to the upper monomer level with increasing temperature (see inset of Fig.8), resulting in the decrease of dimer emission at 290 K. However, this model gives a deviation from the observed spectrum at the low energy side as seen in Fig.8. Although additional conformation such as trimer would be necessary to obtain better fitting, this model neglects the self-absorption effect. It is needed to calculate the line shape taking into account both self-absorption and aggregates.

5. Photoluminescence Spectra of $Ir(ppy)_3$

The PL spectra of $Ir(ppy)_3$ doped in TPD host are shown in Fig.4. Figure 9 shows the PL and excitation spectra of a spin-coated neat $Ir(ppy)_3$ film. As the case of the neat film of TPD, the emission intensity decreases with increasing temperature from 10 K to 290 K. Vibronic structure is also observed, but the intensity ratio of the 0-0 line to the 0-1 line changes drastically for the variation of temperature. The 0-1 line is higher at 140-220 K than the 0-0 line. The peak height of the 0-0 line is a little higher than that of the 0-1 line at 290 K, but almost comparable to each other. This behavior is different from the case of $Ir(ppy)_3$ doped in TPD as seen in Fig.4.

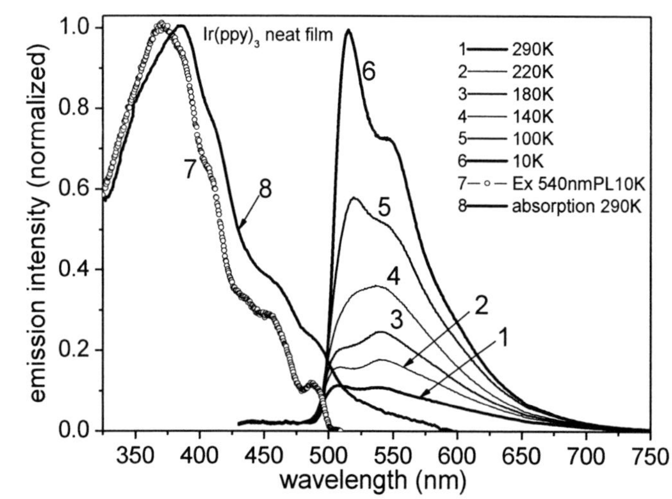

Fig. 9. Photoluminescence (PL) spectra of a 60-nm thick film of $Ir(ppy)_3$ neat film at various temperatures (curves 1-6), which was excited at 360 nm, together with excitation (Ex) spectrum for 540 nm emission at 10 K (curve 7) and absorption spectrum of the neat film at 290 K (curve 8).

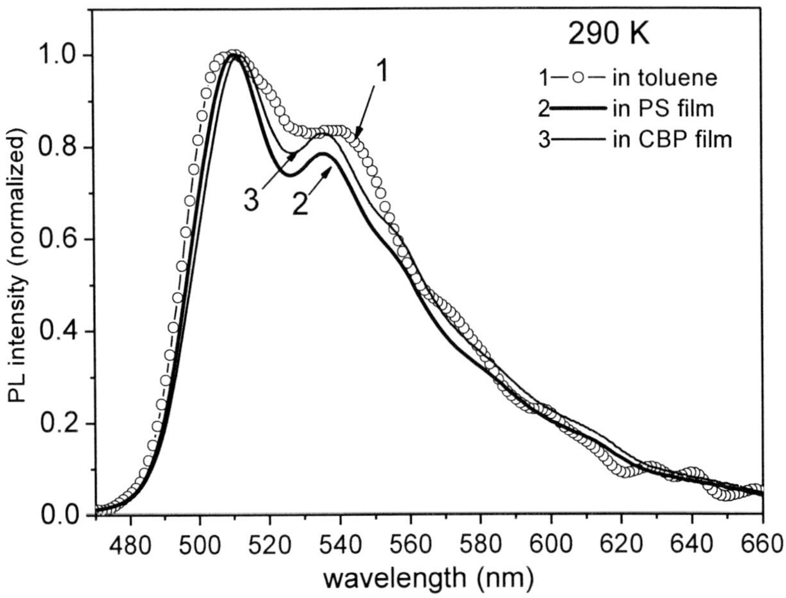

Fig. 10. Photoluminescence spectra of Ir(ppy)$_3$ in toluene solvent (curve 1), in spin-coated polystyrene (PS) film doped with 4 wt% Ir(ppy)$_3$ (curves 2) and CBP film doped with 7.9 wt% Ir(ppy)$_3$ (curves 3), which were measured at room temperature.

The excitation spectrum for the Ir(ppy)$_3$ emission at 10 K is shown in curve 7 in Fig.9, while the absorption spectrum at 290 K is shown in curve 8. From these spectra, it is suggested that the decrease of the 0-0 line intensity, which is observed at high temperatures, is caused by the self-absorption because of strong overlap of the absorption tail with the emission band at high temperatures. In the case of Ir(ppy)$_3$ doped as guest, the 0-0 line never becomes lower than the 0-1 line by changing temperature as shown in Fig.4. This is understood as follows. The absorption intensity of Ir(ppy)$_3$ is too weak in the lightly doped film to give a strong effect of self-absorption on the PL intensity.

Figure 10 shows the PL spectra of Ir(ppy)$_3$ doped in PS and CBP and in toluene solution at 290 K. Unlike the case of the neat film at 290 K, these doped Ir(ppy)$_3$ molecules show the usual vibronic structure, i.e. the 0-1 line intensity is smaller than the 0-0 line one, which is consistent with the case of Ir(ppy)$_3$ doped in TPD (Fig.4). These vibronic structures are similar to each other in the three samples although the intensity ratio of the 0-0 line to the 0-1 line is a little different among them. Only the 0-0 and 0-1 emission lines are observed for Ir(ppy)$_3$ doped in TPD (spin-coated film, at 290 K, see Fig.4), while not only the 0-0 and 0-1 lines but also 0-2, 0-3 and 0-4 lines are observed in Fig.10. This suggests that the resolution in vibronic structure of Ir(ppy)$_3$ depends on the host molecules.

6. Electroluminescence Spectra of Ir(ppy)$_3$

The EL spectra of Ir(ppy)$_3$ were measured using two kinds of OLEDs #Ir1 and #Ir4 with emitting layer of 2.9 and 5.8 wt% Ir(ppy)$_3$-doped CBP, respectively. These devices have the following multi-layer structures.

#Ir1: ITO/α-NPD/2.9wt%Ir(ppy)$_3$:CBP (35nm)/BCP/Alq$_3$ /Al
#Ir4: ITO/α-NPD/ 5.8wt%Ir(ppy)$_3$:CBP(35nm)/BCP/Alq$_3$ /Al.

The layer thickness of the emitting layer is indicated in the parenthesis. Layers of 2,9-dimethyl-4,7-diphenyl -1,10-phenanthroline (BCP) and Alq$_3$ are used as hole-blocking and electron-transport layers, respectively. The devices are sealed by the glass plate to avoid water vapor and oxgen. In Fig.11 the EL and PL spectra of #Ir4 are shown. Similar spectra are also obtained for #Ir1, but the 0-1 intensity relative to the 0-0 line is higher in #Ir4 than in #Ir1. This is understood as follows. The more strong self-absorption effect occurs in #Ir4 than in #Ir1 because the concentration of #Ir4 is higher by two times than that of #Ir1.

When these devices are excited with 380 nm light which was introduced through the ITO transparent anode, it is observed that the PL spectra of Ir(ppy)$_3$ are almost the same as EL spectra. Additionally a very weak and broad PL band with at peak about 450 nm is observed. This is due to N, N'-bis(1-naphtyl)-N, N'-diphenyl-1,1'-biphenyl-4,4'-diamine (α-NPD) used as hole transport layer, whose chemical structure is shown in Fig.12, together with the other organic molecules used in this Chapter. In the EL spectrum of #Ir1 and #Ir4 the α-NPD emission band is also observed although its intensity is much weaker than the case of PL.

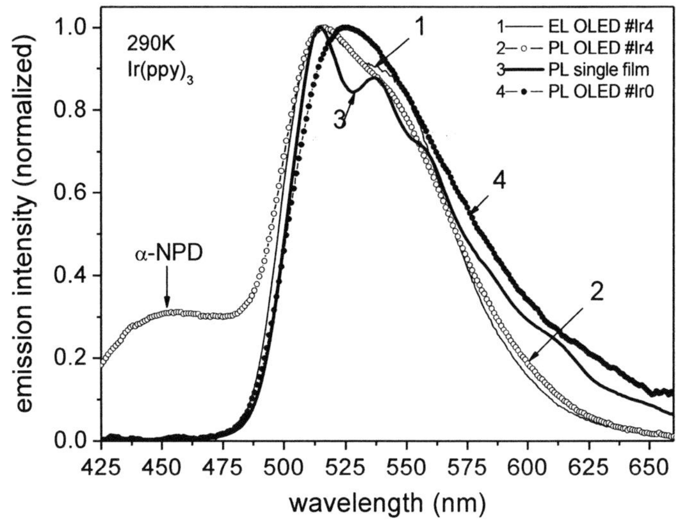

Fig. 11. Electroluminescence (EL, curve 1) and photoluminescence (PL, curve 2) spectra of the OLED #Ir4 with emitting layer of CBP doped with 5.8wt% Ir(ppy)$_3$ at 290 K, compared with PL (curve 3) of a single layer film of CBP doped with 7.9wt% Ir(ppy)$_3$, which was spin-coated on galss plate, at 290 K. Curve 4 shows EL of OLED #Ir0 with emitting layer of 100% Ir(ppy)$_3$. The EL was obtained at 9 V. The PL of OLED and single layer was obtained by excitation at 365 nm. The 450 nm emission band in curve 2 is due to the emission from hole transport layer of α-NPD.

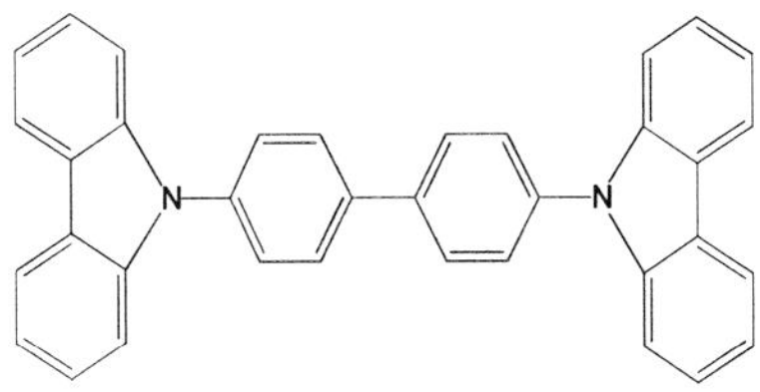

CBP

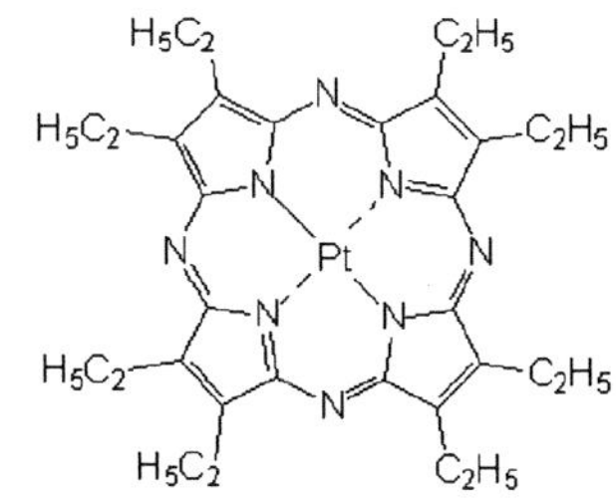

PtOEP

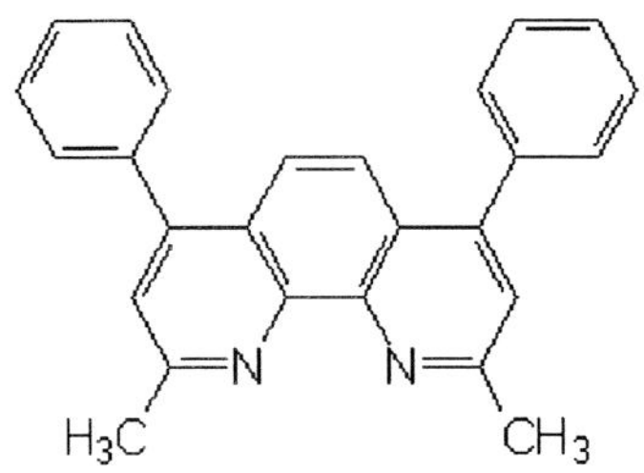

BCP

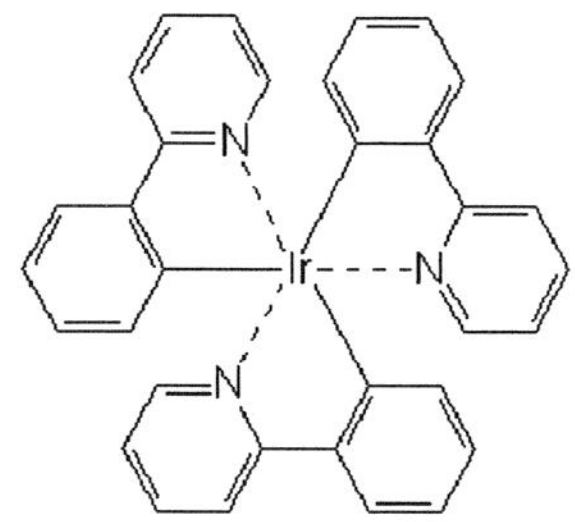

Ir(ppy)3

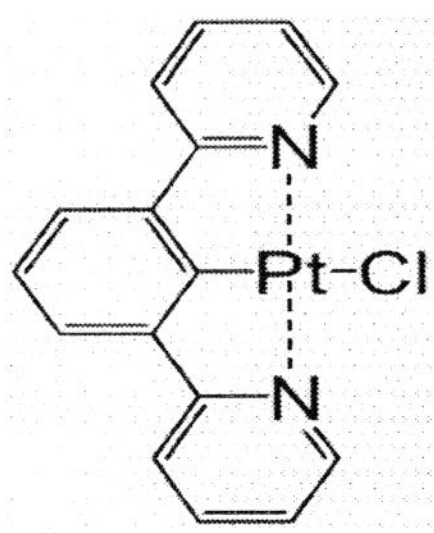

N^C^N-Pt(Cl)

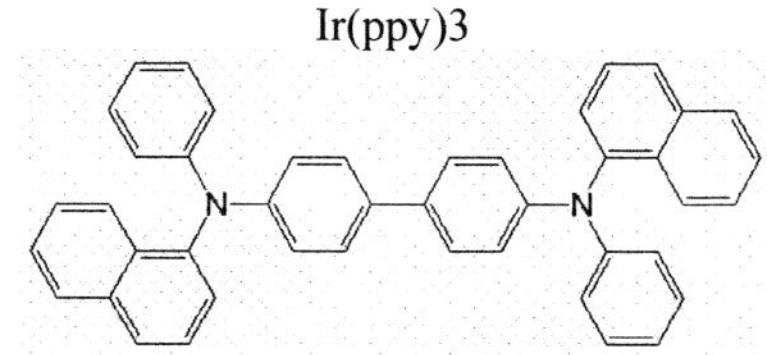

α-NPD

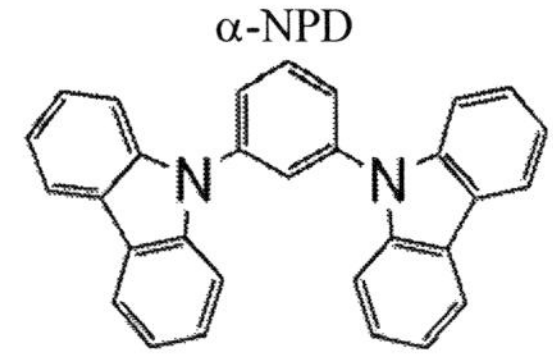

mCP

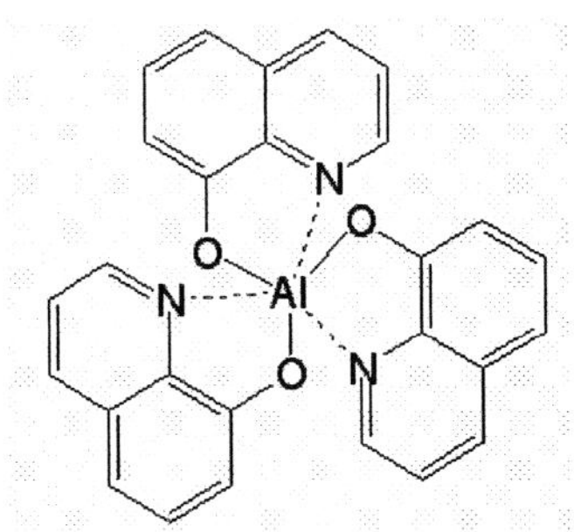

Alq_3

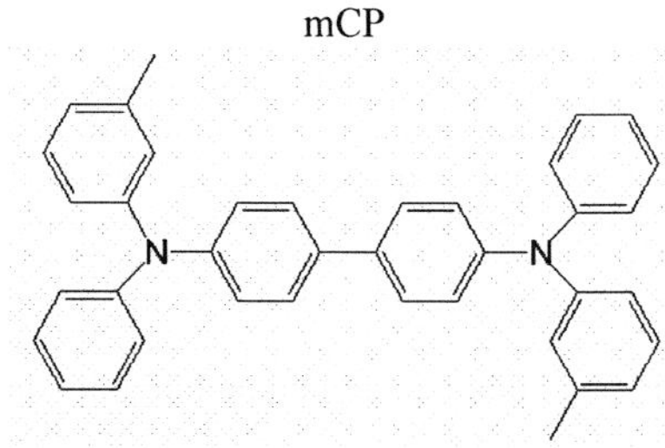

TPD

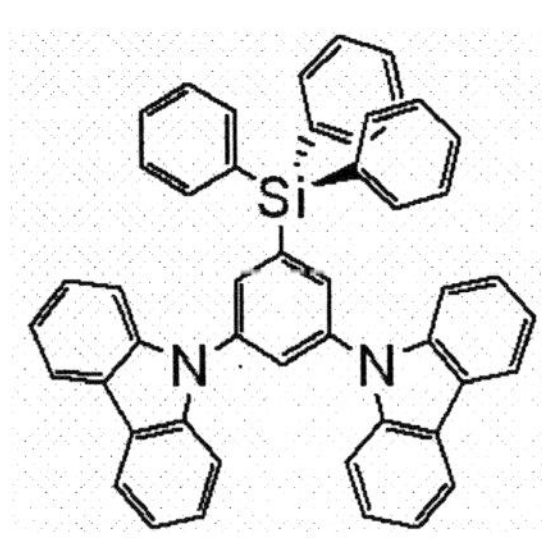

SimCP

Fig. 12. Chemical structures of organic molecules

The room temperature PL spectrum of a single layer film of CBP doped with 7.9 wt% $Ir(ppy)_3$ is also shown in curve 3 of Fig.11 [7]. This film was spin-coated on glass substrate. It is noted that the well-resolved vibronic structure appears in this single layer $Ir(ppy)_3$:CBP film than in the emitting $Ir(ppy)_3$:CBP layer of the multi-layer structured OLEDs (curves 1 and 2 of Fig.11). Regarding this difference between the same $Ir(ppy)_3$-doped CBP films, we discuss in Sec.15.

Curve 4 of Fig.11 shows the EL spectrum of OLED device with emitting layer of 100% $Ir(ppy)_3$ (called #Ir0). The #Ir0 device structure is

#Ir0 : ITO/ $Ir(ppy)_3$(50nm)/LiF/Al.

This device has neither electron transport layer nor hole transport layer. It is noted that EL spectrum of #Ir0 has a single structureless band and its band peak is at 528 nm which is largely shifted from the emission peak of #Ir4. The same 528 nm emission band was observed in #Ir0 by photoexcitation at 380 nm. It is suggested that this red-shifted band is caused by (1) strong self-absorption from neat $Ir(ppy)_3$ film in #Ir0 because the absorption is much intense in 100% $Ir(ppy)_3$ film than in the 5.8wt% $Ir(ppy)_3$ doped film of #Ir4, and (2) solvation effect due to different environments of $Ir(ppy)_3$ molecule between #Ir0 and #Ir4. Regarding the solid state solvation effect, we discuss in Sec.13.

7. Emission Spectra of PtOEP

Figure 13 shows the PL spectra of platinum octaethyl porphyrin (PtOEP) in film of PS doped with 5wt% PtOEP, neat film evaporated on glass plate, and multi-layer structure OLED #Pt3 with emitting layer of CBP doped with 5 wt% PtOEP, together with EL spectra of #Pt0 with a single layer of 100% PtOEP and multi-layer structure OLED #Pt5 with emitting layer of 100% PtOEP. The #Pt0, #Pt3 and #Pt5 OLEDs have the following layer structures.

#Pt0: ITO/PtOEP(50nm)/LiF/Al

#Pt3: ITO/α-NPD/5wt%PtOEP: CBP(33nm)/BAlq/Alq_3 / LiF/Al

#Pt5: ITO/α-NPD/100%PtOEP(45nm)/BAlq/Alq_3 / LiF/Al

In #Pt3 and #Pt5, bis(2-methyl-8-quinolinolato) mono (4-phenylphenolato) aluminum (BAlq) is used as hole blocking layer.

The PL spectrum was obtained by excitation at 360 nm, while the EL spectrum was obtained under the applied voltage of 10 V. The EL spectrum of #Pt3 was the same as the PL spectrum of #Pt3. PtOEP molecules doped in PS and CBP give the same PL band (with a peak at about 645 nm) as the case of film of 5wt% PtOEP doped in TPD. The EL of #Pt3 with CBP doped with 5wt% PtOEP shows the same 645 nm band. Taking into account that the 644.5 nm emission band is observed in PtOEP dissolved in dichloromethane (CH_2Cl_2) solvent [10], the 645 nm emission band is attributable to PtOEP monomer.

The neat film and #Pt0 OLED show additional PL band at 654 nm besides the 645 nm band (curves 3 and 4 of Fig.13). Same emission band is also observed in EL of #Pt0. This indicates that the neat film has an additional emission center (e.g. dimer and trimer) which is different from monomer responsible for the 645 nm emission band. Two EL bands are also obtained from the OLED #Pt5. One band is at 645 nm which is the same as the cases of #Pt0 and doped film. The other is at 651 nm (curve 5) which is different from the cases of #Pt0 and neat film although the EL is generated from layer of 100% PtOEP as #Pt0 and neat film. The 651 nm band peak is shifted from the 654 nm band peak of #Pt0. This peak shift is caused by the different concentration ratio of dimer to monomer between #Pt0 and #Pt5.

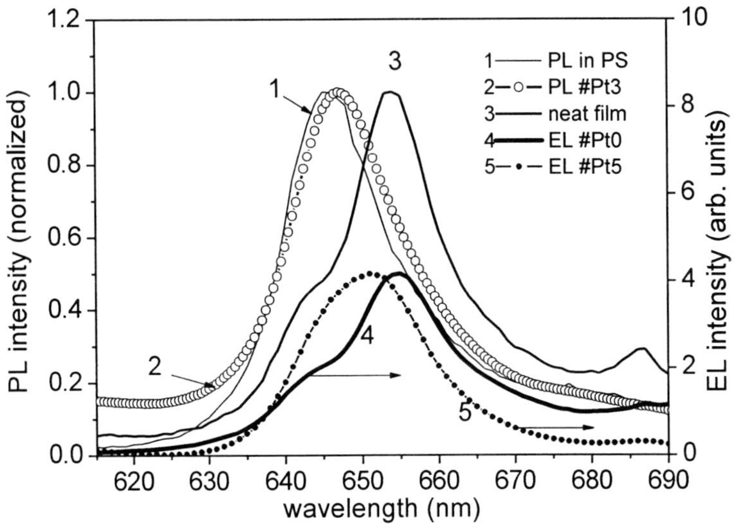

Fig. 13. Photoluminescence (PL) spectra of PtOEP in film of polystyrene (PS) doped with 5wt% PtOEP (curve 1), in the emitting CBP layer doped with 5.8wt% PtOEP of OLED #Pt3 (curve 2), and in PtOEP neat film evaporated on quartz plate (curve 3) at 290 K, together with electroluminescence (EL) spectrum from single 100% PtOEP layer structure OLED #Pt0 (curve 4) and multi-layer structure OLED #Pt5 with 100% PtOEP layer (curve 5).

As mentioned in Sec.6, organic molecules of emitting layer in multi-layer structure OLEDs interact with not only the molecules within the same layer but also molecules in the neighboring layers. Therefore it is reasonable to consider that the environment of 100% PtOEP layer is quite different between #Pt0 and #Pt5, therefore different concentration ratio of dimer to monomer is obtained in the two OLEDs. Of the two OLEDs, the ratio is higher in #Pt0 than in #Pt5.

Figure 14 shows the absorption spectra of 60-nm thick neat PtOEP film (curve 1) and diluted PtOEP in toluene solution (curve 2 [8]), together with the excitation spectra for the 645 nm PtOEP emission from film of PS doped with 5wt% PtOEP (curve 3) and from OLED #Pt3 (curve 4) at 290 K. Kalinowski et al measured the absorption spectra using 70-nm thick PtOEP neat films, film of polycarbonate (PC) doped with 6% PtOEP and PtOEP in toluene solution at room temperature [9]. Their results are consistent with Fig.14, e.g. additional absorption bands were observed at the low energy side of the 645 nm band in the neat films. From Fig.14, it is suggested that PtOEP monomer, which is present in the films doped as guest, has absorption bands at about 535 and 502 nm, while the neat film has additional intense aggregate bands at the longer wavelength side (i.e. at 548 and 510 nm).

Figure 15 shows the low temperature spectra of the neat film and 2wt% PtOEP doped in TPD. The absorption spectrum of PtOEP in CH_2Cl_2 solvent at 290 K is also shown for comparison. The solvent spectrum was obtained from [10]. A well-resolved spectrum is obtained in the neat film at 13 K. Absorption bands are observed at 547.7, 537.5, 517, 508 and 496 nm. Additionally a weak band is also observed at about 532 nm. In the lightly doped film three absorption bands are observed at 532.3, 500 and 495 nm, which is similar to the absorption spectrum of diluted PtOEP in CH_2Cl_2 solvent at room temperature. Red shift is observed for the films. This is due to the solid state solvation effect. Intense PtOEP absorption bands are observed at 529 and 496 nm in CH_2Cl_2 (curve 3 of Fig.15), while at 534.5 and about 500 nm in toluene (curve 2 of Fig.14). The whole spectrum of CH_2Cl_2 is shifted by about 5 nm from the spectrum in toluene. Taking into account that the dielectric constant of CH_2Cl_2 is 9.1 which is largely different from 2.4 for toluene, the peak shift between the two solvents is understood by the solvent effect. When two molecules are in close proximity and form dimer, a low-energy photoluminescence band which is red-shifted from the band of monomer is observed. The same is true for trimer, i.e. the trimer band is red-shifted from the dimer band. There are many reports for the observation of dimer in Pt^{2+}-compounds [11,12,13].

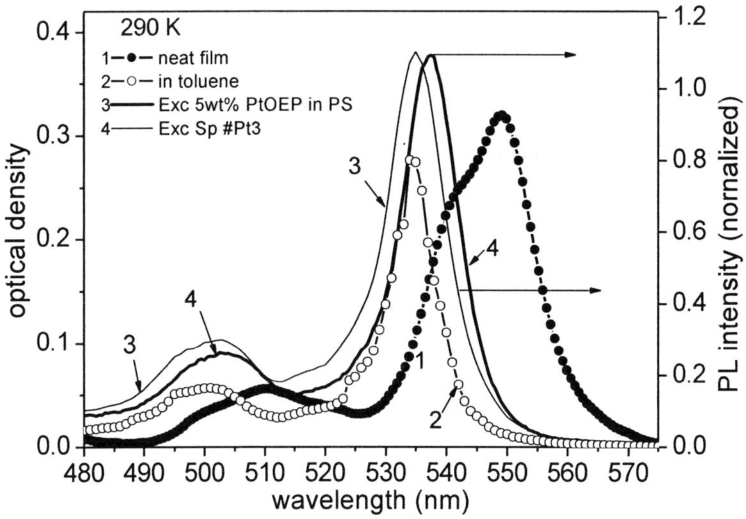

Fig. 14. Absorption spectra of 60-nm thick PtOEP neat films (curve 1) and of PtOEP in toluene solution (curve 2) at 290 K, together with excitation spectra for 645 nm emission in film of polystyrene (PS) doped with 5wt% PtOEP (curve 4)and in the CBP layer with 5wt% PtOEP of OLED #Pt3 (curve 4) at 290 K.

From comparison of the spectra among the neat film at 13 K, lightly doped film at 10 K and solution at room temperature, it is suggested that the neat film contains dimer and trimer (responsible for the 537.5 and 547.5 nm absorption bands, respectively) besides the monomer (responsible for 532 nm absorption band). The 645 nm emission band observed in lightly doped films (Fig.13) is attributed to monomer. Presence of dimer is confirmed by the PL spectra at low temperature. In Fig.16 is shown the excitation spectrum for 645 nm emission of the same film at 10 K. A broad band is observed at about 537 nm. When the film of 5wt% PtOEP doped in TPD is excited with lights of various wavelengths at 10 K, the peak wavelength of the intense PL band changes depending on the excitation wavelength as shown in Fig.17. The excitation wavelength was changed from 520 nm to 550 nm to excite the singlet state of PtOEP directly. No change was observed between 520 and 535 nm where the PL shows a peak at 643 nm. The emission peak shifts to longer wavelength by changing the excitation wavelength from 535 nm to 550 nm as plotted in Fig.16.

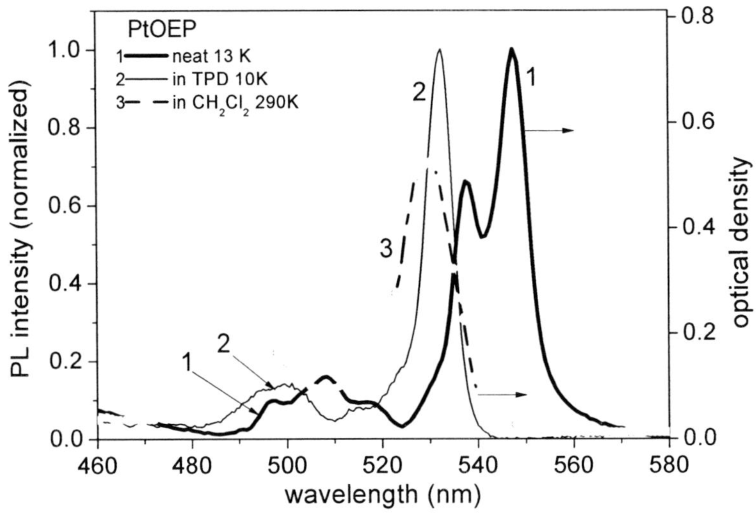

Fig. 15. Absorption spectrum of neat PTOEP film at 13 K (curve 1) and excitation spectrum for 640 nm emission of PtOEP in a film of TPD doped with 2 wt% PtOEP at 10 K (curve 2), together with the absorption spectrum of PtOEP in CH_2Cl_2 solvent at 290 K (curve 3).

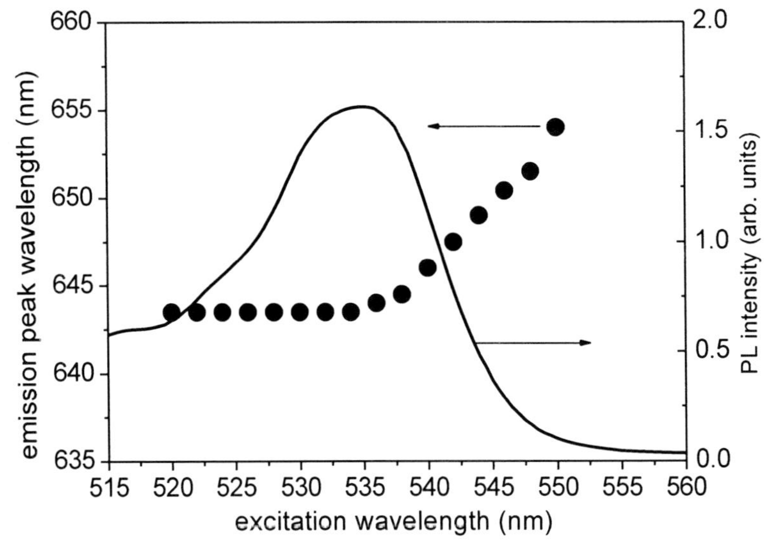

Fig. 16. Excitation spectrum (solid line)for 645 nm emission in film of TPD doped with 5 wt% PtOEP at 10 K. Dependence of the peak wavelength of the intense PtOEP emission band around 645 nm on the excitation wavelength at 10 K (closed circle, left scale).

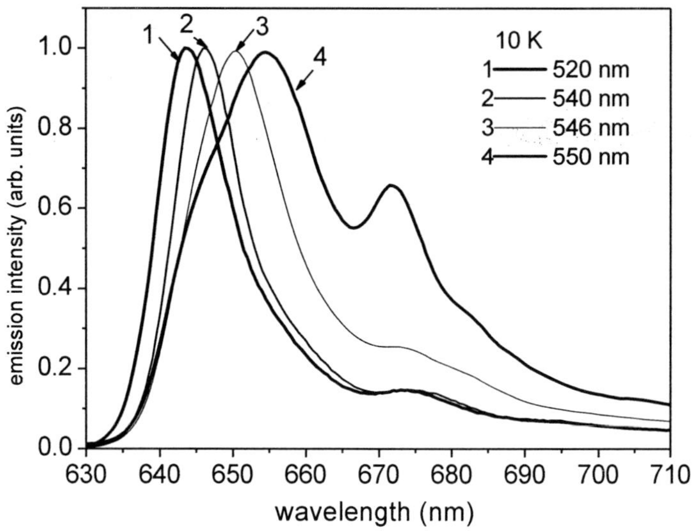

Fig. 17. Photoluminescence spectra of PtOEP in a 60-nm thick film of TPD doped with 5wt% PtOEP, excited at 520, 540, 546 and 550 nm (curves 1, 2, 3 and 4, respectively) at 10 K.

The absorption band of the 5wt% PtOEP film does not show a structured shape at 535-550 nm clearly. However, observation of different emission bands by different excitation wavelengths indicates that the absorption band consists of at least three bands of different origins, one locates at 520-535 nm, the others at 535-550 nm. The first is monomer which is responsible for the emission band with a peak at 643 nm, the second and third are attributed to dimer and trimer which are responsible for the emission bands with peaks at 654 and 672 nm, respectively (see Fig.17).

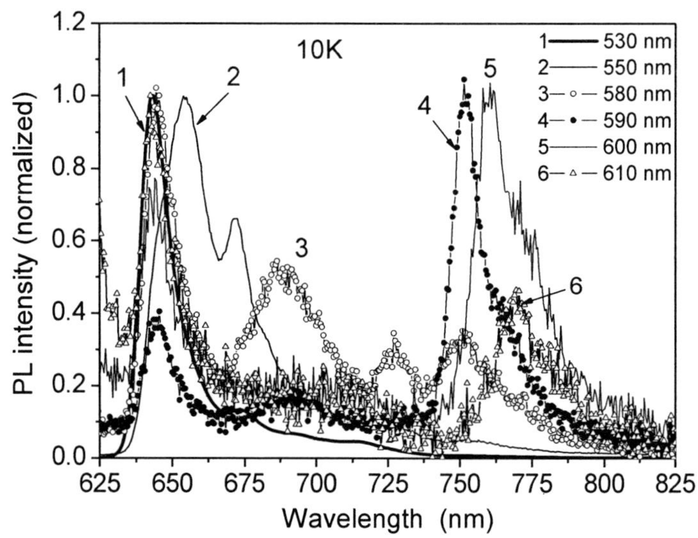

Fig. 18. Photoluminescence spectra of PtOEP in film of TPD doped with 5wt% PtOEP at 10 K, which was excited with lights of various wavelengths.

When the film of 5wt% PtOEP doped in TPD is excited with lights of wavelengths longer than 550 nm at 10 K, it is observed that additional emission bands appear at 750-800 nm as shown in Fig.18. The main emission bands have peaks at 751, 760 and 770 nm. The appearance of these emission bands depends on the excitation wavelength. Figure 19 shows the excitation spectra for 751 and 645 nm emissions at 290 K. The excitation spectra for 650 nm emission in of neat film and for 645 nm emission of 5 wt% PtOEP film are consistent with the absorption spectra shown in Fig.14. The 751 nm emission band is excited at the same absorption bands at 536 and 504 nm as the 645 nm emission band.

Therefore these emission bands are attributed to aggregate, which consists of many PtOEP molecules, because the 750-800 nm emission bands are also observed 100% PtOEP neat films [8,9]. Figures 18 and 19 indicate that such aggregates are present in not only the 100 % neat film but also a lightly doped film like 5 wt % concentration.

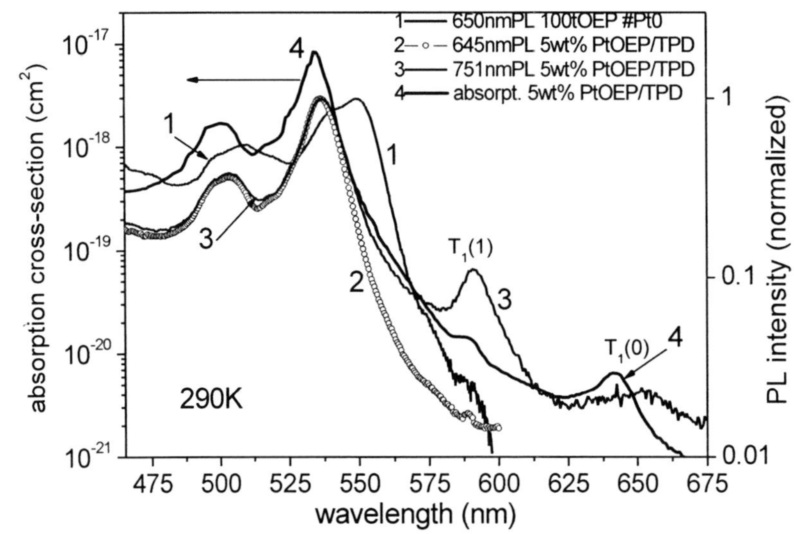

Fig. 19. Excitation spectra for 650 nm emission from 50-nm thick 100% PtOEP layer in #Pt0 OLED device (curve 1), and for 645 and 751 nm emission (curves 2 and 3, respectively) and absorption cross-section spectrum (curve 4, left scale) from 61-nm thick film of TPD doped with 5 wt% PtOEP at 290 K.

Figure 20 shows the PL spectra of a 60-nm thick film of TPD doped with PtOEP which was spin-coated on glass plate. The PtOEP concentration is very low, about 1 wt%. The PL spectra were obtained by excitation with light of various wavelengths at 10 K. The very lightly doped film exhibits a sharp excitation band at 532.5 nm for 640 nm emission. When excited with 520-560 nm light, no peak shift was observed with excitation between 520 and 532 nm light, while the emission band shifts to longer wavelength with increasing the excitation wavelength. The emission band peak is at 640, 643, 645, 650 nm by the 530, 536, 538 and 542 nm excitations, respectively (Fig.20). This indicates the dimer-like center is present even in the very low concentration PtOEP film although there are very few. No emission, however, was observed at 700-850 nm region when excited with light of various wavelengths of 580-650 nm, suggesting that the large aggregate is not created in the very lightly doped film.

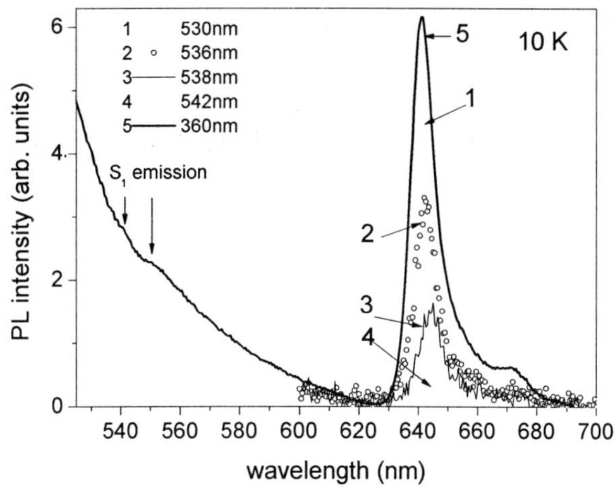

Fig. 20. Photoluminescence (PL) spectra of PtOEP in film of TPD doped with 1wt% PtOEP at 10 K, which was excited with lights of various wavelengths. The vertical arrows indicate the emission from the singlet S_1 state.

8. The S_1 Emission and T_1 Absorption Bands of PtOEP

The emission spectrum by 360 nm excitation is shown in curve 5 of Fig.20. The peak position of the emission band is the same as the band excited with 530 nm. Very weak emission bands are observed at 540 and 552 nm which are superimposed on the tail of the TPD emission band. The peak height of the 552 nm band is about 0.015 of that of the 640 nm intense band. These emission band are attributed to the emission from the singlet state S_1 of PtOEP.

The absorption and PL spectra of PtOEP dissolved in CH_2Cl_2 solution have been studied by Singh and Eftekhari [10]. The absorption spectrum is shown in curve 3 of Fig.15. The absorption bands are observed at 531, 496, 376 and 288 nm, and the intense emission band is observed at 644.5 nm [10]. The 531 and 496 nm absorption bands are close to the 535 and 502 nm absorption bands that are observed in the PtOEP films doped as guest (Fig.19). Besides the 644.5 nm emission band, a weak and broad emission band is observed in the solution at room temperature, which is located at 538-590 nm region. This corresponds to the 540 and 552 nm S_1 emission bands of Fig.20. The peak height of 538-590 nm band is about 0.1 of the 644.5 nm emission band, which is much higher ratio than the case of the film. It is noted that the singlet emission in solution is much enhanced in solution than in solid film.

The absorption band at about 535 nm is attributed to the singlet state of PtOEP monomer as mentioned above. Unlike the case of $Ir(ppy)_3$, the absorption bands due to the triplet state of PtOEP are so weak that it is difficult to observe them by the usual absorption spectroscopic measurement. A very weak excitation band is observed at 588 nm in the excitation spectrum for 650 nm emission in #Pt0 (with a emitting layer of 100% PtOEP) as shown in curve 1 of Fig.19. The lowest-energy absorption band due to the T_1 state is expected to locate near 650 nm from the Stokes shift. The T_1 absorption band was clearly found at 642 nm using the film of 5wt% PtOEP doped in TPD by a highly-resolved absorption measurement as shown by curve 4 of Fig.19 [5]. The 642 and 588 nm bands are attributed to the 0-0 and 0-1 lines associated with the T_1 state due to metal-ligand charge transfer triplet state (^{3}MLCT), respectively. The intensity of the 642 nm band is about three order smaller than the singlet 535 nm absorption band. This means that PtOEP has much weaker spin-orbit interaction than $Ir(ppy)_3$. In the case of $Ir(ppy)_3$, the absorption bands due to ^{3}MLCT are easily observed at 450-500 nm (Fig. 9) [6, 14, 15]. Later we discuss on the reason why the T_1 absorption is much weaker in PtOEP than in $Ir(ppy)_3$.

9. Self-Absorption in Emission Spectra of CBP

Figure 21 shows the PL and excitation spectra of CBP in a 30 nm thick neat film of CBP at various temperatures. The emission spectrum at 290 K consists of three intense bands at 372, 388 and 407 nm and a very weak band at about 430 nm. The middle 388 nm band is the most intense of the four bands. The intensity increases for all the bands with decreasing temperature from 290 K to 10 K. At 10 K the shortest-wavelength emission band at 371 nm is the most intense of the four bands. Therefore the 371, 388 and 430 nm bands observed at 290 K are attributed to the 0-0, 0-1 and 0-2 line, respectively. It is noted that the 0-0 band is much lower than the middle 0-1 at 290 K. Like the case of the TPD neat film (Figs. 3 and 7), this unusual vibronic structure is understood by the self-absorption because the absorption band strongly overlaps with the 0-0 line at 290 K (see curves 5 and 6 of Fig.21).

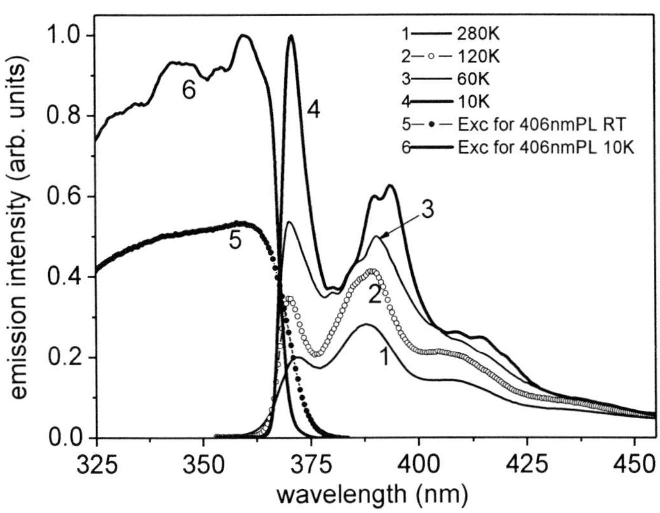

Fig. 21. Temperature dependence of photoluminescence (PL) spectra of CBP in 30-nm thick neat film excited with 300 nm light (curves 1-4), together with the excitation spectra for 406 nm emission at 290 and 10 K (curves 5 and 6, respectively).

Similar unusual vibronic spectrum is observed in a film of CBP doped with 7 wt% iridium bis [(4,6-difluoropheny)-pyridinato-5-(pyridin-2-yl)-1H-tetrazolate (FIrN4) at 290 K as shown in Fig.22. From the absorption spectra monitored by the excitation spectra (curves 6 and 7 of Fig.22), it is also suggested that the absorption gives rise to depression of the 0-0 emission line at high temperatures.

In conclusion, the unusual PL spectra with lower height 0-0 line at 290 K is caused by the self- absorption and small Stokes shift.

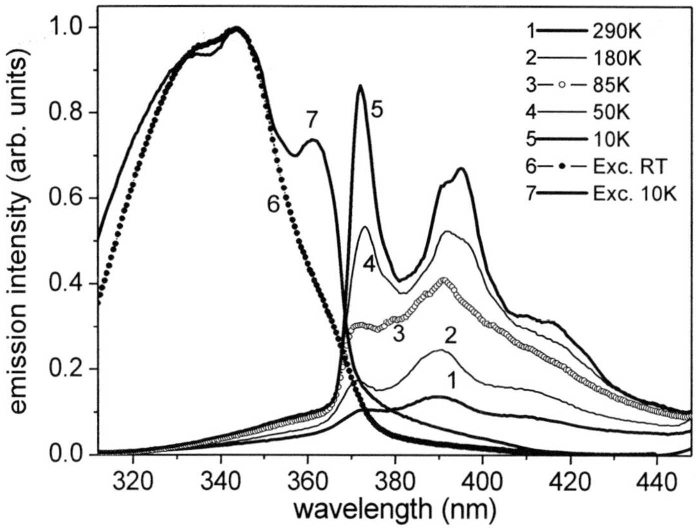

Fig. 22. Temperature dependence of photoluminescence spectra of CBP in 30-nm thick film of CBP doped with 7wt% FirN4, excited with 300 nm light (curves 1-5), together with the excitation spectra for 460 nm FirN4 emission at 290 and 10 K (curves 6 and 7, respectively).

10. Emission Spectra of mCP and SimCP

Figure 23 shows the PL and excitation spectra of 1,3-bis (9-carbazolyl) benzene (mCP) in three films of polystyrene (PS) doped with 1, 5 and 20 wt% mCP and in an mCP neat film at 10 K. The 0-0 line peak position of the PL spectrum shows red-shift due to self-absorption. The amount of shift becomes large with increasing the dopant concentration from 1 to 5 to 20 wt%. No change is observed for the 0-1 line peak position between the 1 and 5 wt% films, but the 0-1 line of 20 wt% film shows red shift, i.e. peak at about 365 nm in 20 wt% film, while at 360 nm in the 1 wt% film. This shift is understood by solid state solvation effect which is more strong in highly doped film than in lightly doped films. This is confirmed from the peak position of the corresponding 0-0 absorption line which is monitored by the excitation spectra of curves 1', 2' and 3' of Fig.23.

In the neat film are observed three triple sharp PL lines with vibronic progression. Their peaks are at 347.5 nm (called a0 line, hereafter, see Fig.23), 351.9 nm (b0), 355 nm (c0), 364.2 nm (a1), 368.6 nm (b1), 373.4 nm (c1), 382.8 nm (a2), 387.5 nm (b2) and 392 nm (c2). The excitation spectrum of the neat film has additional bands at the longer wavelength side in each of the absorption bands due to monomer which are observed in the lightly doped 1 wt% film. Two bands are observed at 342 and 347 nm (called B and C bands, respectively, see Fig.23) at the long wavelength side of the intense band at 340.6 nm (A band). The A absorption band is the 0-0 absorption line of the mCP monomer. Its 0-1 and 0-2 vibrational absorption lines are observed at 324 and 307 nm, respectively.

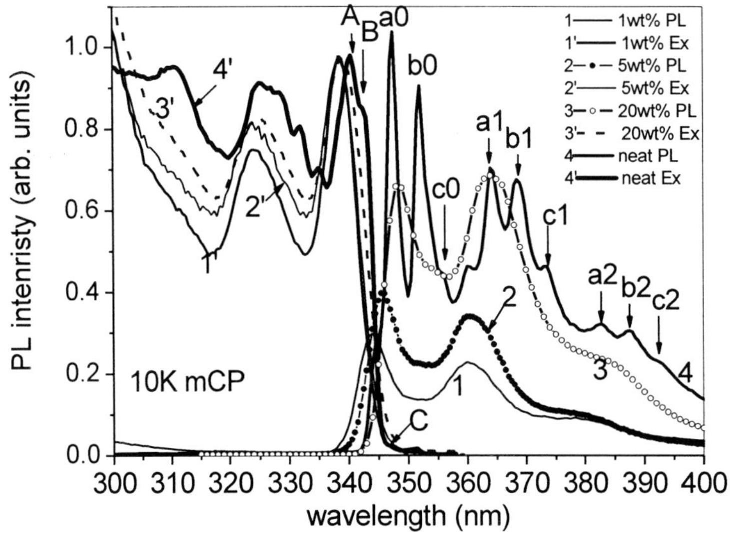

Fig. 23. Photoluminescence (PL) and excitation (Ex) spectra of mCP in films of polystyrene doped with 1, 5 and 20 wt% mCP (curves 1(1'), 2(2') and 3(3'), respectively) and in mCP neat film (curve 4 (4')) at 10 K. The PL and Ex spectra were obtained by 305 nm excitation and for 375 nm emission, respectively.

From characteristics of the peaks observed in Fig.23 (e.g. the a_0-a_1 separation is the same as the b_0-b_1 separation and also same as the c_0-c_1 separation) and from comparison with lightly doped films, it is suggested that there are at least three kinds of conformations in the neat film. One is the PtOEP monomer responsible for the a0, a1 and a2 emission lines and the A absorption band, another is dimer responsible for the b0, b1 and b2 emission lines and the B absorption band, and the other is trimer responsible for the c0, c1 and c2 emission lines and the C band. The a0, a1 and a2 emission lines are attributable to the 0-0, 0-1 and 0-2 vibrational lines associated with the singlet state S_1 of monomer, respectively, the b0, b1 and b2 emission lines are attributable to the 0-0, 0-1 and 0-2 lines associated with the S_1 state of dimer, and the c0, c1 and c2 emission lines are attributable to the 0-0, 0-1 and 0-2 lines associated with the S_1 state of trimer.

Similarly, three triple peaks with vibronic progression are observed at longer wavelength of 440-520 nm in the mCP neat film at 10 K as shown in Fig.24 although they are much weaker than the bands at 345-395 nm. The observed peaks are at 443.2 nm (called m0), 471.7 nm (m1) and 504 nm (m2) which are attributable to the 0-0, 0-1 and 0-2 vibrational lines, respectively, associated with the T_1 state of monomer, at 448.7 nm (d0), 477.0 nm (d1) and 510.5 nm (d2) which are attributable to the 0-0, 0-1 and 0-2 lines associated with the T_1 state of the dimer, and at 455 nm (t0), 482 nm (t1) and 516 nm (t2) which are attributable to the 0-0, 0-1 and 0-2 lines associated with the T_1 state of the trimer.

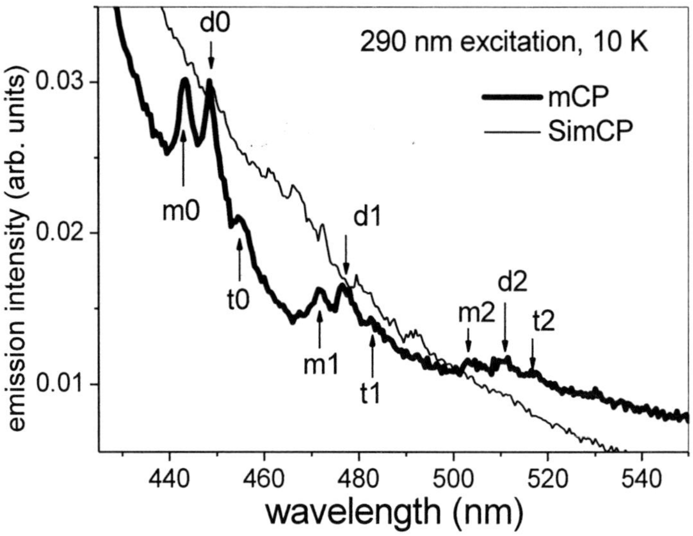

Fig. 24. Photoluminescence spectra of mCP and SimCP neat films in the 425-550 nm spectral range, which are excited at 290 nm at 10 K.

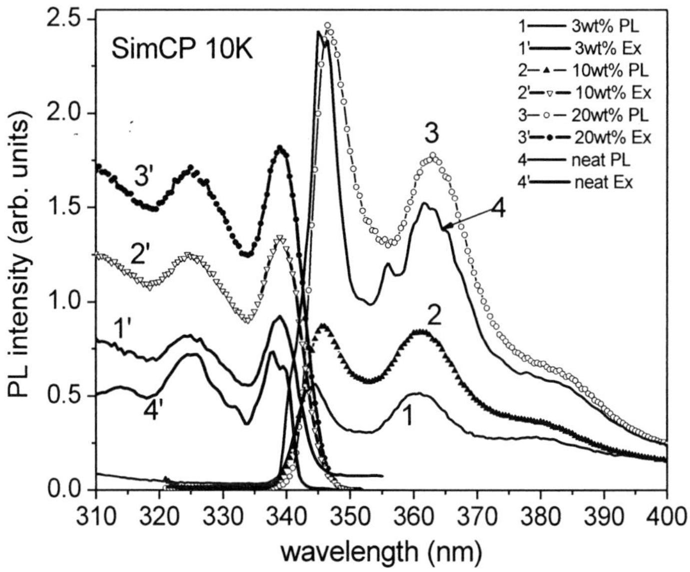

Fig. 25. Photoluminescence (PL) and excitation (Ex) spectra of SimCP in films of polystyrene (PS) doped with 3, 10 and 20 wt% SimCP (curves 1(1'), 2(2') and 3(3'), respectively) and in SimCP neat film (curve 4 (4')) at 10 K. The PL and Ex spectra were obtained by 305 nm excitation and for 360 nm emission, respectively.

Figure 25 shows the PL and excitation spectra of 3,5-bis (9-carbazolyl) tetraphenylsilane (SimCP) in three films of PS doped with 3, 10 and 20 wt% mCP and in SimCP neat film at 10 K. SimCP is a derivative of mCP molecule [16], whose chemical structures shown in Fig.12. No big change is observed in the line shapes among three doped films as the case of mCP. Additional sharp lines are observed in the neat film, indicating the presence of dimer, but not so well resolved from the lines due to monomer. This indicates that, unlike mCP, SimCP has a tendency of avoiding the formation of dimer. This will be understood from the difference of chemical structure between two molecules. It is suggested that Si-ligand in SimCP avoids intermolecular covalent bonding because of the localization of electrons near Si atom. Such an electron distribution induces stronger intramolecular covalent bonding, resulting in more stable chemical structure than mCP without central Si atom. This is confirmed that the glass transition temperature Tg is higher for SimCP (Tg=101 ^{0}C) than for mCP (Tg=55 ^{0}C) [16].

11. Why PtOEP Shows Much Weaker Luminance in OLEDs than $Ir(ppy)_3$?

It is known that OLEDs with PtOEP red emitter show much weaker luminance than those with $Ir(ppy)_3$ green emitter. This is confirmed by comparison of luminance of PtOEP OLED with that of $Ir(ppy)_3$ OLED which has the same multi-layer structure as the PtOEP OLED. For example, we obtained the luminance of 7500 cd/m^2 at applied voltage of 10V in #Ir1, while 40 cd/m^2 in #Pt2 at the same 10 V. The #Pt2 shows lower luminance efficiency by one order than #Ir1 as seen in Fig.26. The organic phosphors PtOEP and $Ir(ppy)_3$ are doped in the same CBP in the emitting layer in the two OLEDs, where no energy transfer from the lowest triplet state T_1 of PtOEP and $Ir(ppy)_3$ to the CBP host because the T_1 state locates below the T_1 state of CBP at 2.56 eV. Why such a large difference occurs between PtOEP and $Ir(ppy)_3$?

The luminescence intensity P is given by

$$P = N\, k_r\, \Phi\ ,$$

where N is the population at the emitting state of the excited molecules, k_r is the radiative transition probability which is the inverse of radiative lifetime τ_r, Φ is the quantum efficiency which is given by $\Phi=k_r/(k_r + k_{nr})$, k_{nr} is non-radiative transition probability from the emitting state. Here the emitting state means the lowest triplet state T_1 in the case of phosphorescent materials.

The photoluminescence quantum efficiency (i.e. internal quantum efficiency in OLED device) is nearly 100% for $Ir(ppy)_3$ doped in CBP at room temperature [17, 18], while 40 % for PtOEP doped in CBP [19] and 41.5 % for PtOEP in both de-aerated toluene and chloroform [8]. The radiative lifetime of the T_1 state is estimated as approximately 2 μs at room temperature for $Ir(ppy)_3$ [7, 20], while about 120 μs for PtOEP at room temperature [8]. This leads to about 0.004 as the luminescence intensity ratio of PtOEP to $Ir(ppy)_3$, i.e. a much weaker PtOEP emission than the $Ir(ppy)_3$ emission. Therefore we understand that the radiative transition probability of the T_1 state is strongly responsible for the difference of emission intensity between the two phosphors. The probability depends on the spin-orbit coupling, which becomes strong with increasing the atomic number of the central metal ion in the organic compounds (called heavy atom effect). The atomic number of Pt^{2+} and Ir^{3+} ions is 78 and 77, respectively, almost the same for the two ions. This means that, besides the heavy atom effect, another factor is responsible for difference of spin-orbit coupling between PtOEP and $Ir(ppy)_3$.

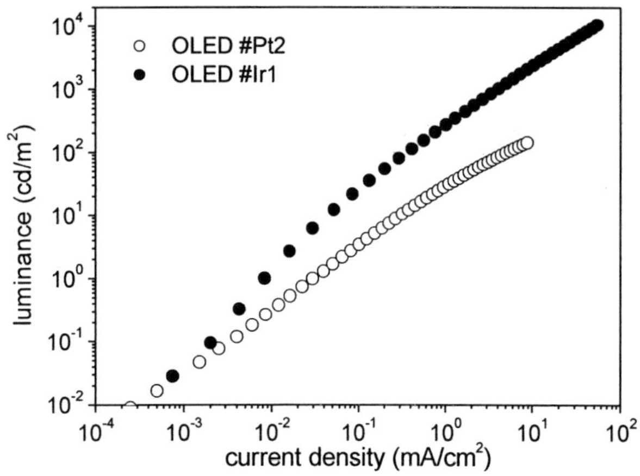

Fig. 26. Luminance of the #Pt2 and #Ir1 OLEDs plotted against the current density.

Figure 27 shows the absorption and EL spectra of PtOEP, which are compared with those of Ir(ppy)$_3$ at 290 K. These spectra are plotted against photon energy. It is noted that the emission band width is much wider in Ir(ppy)$_3$ than in PtOEP. The same is true for the low-energy absorption bands. This suggests that the mixing of the electronic states of central Ir^{3+} metal ion with the ligand states is much higher than the case of PtOEP, i.e. the orbital overlap between the metal ion and ligand molecules is much larger in Ir(ppy)$_3$ than in PtOEP. Strong overlap gives rise to intense spin-orbit coupling. The next question is why PtOEP has much weaker orbital overlap.

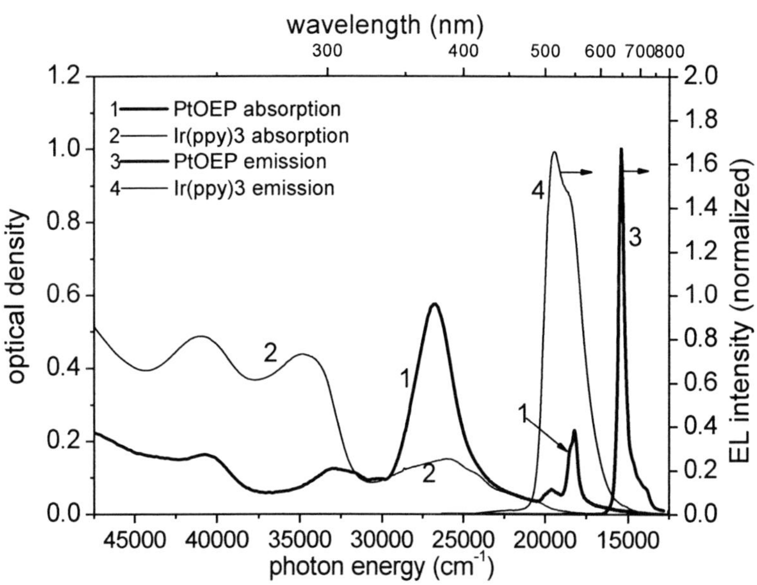

Fig. 27. Absorption spectra of 60 nm-thick films of PtOEP and Ir(ppy)$_3$ at 290 K and electroluminescence (EL) spectra of #Pt2 and #Ir1 OLEDs, which are plotted against photon energy.

The electronic states of the highest occupied molecular orbital (HOMO) and the lowest unoccupied molecular orbital (LUMO) have been clarified for Ir(ppy)$_3$ using the time-dependent density function theory TDDFT [14,21]. According to the TDDFT calculation, the electronic population of d-electrons of Ir^{3+} with $(5d)^6$ outer electron configuration is about 50 % in the HOMO level and the rest is π electrons of ligand phenyl pyridine (ppy) molecule.

This means that an efficient electron overlapping is achieved in $Ir(ppy)_3$, leading to delocalization of the excited triplet state over the ligands and Ir^{3+} metal and to formation of the metal-ligand charge transfer (MLCT) state. It is suggested that such an overlapping is possible by the following reason. The d orbital of Ir^{3+} splits to two levels of six-fold degenerate t_{2g} and four-fold degenerate e_g in octahedral symmetry. The t_{2g} level fully occupied by six d-electrons is located close to the π levels of ligand. Therefore the d- and π electrons are mixed to each other easily, leading to the strong spin-orbit interaction.

The symmetry of Pt^{2+} ion in PtOEP is D_{4h}. In this symmetry the lower-energy t_{2g} level splits into two fold degenerate e_g and non-degenerate b_{2g}, while the higher-energy e_g level splits into two non-degenerate a_{1g} and b_{1g}, where the a_{1g} level is lower than the b_{1g} level and also the b_{2g} level. Pt^{2+} ion has $(5d)^8$ outer electron configuration. The HOMO level is formed by the fully occupied b_{2g} level. The π electrons of porphyrin ligand surrounding the central Pt^{2+} ion has localized in porphyrin itself, and very weak overlapping with Pt^{2+} ion. This suggests that the π level is energetically far from the b_{2g} level of Pt^{2+}. As the result, the spin-orbit interaction becomes much weaker in PtOEP than in $Ir(ppy)_3$.

The radiative lifetime of the T_1 state is two order longer in PtOEP than in $Ir(ppy)_3$. The long lifetime causes the energy dissipation between the neighboring PtOEP molecules. In the case of photoluminescence of organic molecules excited with high power laser, triplet-triplet annihilation occurs, leading to enhancement of the non-radiative probability [3,7,8]. Such a nonlinear phenomenon, however, does not occur when molecules are excited with non-coherent light like radiation from Xe-lamp. In the case of non-coherent light excitation, PtOEP with long radiative lifetime of 120 μs gives rise to the energy dissipation from the excited PtOEP to unexcited neighboring PtOEP molecule by overcome of potential barrier between two neighboring molecules by thermal activation energy [22] (see Sec.12). This is confirmed from the observation that the PtOEP emission intensity decreases with increasing temperature from 10 K to room temperature [5], because the energy dissipation probability increases with increasing temperature.

There is additional reason for the very weak luminance in PtOEP OLEDs. Dimer of PtOEP are easily formed even in lightly PtOEP-doped films. The concentration of PtOEP is usually 3-6 wt % in OLEDs, where small aggregates like dimer are present as mentioned in Sec.7. The non-radiative processes from the monomer PtOEP to the dime occur because the T_1 state of dime is lower than that of monomer, giving rise to weak EL luminance intensity. This process is not negligible in the actual OLED devices.

12. What Pt^{2+}-Compounds Emit Intense Phosphorescence ?

If the b_{2g} level of Pt^{2+} locates energetically close to the π level of ligand, one can expect strong spin-orbit interaction. Such a situation is produced by the strong ligand field for the central Pt^{2+} ion. There is a Pt^{2+}-compound, terdentate N^C^N-coordinating 1,3-di(pyridyl) benzene Pt(Cl) with Cl at one of the four coordinate of Pt, which is called N^C^N-Pt(Cl) [23] or Pt(dpb)Cl [24]. The chemical structure is shown in Fig.12. The symmetry of Pt^{2+} is C_{2v}, which is lower than that of PtOEP. Therefore the ligand field splitting of d electrons of Pt^{2+} is wider than the case of PtOEP. It is expected that, unlike the case of PtOEP, the b_{2g} level of Pt^{2+} is located close to the π level of ligand, leading to the mixing of the two orbitals. In fact the strong mixing of the d orbital and π orbital is found by the TDDFT calculation for N^C^N-Pt(Cl) [24]. As a result we can expect high phosphorescence intensity for this compound.

Intense EL luminance is certainly obtained in the OLED with N^C^N-Pt(Cl) doped in CBP, e.g. about 17000 cd/m^2 at applied voltage of 10 V. The phosphorescence lifetime of N^C^N-Pt(Cl) is about 6.8 μs and the photoluminescence quantum efficiency Φ is 60 % at room temperature [23]. These values gives 11 μs as the radiative lifetime τ_r, which is one order shorter than τ_r of PtOEP, indicating that the spin-orbit interaction of the emitting triplet T_1 state with the singlet state is much stronger in N^C^N-Pt(Cl) than in PtOEP. In this way we understand the reason why N^C^N-Pt(Cl) gives much higher EL luminance than PtOEP doped in CBP (which gives 40 cd/m^2 for at the same 10 V).

In conclusion, it is suggested that the energy levels of d-electrons of Pt^{2+} ion relative to the energy levels of ligand orbitals determine the spin-orbit coupling of the molecules. The relative positions of energy levels are determined by the local symmetry of Pt^{2+}. The lower the symmetry of metal ion, the stronger the spin-orbit coupling, leading to high phosphorescence intensity.

The integrated PL intensity of N^C^N-Pt(Cl) decreases with increasing temperature from 8 K, e.g. the intensity at 300 K reduces to 68 % of the intensity at 8 K in thin film of N^C^N-Pt(Cl)-doped polycarbonate (PC) [23]. This ratio is close to the quantum efficiency of 60 % which was measured using dichloromethane solution at room temperature. Such a reduction is explained by dissipation (i.e. diffusion) of excited energy from the T_1 state to the neighboring N^C^N-Pt(Cl) molecule. The dissipation rate k_D from the T_1 state of the excited guest molecule M_1 to the T_1 state of unexcited neighboring guest molecule M_2 is given by

$$k_D = K_D[A_T + \exp(-E_D/kT)],$$

where K_D is coupling parameter between two guest molecules, A_T is parameter due to tunneling of the excited energy from the excited to unexcited molecules, and E_D is thermal activation energy to overcome the potential barrier between the neighboring molecules [18]. Back transfer from the M_2 molecule to M_1 might be possible at high temperatures like room temperature. However it is negligible because the M_2 molecule has many processes after excitation, e.g. energy dissipation and diffusion to neighboring molecules. The probability of back transfer to the original M_1 molecule is very small.

The parameter K_D depends on the radiative lifetime of the T_1 state, i.e. the radiative lifetime is important factor to make the dissipation possible. If the lifetime is short, the rate of radiative transition from the T_1 state to the ground state S_0 becomes much higher than the rate of energy transfer to the neighboring molecule, resulting in depression of dissipation process. The τ_r=11 μs of N^C^N-Pt(Cl) is rather long, therefore the dissipation process occurs simultaneously with the radiative process to the S_0 state, giving rise to decrease of PL intensity with increasing temperature. If τ_r is more shorter than 11 μs, we would obtain higher quantum efficiency than 60 %. To shorten the radiative lifetime of N^C^N-Pt(Cl), one has to modify the chemical structure so as to reduce the local symmetry around Pt^{2+} ion.

The parameters K_D and A_T are temperature independent, therefore the temperature dependence of PL intensity is determined by the thermal activation. Since the quantum efficiency (0.60) at room temperature is close to the intensity ratio (0.68) of the RT intensity of PL to the 8K intensity [23], we predict the PL quantum efficiency at 8 K is nearly 100 % for N^C^N-Pt(Cl). The following three reasons are conceivable for the 100% efficiency. (1) The non-radiative process of dissipation to neighboring molecule is negligible at such a low temperature because the thermal activation does not occur and because the tunneling process is negligible in very lightly doped PC film. (2) Energy transfer from the T_1 state of N^C^N-Pt(Cl) to the PC host is impossible because the T_1 state of host locates far above that of guest. (3) The non-radiative multi-phonon transition from the T_1 state to the ground state does not occur by the energy gap law because of wide energy gap (2.48 eV) between the T_1 and S_0 states although the higher-energy singlet states including the S_1 state mix with the T_1 state by the spin-orbit interaction.

13. Solid State Solvation Effect

Absorption and PL bands of organic solute molecule in liquid solutions change the peak wavelength without changing their line shapes, depending on molecule of solvent. The shift becomes large in the solvent with higher dielectric constant because the permittivity of solvent changes the local electric field around the solute and then the energy states of the solute is changed by the interaction of dipole of solute molecule with the electric field due to solvent. This is called solvent effect or solvation effect [25].

This idea of solvation effect was applied to the case of solid films since the similar phenomena were observed for dye in films [26, 27, 28]. For example, when Alq_3 is doped in TPD film, it was observed that the green PL band of Alq_3 shows red shift from 498 to 522 nm, without change in spectral shape and peak width, as the concentration of Alq_3 is increased from 2.5 to 100 % [1,26]. This is understood as follows. When the DCM2 concentration is increased in the relatively less-polar Alq_3 host (dipole moment: 5.5 D for Alq_3 and 11.3 D for DCM2 [27]), the distance between nearest-neighbor DCM2 molecules decreases, resulting in increase of the local polarization field and then leading to red shift with increasing DCM2 concentration [28]. A theory to explain this sold state solvation effect was suggested by Baldo et al [1]. They considered that the polar Alq_3 molecules form ordered domain in TPD host, which consists of many Alq_3 molecules, by the dipole-dipole interaction among the neighboring molecules in less polar molecular TPD host (dipole moment: 1.5 D [1]). The energy gap between the excited and ground states of the central molecule in the domain decreases with increasing the domain size. It was assumed that the observed emission is caused by the central molecule in the maximum-size domain although domains with various sizes are present.

The solvation effect is also observed for TPD, $Ir(ppy)_3$ and PtOEP when we measure the absorption and PL spectra by changing solvent molecule in solution [7, 8] and by changing dopant concentration and host molecule of film variously (see e.g. Figs. 3 and 5 for TPD). The peak shift in the emission band is observed not only between films but also between liquid solution and sold film as seen in e.g. Figs. 10 and 15 for PtOEP. The amount of the relative peak shift is mainly understood by the relative values of dielectric constant and polarizability between host and guest molecules.

14. UNUSUAL Alq_3 EMISSION SPECTRUM

As mentioned in Sec.1, the emission spectrum of Alq_3 is quite different from the other organic molecules. For example, large Stokes shift and structureless broad emission band are pointed out. We shall examine the reasons.

There are many reports on the EL spectra of Alq_3 from 100% Alq_3 neat layer since the first report of EL band with a peak at 550 nm with a 60-nm thick film by Tang and VanSlyke [29]. We notice that the EL peak wavelength is different among the reports although a Gaussian-shape emission band is observed. For example, the peak wavelength is about 532 nm [30], 528 nm [31], 525 nm [32], 520 nm [33], and 510 nm [34]. Similarly the PL spectrum of neat film at room temperature is different among the investigators although the difference is much smaller than the case of EL. For example, the PL peak is at 530 nm in 50-nm thick film [34], about 530 nm in 100-nm thick film [30], 525 nm in 60-nm thick film [32], 522 nm in 200-nm thick film [1]. We obtained a peak at 522 nm at room temperature in 10-nm thick film (Fig.1). Observation of different peak positions in neat films indicates that the morphology of neat film is different among the films fabricated. One of these reasons is a different mixture ratio of two isomers of *mer*- and *fac*-Alq_3 among the films [2,35].

The PL spectrum of Alq_3 in solution shows a single Gaussian-shape band, i.e. the peak at 540 nm in N,N-dimethylformamide (DMF) solution at 298 K, at about 510 nm in the solution of the same DMF at room temperature [30] and at 475 nm in ethanol at 77 K [36]. No vibronic structure is observed even for isolated single Alq_3 molecule in liquid solution. The same is true for the PL spectrum from a film of lightly doped TPD (2.5% Alq_3) at room temperature [1]. Why Alq_3 molecule does not show the vibronic structure?

The large Stokes shift means a large Huang-Rhys parameter (or called Huang-Rhys factor) S [4,37], indicating a large Franck-Condon shift. The Franck-Condon shift is caused by difference of the bottom of adiabatic potential parabola (i.e. equilibrium position in the configurational coordinate) between the excited and ground states. The spectral distribution of emission band involving 0-n (n=0, 1,2,3, , ,) vibrational transitions is proportional to the Franck-Condon factor F_{0n} :

$$F_{0n}= \exp(-S)S^n/n!$$

Therefore the S factor is approximately estimated from the intensity ratio of the 0-1 line to the 0-0 line. Since no vibronic structure appears in Alq_3 thin films even at 10 K (Fig.1), we cannot estimate the S factor for Alq_3. Brinkmann et al found the vibronic structure with progression of 525 cm^{-1} in polycrystalline α- and β-Alq_3 at 4.2 K [2], then estimated S=2.6. Usually the S factor is less than 1 for most organic molecules. Although we cannot estimate the S factor in thin film of Alq_3, it is suggested from the large Stokes shift that the S factor is more than 2.6 in thin film.

According to calculation with TDDFT, the electron density in HOMO is localized on different ligand from that in LUMO [30, 38]. Different charge distribution between HOMO and LUMO means the different charge distribution (and different orientation of dipole moment) between the ground state S_0 and the emitting singlet state S_1, resulting in the large shift of the equilibrium coordinate between the ground and excited state adiabatic potential parabolas. In this way we understand the reason why large Stokes shift occurs in Alq_3.

Broad emission band is observed in Alq_3. This is also understood by the large Huang-Rhys factor S. Simulation of spectral distribution of emission band using various S values indicates that the much more vibrational states are involved in the emission spectrum with increasing the S number [37, 39]. For example, only four or five vibrational transitions are involved for S=0.4, while at least twenty vibrational transitions are involved for S=20. This means that Alq_3 with large S value gives a broad emission band.

15. Appearance of Resolved Vibronic Structure

Organic molecular complexes exhibit a more or less vibronic structure in the emission and absorption spectra. The resolution in the vibronic structure becomes high with decreasing temperature from room temperature to low temperature near 10 K as shown in Fig.5. There are three molecule groups regarding the PL line shape. The first group is molecules with a well resolved vibronic structure, the second is molecules with a less resolved structure, and the last is molecules with no structure even at 10 K. For example, CBP, TPD, PtOEP, mCP and SimCP belong to the first group, and Alq_3 to the last group. $Ir(ppy)_3$ in multi-layer OLED device belongs to the second group, while $Ir(ppy)_3$ doped in CBP film, which is spin-coated on glass plate, belongs to the first group (Fig.11).

It has been suggested that the highly resolved vibronic structure is due to the transition from the ligand-centered (LC) $\pi-\pi^*$ state, while poorly resolved one from the metal-ligand-charge-transfer (MLCT) state [15, 40, 41, 42]. From this suggestion, the less resolved green emission of $Ir(ppy)_3$ is attributed to the MLCT state, which is consistent with the TDDFT calculation [14, 21]. However, it should be reminded that the vibronic resolution in PL spectrum becomes high with decreasing temperature from room temperature (Figs. 4 and 5). This indicates that the appearance of resolved vibronic structure is caused by not only the electronic state (either LC or MLCT) but also the electron-vibration interaction.

The vibronic structure of the emission spectrum is caused by the vibrational transitions from the vibrational quantum number m=0 in the excited state to the vibrational quantum numbers n=0, 1, 2, 3, , , in the ground state as mentioned in Sec.2. The vibrational mode involved in the ground state determines the vibronic progression, i.e. the separation of the 0-0 and 0-1 lines, etc. If only one mode is involved, we expect to observe highly resolved vibronic structure. This means that if many vibrational modes with different vibrational energies are involved, we observe a less structured spectrum by overlapping among various vibrational lines. Similarly if there is a large distribution of molecular environments in thin film, inhomogeneous broadening occurs, resulting in the smearing out of vibronic structure in the emission spectra. Decrease of temperature induces to depress the distribution due to thermal movement, giving rise to well-resolved vibronic spectra.

The LC $\pi-\pi^*$ transition is responsible for the green emission of Alq_3 and then we expect a resolved spectrum. But no vibronic structure is observed for Alq_3. This is explained as follows. In Alq_3, no long-range order is present in thin films, which was clarified by the absence of X-ray diffraction peaks, while there is not only a large disordered distribution of molecular packings [2] but also mixture of two isomers of *mer-* and *fac*-Alq_3 in the amorphous films. As a result Alq_3 does not show any resolved spectrum.

In Pt^{2+}–compounds such as PtOEP with porphyrin-plane, intermolecular stacking interactions occurs between the planar cations, leading to formation of planar-parallel orientation in the constituent molecules, i.e. ordered stacking structure. As a result, unlike the case of Alq_3 with random distributions of various conformations, resolved vibronic spectrum is observed. In conclusion it is suggested that the spectral line shape of whether it is well resolved or not depends on the chemical structure of molecule and the distribution of conformations in amorphous thin films.

Regarding $Ir(ppy)_3$-doped CBP film, it shows different resolved spectrum depending on environment of the film. A well-resolved vibronic structure appears in this single layer $Ir(ppy)_3$:CBP film than in the emitting $Ir(ppy)_3$:CBP layer of the multi-layer structured OLEDs (see Sec.6). This difference is understood from the above-mentioned model. The $Ir(ppy)_3$ molecules in OLEDs interact with not only the molecules within the same layer but also molecules in the neighboring layers. Diffusion of the components from the neighboring layers and the interface is not negligible in the multi-layer structure OLEDs. Such various interactions induce a large distribution of environments and geometrical strutures for $Ir(ppy)_3$ molecules. As a result, the $Ir(ppy)_3$ film of OLED shows less-resolved vibronic structure and disappearance of weak 0-2 and 0-3 vibrational lines.

Watanabe et al observed EL spectra with different intensity ratio between the 0-0 and 0-1 vibronic lines in the EL spectra of $Ir(ppy)_3$ in 6.5 wt% $Ir(ppy)_3$-doped CBP layer in multi-layer OLEDs by changing the thickness of electron transport layer of Alq_3 variously between 20 and 60 nm [43]. The structure of OLED device is quite similar to that of #Ir4. When the Alq_3 layer becomes thick, the lower-height 0-1 line becomes close to the 0-0 line height. When the thickness of emitting layer was changed from 10 to 60 nm but the Alq_3 layer was fixed at 30 nm, no change was observed in the intensity ratio between the 0-0 and 0-1 lines [43]. These experimental results indicate that the conformation of $Ir(ppy)_3$ molecules in the emitting layer is changed by not only the nearest-neighboring layer but also next-neighboring layer, suggesting inter-layer interaction which is due to dipole-dipole interaction of molecules among the neighboring layers.

16. Formation of Aggregates in Thin Films

As mentioned in Sec.7, the PL and absorption spectra indicate that PtOEP molecules forms not only small aggregate such as dimer in neat film but also large aggregate besides monomer. PtOEP dimer was formed even in lightly PtOEP-doped TPD, polystyrene and polycabonate films. Dimer and trimer of mCP are also formed in mCP neat film. TPD dimer is also formed in the TPD neat film. On the other hand, dimer is not found in Alq_3 and $Ir(ppy)_3$ neat films, which show peak shift in the emission bands, by the solid state solvation effect, with increasing concentration of the dopant. Thus, there are two groups in the doped and neat films. The first group is that the constituent molecules form aggregates as the cases of PtOEP and mCP, the second group is that the constituent molecules do not form aggregate as the cases of Alq_3 and $Ir(ppy)_3$.

Formation of aggregate is due to the intermolecular covalent bonding. PtOEP has plane of porphyrin which has a strong tendency towards π–π stacking as the case of square planar Pt^{2+}-compounds [11]. Therefore dimer is easily formed, but monomer PtOEP also exists together with the aggregates even in heavily doped films.

Alq_3 and $Ir(ppy)_3$ molecules do not have such a planar structure, resulting in poor stacking. The TDDFT calculation for Alq_3 shows absence of effective HOMO overlap between neighboring molecules [38]. This indicates that formation of stable dimer and trimer are difficult in Alq_3 film although each of Alq_3 molecules has dipole-dipole interaction with neighboring molecules.

Molecules confined in thin film take their stable positions and conformation finally after they are randomly distributed by evaporation or spin-coating. When the concentration of dopant is increased and the intermolecular distance becomes short, molecules like PtOEP and mCP form aggregates by strong intermolecular covalent bonding, while molecules like Alq_3 and $Ir(ppy)_3$ do not. Like this the strength of intermolecular bonding depends on the chemical structure of each molecule.

17. Microcrystals in Organic Thin Films

Atomic force microscopy (AFM) and friction force microscopy (FFM) were used to study the morphology of TPD and Alq_3 film surfaces [44]. The images of AFM and FFM indicate that partial crystallization appears in TPD single layer film when it is kept at room temperature in a week after evaporation on glass, while Alq_3 film is extremely stable even for a few years and such microcrystals are not formed even at high temperature of 80 °C [45]. This means that, unlike Alq_3, TPD molecules form microcrystals easily.

We observed the almost the same PL line shapes of TPD at 10 K between the TPD neat film and 5wt% TPD doped polystyrene (PS) film (Figs. 3 and 5) although the peak shift is observed in the two films by solvation effect. The spectra of Figs.3 and 5 were measured by photoexcitation in the absorption bands of TPD dopant and PS host several weeks after fabrication of films. Therefore it is confirmed that the amorphous molecules are still present for a long time.

As shown in Fig.6, TPD doped as host for $Ir(ppy)_3$ guest and TPD in the neat film give emission band at about 404 nm in addition to the monomer emission band at 397 nm. It is suggested the 404 nm band is due to not excimer but aggregate such as dimer, which consists of small numbers of TPD molecules, because it locates close to the monomer band. Where is the emission band due to microcrystals suggested from the AFM and FFM images ?

When the neat film and 5 wt% TPD-doped PS film are excited at light of 260-295 nm, additional PL band is observed at 690 nm at 10 K as shown in curves 1 and 2 of Fig.28. The same PL band is also observed in 5wt% TPD-doped polycarbonate (PC) film (curve 3 of Fig.28). The 690 nm emission band is excited by the absorption bands different from the bands of isolated TPD and PS molecules (curve 4 of Fig.28). This structureless Gaussian-shape 690 nm emission band was also observed in TPD film doped with 5wt% $Ir(ppy)_3$. From these results, it would be suggested that the 690 nm emission band is attributed to the microcrystals.

A very weak PL band is observed at 700 nm in CBP neat film besides the PL band of isolated CBP molecule. The 700 nm band is also observed in films of CBP doped with 7 wt% FIrpic and doped with 7 wt% FIrN4. The excitation spectrum for this emission is different from the excitation spectrum for emission of the isolated CBP molecule. The 700 nm band enhances with decreasing temperature to 10 K. This emission band is quite similar to the 690 nm emission band observed in films containing TPD. Therefore, like the case of TPD, the 700 nm emission band is also attributed to the microcrystals of CBP.

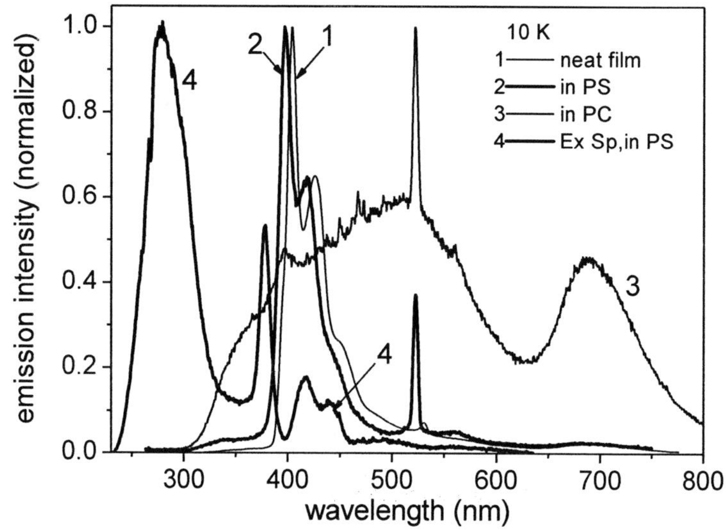

Fig. 28. Photoluminescence spectra of TPD neat film (curve 1) excited at 265 nm and films of 5wt% TPD doped polystyrene (PS) and polycarbonate (PC) (curves 2 and 3, respectively), which are excited at 260 nm, together with excitation spectrum for 690 nm emission in the 5wt% TPD doped PS (curve 4). The measurements were made at 10 K. Sharp lines at 530 nm of curve 1 and at 530 nm of curves 2 and 3 are due to the harmonic of the excitation light.

Presence of domain in films has been suggested for Alq_3 films [1]. Unlike the case of TPD neat film, however, the AFM image does not show such a formation in Alq_3 neat films [45]. The PL spectral line shape is the same between the neat film and lightly doped film. We could not find any additional emission band in the Alq_3 neat film clearly (see Fig.1). This is understood as follows. Each molecule of Alq_3 does not have strong interaction with neighboring molecules in the domain but weak dipole-dipole interaction that induces small energy shift from the isolated molecule by the perturbation from molecules in domain. This means that the domain is not practically present in film and not detected by physical measurements like AFM. We believe that the domain means presence of sphere with maximum radius where each molecule in film has intermolecular interaction.

ACKNOWLEDGMENTS

We thank Profs. Alfons Penzkofer and Chin-Ti Chen for kindly supplying the spin-coated and evaporated organic films that are used in the present study. We also thank Mr. Hideyuki Murayama for assistance in photoluminescence measurements and Mr. Nadeer Aljaroudi for assistance in numerical calculation. The present work was partly supported by the Grant-in-Aid for the Scientific Research from the Japan Society for Science Promotion.

REFERENCES

[1] M.A. Baldo, Z.G. Soos and S.R. Forrest, *Chem. Phys. Lett.* 347 (2001) 297.
[2] M. Brinkmann, G. Gadret, M. Muccini, C. Taliani, N. Masciocchi and A. Sironi, *J. Am. Chem. Soc.* 122 (2000) 5147.
[3] N.J. Turro, *Modern Molecular Photochemistry* (University Sci. Books, Sausalito, 1991).
[4] M. Pope and C.E. Swenberg, *Electronic Processes in Organic Crystals*, (Clarendon Press, Oxford, 1982).
[5] T. Tsuboi, H. Murayama and A. Penzkofer, *Thin Solid Films* 499(2006)306.
[6] T. Tsuboi and M. Tanigawa, Thin Solid Films 438-439 (2003) 301.
[7] W. Holzer, A. Penzkofer and T. Tsuboi, *Chem. Phys.* 308 (2005) 93.
[8] A.K. Banzal, W. Holzer, A. Penzkofer and T. Tsuboi, Chem. Phys. (2006), in press.
[9] J. Kalinowski, W. Stampor, J. Szmythowski, M. Cocchi, D. Vrgili, P. Di Marco, *J. Chem. Phys.* 122 (2005) 154710.
[10] J.J. Singh and A. Eftekhari, *NASA Technical Paper* 3619 (1996).
[11] S.W. Lai, M.C.W. Chan, T.C. Cheung, S.M. Peng and C.M. *Che, Inorg. Chem.* 38 (1999) 4046.,
[12] V.M. Miskowski and V.H. Houlding, *Inorg. Chem.* 28 (1989) 1529.
[13] T. Dienel, H. Proehl, T. Fritz and K. Leo, *J. Lumin.* 110(2004) 253.
[14] P. J. Hay, *J. Phys. Chem.* A 106 (2002) 1634.
[15] M.G. Colombo, T.C. Brunold, T. Riedener, H.U. Gudel, M. Fortsch and H.B. Burgi, *Inorg. Chem.* 33 (1994) 545.
[16] S.-J. Yeh, M.-F. Wu, C.-T. Chen, Y.-H. Song, Y. Chi, M.-H. Ho, S.-F. Hsu and C. H. Chen, *Advanced Mat.* 17(2005) 285.
[17] K. Goushi, R. Kwong, J.J. Brown, H. Sasabe and C. Adachi, *J. Appl. Phys.* 95 (2004)7798.
[18] T. Tsuboi, *J. Lumin.* 119-120 (2006) 127.
[19] C. Adachi, M.A. Baldo, S.R. Forrest, S. Lamansky, M.E. Thompson and R.C. Kwong, *Appl. Phys. Lett.* 78 (2001) 1622.
[20] W.J. Finkenzeller and H. Yersin: *Chem. Phys. Lett.* 377 (2003) 299.
[21] N.G. Park, G.C. Choi, J.E. Lee and Y.S. Kim, *Current Appl. Phys.* 5 (2005) 79.
[22] T. Tsuboi and N. Aljaroudi, unpublished.
[23] T. Tsuzuki and S. Tokito, *Proc. Int. Super-Functionality Organic Devices*, IPAP Conf. Series 6 (2005) 99.
[24] W. Sotoyama, T. Satoh, H. Sato, A. Matsuura and N. Sawatari, *J. Phys. Chem.* A 109 (2005) 9760.
[25] A.P. Wilde and R.J. Watts, *J. Phys. Chem.* 95 (1991) 622.
[26] V. Bulovic R. Deshpande, M.E. Thompson and S.R. Forrest, *Chem. Phys. Lett.* 308 (1999) 317.

[27] V. Bulovic, A. Shoustikov, M.A. Baldo, E. Bose, V. Kozlova, M.E. Thompson and S.R. Forrest, *Chem. Phys. Lett.* 287(1998) 455.

[28] C.F. Madigan and V. Bulovic, *Phys. Rev. Lett.* 91 (2003)247403.

[29] C.W. Tang and S.A. VanSlyke, *Appl. Phys. Lett.* 51 (1987) 913.

[30] P.E. Burrows, Z. Shen, V. Bulovic, D.M. McCarty, S.R. Forrest, J.A. Cronin and M.E. Thompson, *J. Appl. Phys.* 79 (1996)7991.

[31] S.A. VanSlyke, C.H. Chen and C.W. Tang, *Appl. Phys. Lett.* 69 (1996) 2160.

[32] H.Q. Tang, H.X. Liao, K, Xu, Z.X. Zhou and L.H. Zhu, *J. Lumin.* 118 (2006) 39.

[33] J. Kalinowski, N. Camaioni, P.D. Marco, V. Fattori and G. Giro, *Int. J. Electron.* 81 (1996) 377.

[34] R.J. Curry and W.P. Gillin, *J. Appl. Phys.* 88 (2000) 781.

[35] M. Muccini, M.A. Loi, K. Kenevey, R. Zamboni, N. Masciocchi and A. Sironi, *Adv. Materials* 16 (2004) 861.

[36] R. Ballardini, G. Varani, M.T. Indelli and F. Scandola, *Inorg. Chem.* 25 (1986) 3858.

[37] B. Henderson and G.F. Imbusch, *Optical Spectroscopy of Inorganic Solids* (Oxford Sci. Publ., Oxford, 1989).

[38] B.C. Lin, C.P. Cheng, Z.Q. You and C.P. Hsu, *J. Am. Chem. Soc.* 127 (2005) 66.

[39] Y. Farge and M.P. Fontana, *Electronic and Vibrational Properties of Point Defects in Ionic Crystals* (Amsterdam, North-Holland, 1974)

[40] M. Maestri, D. Sandrini, V. Balzani, A. von Zelewsky and P. Jolliet, *Helv. Chim. Acta* 71 (1988) 1053.

[41] S. Lamansky, P. Djurovich, D. Murphy, F. Abdel-Razzaq, S.R. Forrest and M.E. Thompson, *J. Am. Chem. Soc*, 123 (2001) 4304.

[42] C.B. Blanton, Z. Murtaza, R.J. Shaver and D.P. Rillema, *Inorg. Chem.* 31 (1992) 3230.

[43] T. Watanabe, K. Nakamura, S. Kawami, Y. Fukuda, T. Tsuji, T. Wakimoto,. S. Miyaguchi, M. Yahiro, M. Yang and T. Tsutsui, *Synth. Met.* 122 (2001) 203.

[44] H. Mu, H. Shen and D. Klotzkin, Solid-St. Electro. 48 (2004) 2085.

[45] E.M. Han, L.M. Do, N. Yamamoto and M. Fujihira, *Thin Solid Films* 273 (1996) 202.

In: Trends in Optical Materials Research
Editor: Owen G. Gardens, pp. 91-112

ISBN: 1-60021-290-5

Chapter 3

UNDERSTANDING THE UPCONVERSION LUMINESCENCE PROPERTIES OF RARE-EARTH DOPED OXIDE NANOCRYSTALS

Amitava Patra*

Department of Materials Science, Indian Association for the Cultivation of Science, Jadavpur, Kolkata 700 032, India

ABSTRACT

Here, we report the role of dopant and co-dopant concentration on the modification of crystal structure and the upconversion properties (tuning and efficiency) of rare-earth doped oxide nanocrystals. We have synthesized rare-earth doped oxide nanocrystals using non-ionic surfactant-stabilized reverse emulsions. The non-ionic surfactant in the Span series, i.e. sorbitan monooleate (Span 80) of hydrophilic-lipophilic balance (HLB) values of 4.3 has been used for the stabilization of emulsions. Bright green (550 nm) and red (670 nm) upconversion emission are observed at room temperature with 980 nm excitation from the $^4S_{3/2}$ and $^4F_{9/2}$ levels of ZrO_2:Er^{3+} nanocrystals. The total luminescence values are 11960, 7132 and 302 Cd/m^2 for 1000°C heated 0.05,0.25 and 1.0 mol% Er doped ZrO_2 nanocrystals, using 850 mW of cw excitation power source. The effects of co-dopants (Yb^{3+}, Tm^{3+} and Y^{3+}) in the ZrO_2: Er matrix on the upconverted emission are also reported. The emission spectra and the pump intensity dependence of the luminescence intensities are used to understand the excitation mechanisms,i.e. excited-state absorption (ESA) or energy transfer upconversion (ETU) processes. It is observed that the upconversion intensity increased as the excitation power increased under lower excitation power, but saturated at higher excitation power at high concentration of dopant ions.

*Author to whom correspondence should be addressed; electronic mail: msap@iacs.res.in Fax: (91)-33-2473-2805

1. INTRODUCTION

Towards the end of the 20th century and into new millennium, there has been a renaissance in the study of pure and doped nanomaterials for photonic applications [1]. One class of such materials is represented by rare-earth doped nanocrystals that have been investigated as candidates to be used as phosphors in luminescent displays, cold lamps, lasers, amplifiers, upconversion and imaging of biological systems [1-4]. Combining the promising optical properties of rare-earth ions and nanoparticles, the study of frequency upconversion (UC) processes of rare-earth ions in oxide nanoparticles is important. As a matter of fact, upconversion emissions of rare-eath ions -doped nanocrystals have already been achieved in different crystal structure hosts, such as ZrO_2, Y_2O_3, Lu_2O_3, $BaTiO_3$, TiO_2, ZnO, LaF_3, and Gd_2O_3 [5-20]. As these potential applications are still very much in the design-phase, further fundamental research in the field of upconversion spectroscopy remains a challenge. The process of photon upconversion is a way to convert long-wavelength excitation radiation into shorter wavelength output radiation. Upconversion is one of the most studied nonlinear optical processes compared to second harmonic generation and two-photon absorption. The upconversion is based on the existence of at least two real, metastable emitting states, which is the basic difference from second harmonic generation and two-photon absorption. As a result, upconversion has the potential to be a very efficient process. The predominant mechanisms of upconversion processes are excited state absorption and energy transfer upconversion. In both of these cases, a simple cw diode laser can be used for the upconversion process, in contrast to expensive mode locked lasers used in two-photon upconversion in organic fluorophore. Therefore, efficient upconverting rare-earth doped nanocrystals could be good replacement of organic dyes in two-photon confocal microscope imaging. Presently it is recognized that the UC efficiency depends on the concentration of dopant, crystal structure and crystal host [5, 8 and13]. Also, the process of "miniaturization" of materials to nanometer scale revealed that the radiative electronic relaxation probabilities of rare-earth ions doped nanoparticles may be significantly different from their bulk counterparts [21-23]. It is well established that in the luminescence of rare-earth ions the highest phonon frequencies of the host lattice are responsible for non-radiative relaxations [24]. In accordance with the energy law, the presence of a large gap between emitting and terminal levels reduces the probability of nonradiative decay. Therefore, lower host phonon energy has a greater number of phonons connecting the emitting level with the next lower level. This will decrease the non-radiative relaxation probability and increase the quantum yield of luminescence. To overcome the phonon decay problem it is necessary to choose a lattice which has much lower phonon energy. The zirconia matrix seems to be an ideal medium for preparation of highly luminescent materials because it is chemically and photochemically stable, has a high refractive index and low phonon energy [8]. In case of rare-earth ions, the electronic f-f transitions involve electrons that are localized in atomic orbital of the ions. Therefore, no size-dependent quantization effect (due to confinement of delocalized electrons) is found in these transitions. However, a nanostructure can be utilized to manipulate the excitation dynamics of rare-earth ions. A very broad range of unprecedented optical properties can be observed by changing the host lattice and dopant concentration. The main goal of this work is to understand the role of dopant, co-dopant concentration, on the upconversion luminescence efficiency of rare-earth ion doped oxide nanocrystals.

Various chemical methods have been developed to control the size and distribution of nanoparticles during synthesis. Due to the high surface area and the surface energy, the nanoparticles undergo aggregation into larger particles. In conventional methods, uncontrolled nucleation and subsequent growth of the precipitated particles in a bulk aqueous medium finally generates larger particles with a wide size distribution. To overcome these problems, many recent studies have used emulsions to control the size and morphology of the nanoparticles [25]. Generally the reverse micelles formed in water-in-oil (w/o) type emulsions act as the micro- or nano- reactors in which reactions are carried out [26, 27]. It is known that a minimum critical micelle concentration (CMC) is needed for the surfactant molecules to self-aggregate, giving rise to "reverse micelles" in organic solvents. Here, we use a sol-emulsion-gel process to prepare rare-earth doped oxide nanoparticles. In the present investigation, the two immiscible liquids, i.e. for the w/o in oil type emulsions (reverse emulsion) where dispersion of the water phase in the oil phase are used. It is known that thermodynamically, the increase in surface area (ΔA) of the dispersed phase associated with a free energy (ΔG) can be represented as [28]

$$\Delta G = \gamma \Delta A \qquad (1)$$

where γ is the interfacial tension. The equation indicates that a low interfacial tension favors droplet disruption. In fact, the high interfacial tension between the dispersed water phase and continuous oil phase is reduced by the addition of a system compatible amphiphilic surface active agent or surfactant (emulsifier in the present case), i.e. the molecule with a polar head and a non-polar tail [28]. The surfactant molecules orient themselves according to the polarities of the involved chemical constituents. Thus, due to the high polarity of water, the polar heads of the surfactant molecules at the water-oil interface are oriented towards the water droplets and get adsorbed on their surfaces and prevent the coalescence of the droplets by steric hindrance occurring between two droplets (steric stabilization). Therefore, the emulsifier added to the system: (i) lowers the interfacial tension between two immiscible liquids followed by droplet disruption and (ii) prevents re-coalescence of droplets by adsorbing on their surfaces, thus producing a stable emulsion. In the present investigation, non-ionic surfactant Span 80 is used to stabilize the reverse emulsions. The surfactants in the Span series, i.e. Span 80, in the present case, are basically the fatty acid esters of anhydro sorbitols which are good oil soluble emulsifying agents. A high (hydrophilic-lipophilic balance) HLB value of the surfactant indicates strongly hydrophilic character while a low value is an indication of a strong hydrophobic nature. Considering these points, non-ionic surfactants, i.e. Span 80 with HLB values of 4.3 was selected for the present study. Scheme1 depicts the molecular structures of the amphiphilic Span 80, where the hydrophilic sorbitan group acts as a 'polar head' and the hydrophobic fatty acid group acts as the 'non-polar tail'.

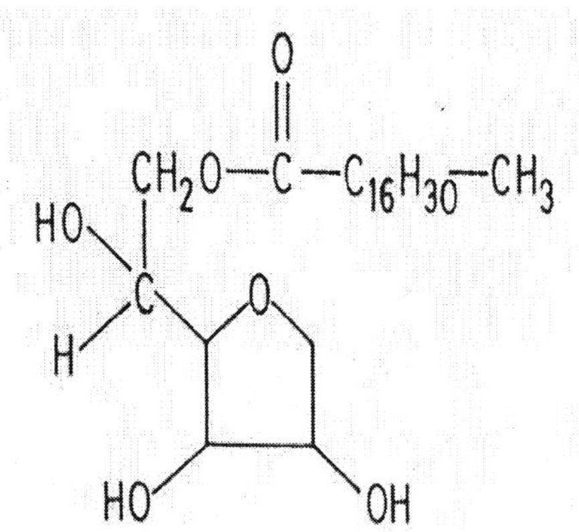

Span-80

Scheme 1. Molecular structures of Span -80

2. EXPERIMENTAL PROCEDURE

The sol-emulsion-gel method was used for the preparation of Er^{3+} doped oxide nanoparticles. A flow chart shown in scheme 2 summarizes the synthetic approach used for the preparation of Er^{3+}-doped ZrO_2 nanoparticles.

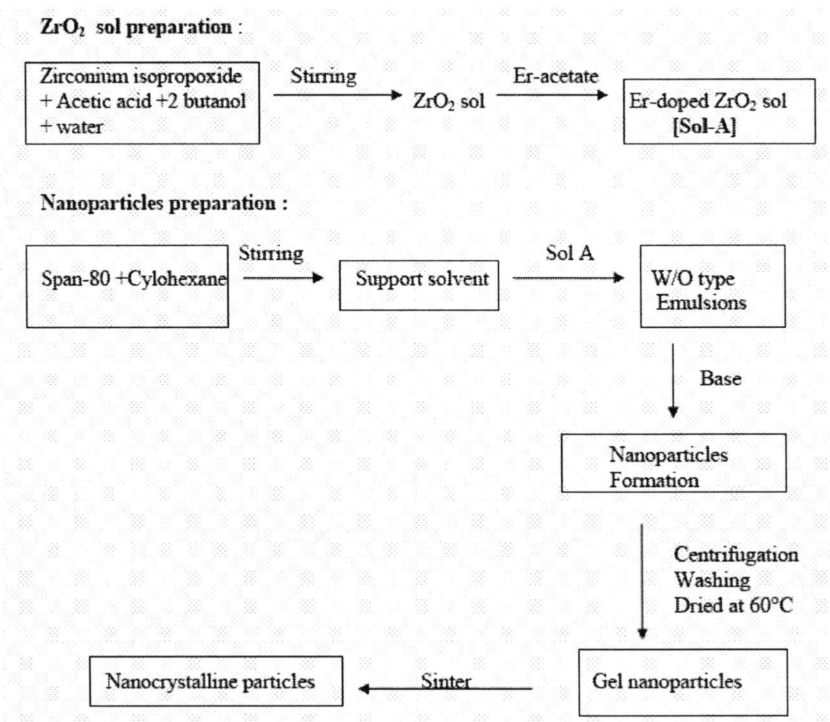

Scheme 2. Schematic flow chat of nanocrystalline particles preparation

Zirconia isopropoxide (Fluka) and erbium acetate were used as the starting materials. The ZrO_2 sol preparation technique is described below. First, three milliliters of glacial acetic acid was slowly added to 10 ml of zirconia isopropoxide and stirred for 30 min. Then 20 ml of n-propanol was added to the solution, which was further stirred for 15 min at room temperature. 4ml of 50% aqueous acetic acid was slowly added to the above solution under stirring which resulted in a clear transparent sol. Then, a stoichiometric amount of erbium acetate was added to this sol. The emulsified sol droplets were obtained through water-in-oil (w/o) type emulsions with cyclohexane and sorbitan monooleate (Span 80, fluka) as the organic liquid (oil phase) and non-ionic surfactant, respectively. The principle of the process involves the dispersion of an aqueous sol containing the desired constituents under agitation into a water-immiscible organic liquid for low dielectric constant. The support solvent containing 5 vol% of span 80 in cyclohexane was used for emulsification, i.e. for the preparation of water – in – oil (w/o) type emulsion in the present study. The volume ratio of the sol and cyclohexane was 1: 4. A measured amount of erbium doped zirconia sol was then dispersed in the solvent under stirring condition. The sol droplets formed in the process were then gelled by the controlled addition of triethylamine (TEA). The gel particles were separated by centrifugation followed by washing with acetone and methanol. The product was dried at 60°C in air for 12 hrs. The dried materials were heated at different temperatures up to 1000°C for 1 hr. Er doped $BaTiO_3$ nanoparticles preparation technique is already described [13].

Transmission electron microscopy (TEM) was employed to determine the morphology and the particle size of the resulting powders; the crystalline phases of calcined powders were identified by X-ray diffraction (XRD). The crystallite sizes of the nanoparticles were calculated following the Scherrer's equation:

$$D = K\lambda/\beta\cos\Theta \qquad (2)$$

where K = 0.9, D represents crystallite size (Å), λ, the wavelength of Cu $k\alpha$ radiation and β the corrected half width of the diffraction peak. We pressed the particles to form a smooth, opaque flat disk for optical study. The samples were irradiated with a diode laser tuned to 980 nm. A CCD–coupled spectrometer recorded the fluorescence spectra. The absolute fluorescence intensity was measured with a Minolta LS-110 Luminance meter.

3. Results and Discussion

3.1. Formation of ZrO_2 Nanoparticles

In this study, ZrO_2 nanoparticles were obtained by sol-emulsion-gel technique. The proposed mechanism for the separation of ZrO_2 nanoparticles in presence of TEA is presented in Scheme 3. In presence of TEA, at pH 8, the above-mentioned ZrO_2 complex became destabilized due to the basic hydrolysis of the surfactant molecules (the fatty acid esters) attached to the nanoparticles and formed new complex (capped) with the nitrogen atom of TEA molecules.

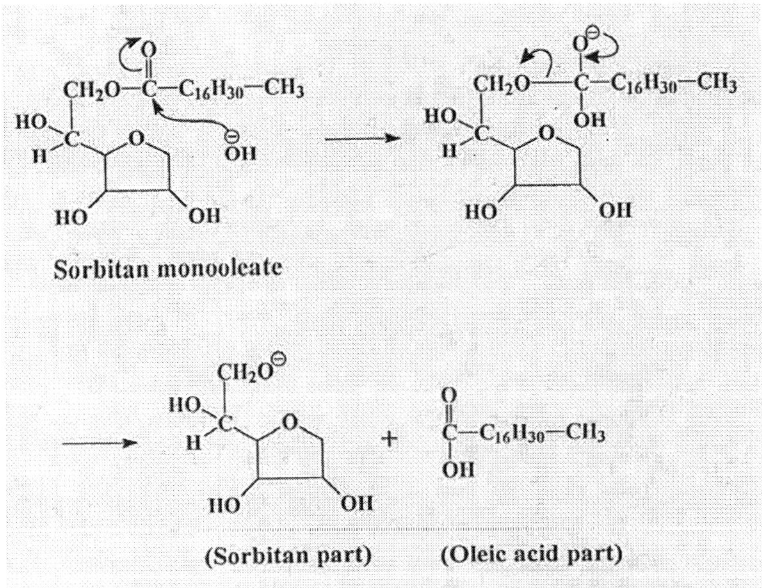

Scheme 3. Proposed mechanism for the separation of TEA capped ZrO_2 nanoparticles

The solubility of TEA-capped ZrO_2 nanoparticles decreased in the acetone medium, leading to the flocculation as free-standing particles. The basic hydrolysis of the surfactant molecules is considered to be consisted of the following essential steps:

(i) The TEA molecule in presence of H_2O form triethylammonium hydroxide $(C_2H_5)_3N + H_2O \rightarrow (C_2H_5)_3N^+HOH^-$

(ii) The triethylammonium hydroxide increases the pH of the system by liberating hydroxide (OH^-) ions

(iii) The (OH^-) ion attacks the carbon (C) atom of the ester carbonyl (>C=O) group in the surfactant molecule (sorbitan monooleate as a typical example), causing hydrolysis into the sorbitan and oleic acid parts (Scheme 3).

(iv) Hydrolysis of the sorbitan monooleate destabilizes the emulsion system followed by the flocculation of the generated particles.

3.2. Structural Investigations

Figure 1 shows the XRD patterns of 0.05 mol% Er^{3+} doped ZrO_2 nanocrystals obtained after heating at 900, 1000 and 1100^0C. Both monoclinic and tetragonal phases are present here. Two lines at 28.2° ($11\bar{1}$) and 31.4° (111) for the monoclinic phase (m) (JCPDS 37-1484) and the line at 30.17° (111) are observed for the meta-stable tetragonal phase (JCPDS 17- 0923) of ZrO_2. The volume fractions of monoclinic and tetragonal phases are 90.62% and 9.38%, respectively for 900^0C heated samples. The volume fraction of monoclinic is increased from 90.62% to 98.84 % with increasing the temperature from 900 to 1000°C. Finally, the monoclinic phase is reached to 100% at 1100°C. It is clearly seen that with increasing the temperature of heating the volume fraction of the monoclinic phase is increased. The shifting of the lines with changing the temperature of heating indicates structural change with temperature of heating, indicating Er^{3+} ions substitute the Zr^{4+} ions. The calculated crystallite sizes from the (11 $\bar{1}$) peak are 31.49, 42 and 42.06 nm for 900°C, 1000°C and 1100°C, respectively. Therefore, crystallite size increases with increasing the temperature of heating. It reveals that crystal phase can be tuned by monitoring the temperature of heating.

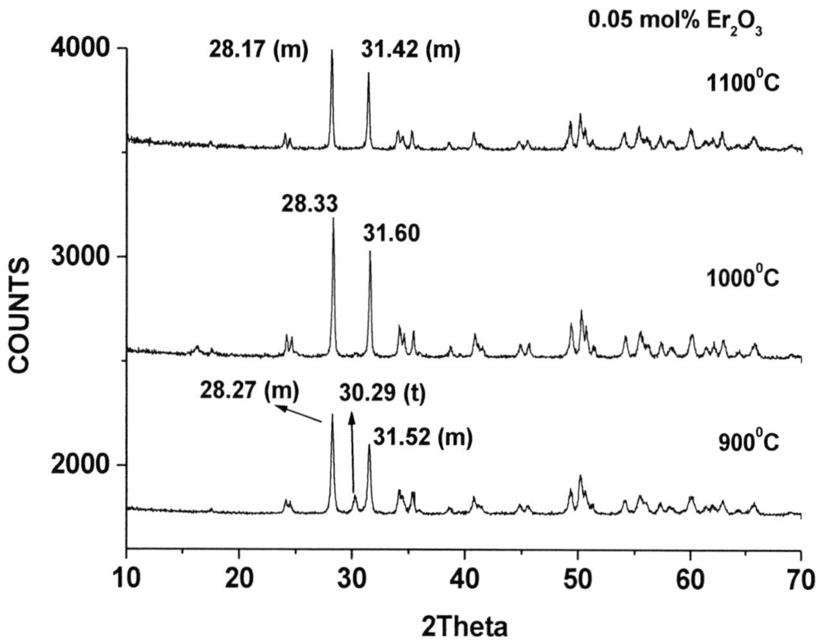

Figure 1. X-ray powder diffraction patterns of 0.05 mol% Er doped ZrO_2 nanocrystals heated at different temperatures for 1 h

Figure 2 presents the XRD patterns of different concentration of Er doped ZrO_2 nanocrystals heated at 1000°C. The change of crystal phase with changing the dopant concentration is observed at fixed temperature. The volume fraction of the monoclinic phase is 98.99% for the undoped ZrO_2 sample. However, the volume fraction of monoclinic phase decreases from 98.8 to 71.92% with increasing the dopant concentration from 0.05 to 1.0 mol% Er-ion. On the other hand, the volume fraction of tetragonal phase increases with increasing the dopant concentration. Therefore, crystal phase composition can be tuned by changing the dopant concentration. At low dopant concentrations, monoclinic structure of zirconia is observed with space group of $P2_1/c$, where each Zr atom is in seven-fold coordination with oxygens [29]. The crystal structure of tetragonal ZrO_2 is a body centered lattice with the space group $P4_2/nmc$. The symmetries of lattice are 2/m (C_{2h}) and 4/mmm (D_{4h}) for monoclinic and tetragonal, respectively. It is well known that doping of trivalent ions in zirconia has been traditional approach for the metastable tetragonal phase stabilization [30]. The calculated crystallite sizes from the (11 $\bar{1}$) peak are 41.57, 42.06, 30.72 and 42.28 nm for undoped, 0.05, 0.2 and 1.0 mol% Er doped ZrO_2 nanocrystals at 1000°C, respectively.

The tetragonal content increases from 1.2 to 28.08% for 0.05 to 1.0 mol% Er^{3+} ions concentration, respectively.

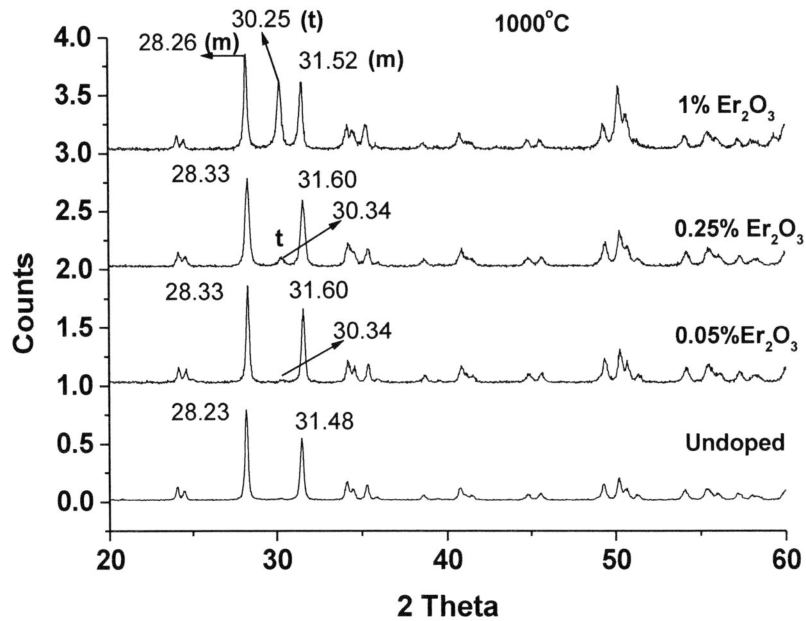

Figure 2. X-ray powder diffraction patterns of different concentrations of Er doped ZrO_2 nanocrystals prepared at 1000°C for 1 h

However, the crystallite size is not dependent on dopant concentration in this study. Further increase in temperature to 1100°C, 100% monoclinic phase is obtained for undoped and 0.05 mol% Er-doped ZrO_2 samples. The volume fractions of monoclinic phase are 95.7% and 88.67% for 1000°C heated 0.2 and 1.0 mol% Er-doped ZrO_2 nanocrystals, respectively. We did not observe any significant change in the crystallite sizes with changing the concentration of dopant. The sizes are 41.9, 42, 34.8 and 42.7 nm for undoped, 0.05, 0.2 and 1.0 mol% Er doped ZrO_2 nanocrystals prepared at 1100°C, respectively. Results reveal that crystal phase changes with changing the dopant concentration. The increase in the tetragonal phase is may be due to the removal of oxygen vacancies. The oxygen vacancies are considered to be responsible for the formation of tetragonal phase instead of forming the thermodynamically stable monoclinic phase upon crystallization. Usually, oxygen vacancies are introduced by incorporation of allo-valent ions, including rare-earth ions, into ZrO_2 lattice [30]. Monoclinic phase is most stable when the composition is close to the stiochiometry value. The symmetries of lattice are C_{2h}, D_{4h} and O_h for monoclinic, tetragonal and cubic, respectively. It is known that monoclinic phase is less symmetric than tetragonal phase and cubic phase. These results clearly indicate the crystal structure and symmetry change with changing the temperature of heating and dopant concentration.

Figure 3 shows the TEM micrographs of 1000°C heated 1.0 mol% Er doped and 0.25Yb-0.25Er co-doped ZrO_2 and selected area electron diffraction of 1000°C heated 1.0 mol % Er doped ZrO_2 nanocrystals. The average size of the particles is ~ 42 nm for Er- doped nanocrystals in agreement with the results obtained from X-ray diffraction studies. The figure of 3c shows the selected area of electron diffraction pattern. The diffraction spots are observed in the selected area electron diffraction (SAED) pattern and this diffraction pattern confirms the (002) plane of monoclinic crystal phase of ZrO_2.

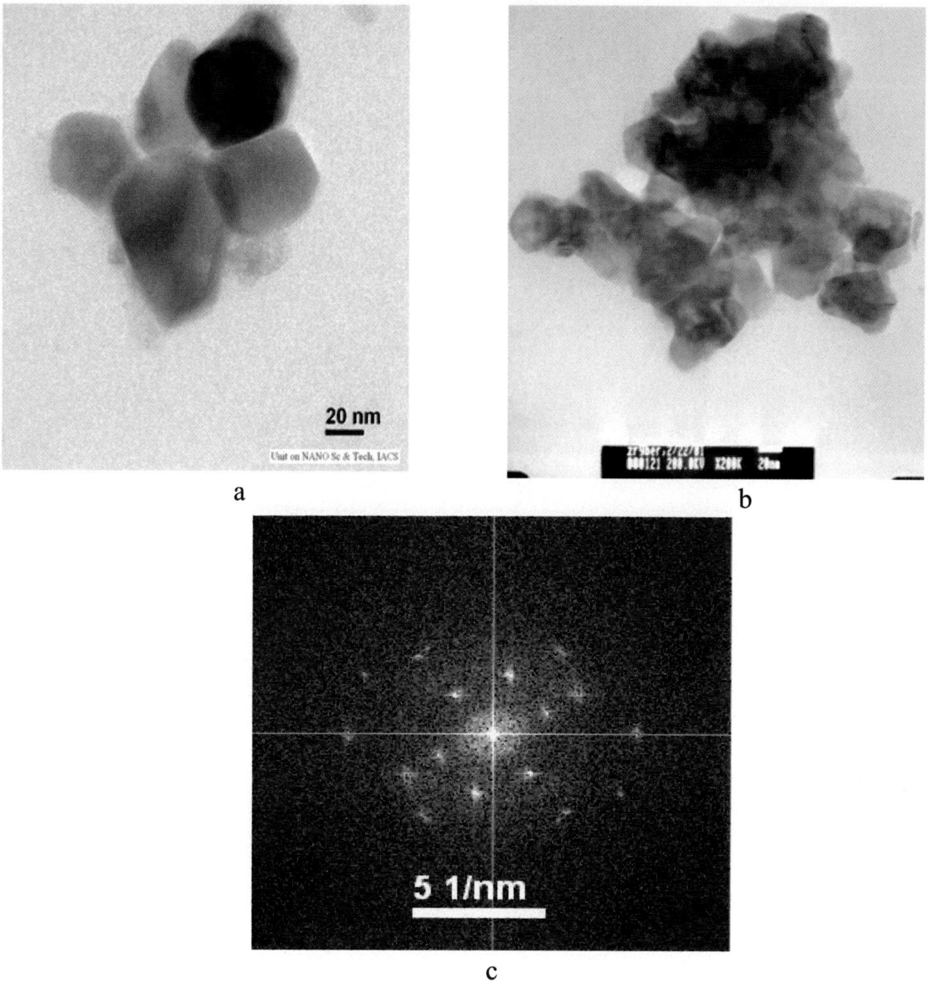

Figure 3. TEM micrographs Er doped (a) and Yb-Er co-doped (b) and selected area electron diffraction (c) 1.0 mol % Er doped ZrO_2 nanocrystals annealed at 1000°C

3.2. Upconversion Emission Properties

Effect of Concentration

The upconverted emission spectra of the ZrO_2:Er^{3+} nanocrystals in different concentrations of Er^{3+}-ions under infrared excitation are shown in Figs. 4(a-d). Generally, the f-f transition of rare-earth ion is very narrow and exhibit multiple structures derived from electronic interactions, as well as spin-orbit coupling. The distinct splitting of emission bands is observed for the Er^{3+}-ions due to the variation of the local structure (crystal field) around Er^{3+} ions in the ZrO_2 nanocrystals.

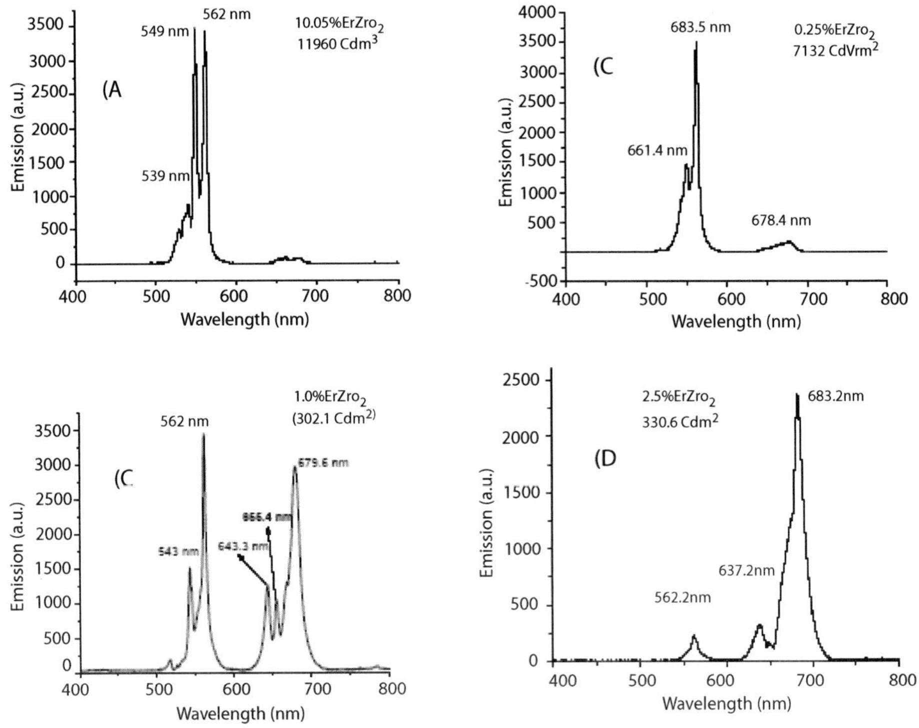

Figure 4. Upconverted emission spectra of different concentrations of Er^{3+} in ZrO_2 nanoparticles heated at 1000 °C

Therefore, the different spectral shape of green upconversion emission is due to the variation of the crystal structures of ZrO_2 nanocrystals. A significant change of upconversion emission wavelength from green to red is observed with increasing the concentration of dopant. It is clearly seen in the figures 4(a-d) that the relative increase in intensity of red emission with respect to the intensity of green emission is observed with an increasing concentration of erbium. A significant drop in the overall intensity is observed with an increasing concentration of Er^{3+} ions. The upconversion luminance values are 11960, 7132, 302.1 and 330.6 Cd/m^2 for 0.05, 0.25, 1.0 and 2.0 mol% Er doped samples, respectively [8]. It clearly reveals that concentration of ions plays an important role on the tuning of upconversion emission spectra. Figure 5 shows the variation of luminance value with changing the concentration of Er^{3+} ions in $BaTiO_3$ nanocrystals. Here, we also observed that there is a significant drop in the overall intensity with increasing concentration of Er^{3+} ions. The upconversion luminance values are 17130 Cd/m^2, 9487 Cd/m^2, 1653 Cd/m^2 and 540 Cd/m^2 for 0.25, 1.0, 2.5 and 10-mol% Er^{3+}doped $BaTiO_3$ nanocrystals, respectively [13]. The intensity ratio of the green emission to that of the red emission decreases with increasing the concentration of Er^{3+} ions. For example, in the 0.25-mol% sample, the ratio becomes 3:1, while in the 2.5 mol% sample we obtain a ratio of 0.888:1. On the other hand, higher concentration favors the red emission. Since the size of the Er^{3+} ion (~ 0.10 nm) is almost intermediate between those of the Ba^{2+} ion (~0.14 nm) and Ti^{4+} ion (~0.06 nm), it seems equally likely that the Er^{3+} ions occupy the site of either ion in the $BaTiO_3$ lattice.

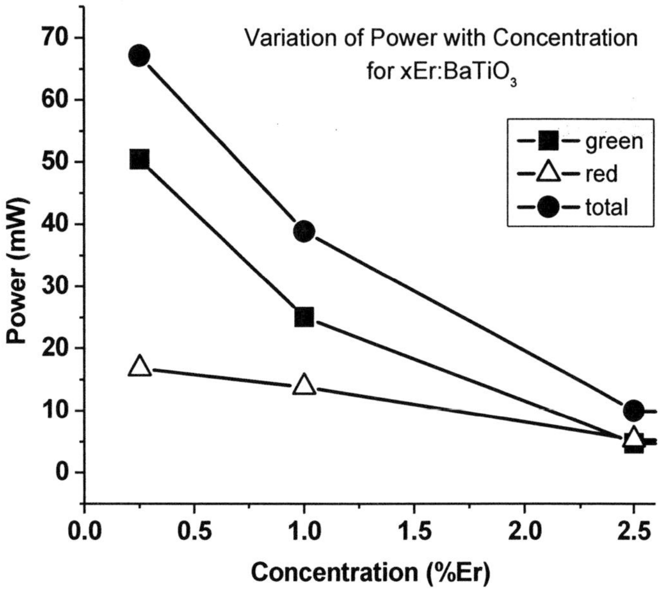

Figure 5. Variation of the luminescence value with changing the concentration of Er^{3+} ions in $BaTiO_3$ nanocrystals prepared at 1000°C

At low concentration (0.25 mol%) the mean distance between the Er^{3+} ions is estimated by $R = 0.62/ (N)^{1/3}$ to be 1.81 nm but for high concentration (2.5 mol%) the distance is 0.84 nm [13]. Therefore, the erbium ions in low concentration are usually randomly distributed in the host lattice and the Er^{3+} - Er^{3+} distances are too far apart. However, at higher concentration, the distances between two Er^{3+} ions are shortening [31], thus leading to formation of Er^{3+} clusters. As a result concentration-quenching processes will be the predominant nonradiative decay process at higher concentrations as a result the emission intensity sharply decreases.

The mechanism of the upconverted emission of Er^{3+} has been well established in the literature [2]. To explain this phenomenon, we use the typical energy level diagram of Er^{3+} ions as shown in Fig. 6. The diode-laser wavelength (980 nm) matches the absorption transition between the ground state, $^4I_{15/2}$, and the excited level $^4I_{11/2}$. After first-level excitation, the same laser pumps the excited atom from the $^4I_{11/2}$ to the $^4F_{7/2}$ level. Subsequent nonradiative relaxation populates the $^4S_{3/2}/\ ^2H_{11/2}$ and the $^4F_{9/2}$ levels. Bright green (550 nm) and red (670 nm) emission are observed due to the transitions $^4S_{3/2} \rightarrow\ ^4I_{15/2}$ and $^4F_{9/2} \rightarrow\ ^4I_{15/2}$, respectively. Usually, the erbium ions are randomly distributed in the host lattice at low concentration, but at higher concentration concentration quenching occurs due to closer nearest neighbor of the Er^{3+} ions. At low dopant concentration, the $^4S_{3/2}/\ ^2H_{11/2}$ levels decay radiatively to $^4I_{15/2}$. Therefore, the green emission has a higher intensity and this emission will be mainly contributed by the excited-state absorption (ESA) process. At a higher concentration, the luminescent lifetime of $^4S_{3/2}/\ ^2H_{11/2}$ levels are shortened as a result of the cross-relaxation processes between ($^2H_{11/2} \rightarrow\ ^4I_{9/2}$) and ($^4I_{15/2} \rightarrow ^4I_{13/2}$) transitions. At high dopant concentration, the energy transfer upconverting process will becomes dominant at this time. As a result, the green emission disappears and the red emission is observed upon excitation at 980 nm.

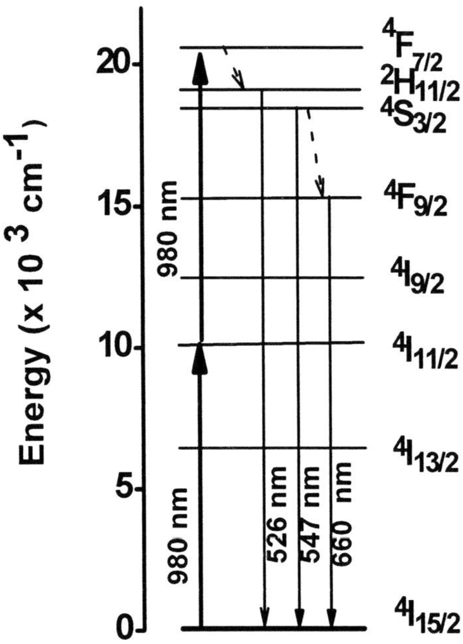

Figure 6. The energy level diagram of Er^{3+} ion under infrared excitation

The enhanced interactions between the ions induce increased relaxation of $^4I_{11/2}$ – $^4I_{13/2}$ at higher dopant concentration (Fig.6). As the long lifetime of $^4I_{13/2}$ level, the electrons populated in this level will reabsorb the excited light of 980 nm to arrive at $^4F_{9/2}$ level, and then go back to the ground state to emit the red light (670 nm). It is reported [8] that at a higher concentration, the luminescent lifetime of $^4S_{3/2}$/ $^2H_{11/2}$ levels are shortened due to the cross-relaxation (CR) processes between ($^2H_{11/2} \rightarrow {}^4I_{9/2}$) and ($^4I_{15/2} \rightarrow {}^4I_{13/2}$) transitions. This cross - relaxation process is dominant at higher concentration of Er^{3+} ions. Therefore, concentration quenching process takes place at higher concentrations of dopant ions.

Effect of Co-Dopant

Figure 7 shows the upconverted spectra of Y^{3+} co-doped Er^{3+} in ZrO_2 nanocrystals prepared at 1000°C. The upconversion luminance values are 1199 and 905 Cd/m^2 for 1.0Er1.5Y and 2.5Er1.5Y doped ZrO_2 nanocrystals prepared at 1000°C, respectively. Here, co-dopant concentration is fixed (1.5 mol%) and concentration of Er^{3+} has changed from 1.0 to 2.5 mol%. However, the upconversion luminance values are 302 and 330 Cd/m^2 for 1.0 and 2.5 mol% Er doped ZrO_2 nanocrystals in absence of co-dopant (Y^{3+}-ion). This result clearly indicates that co-dopant plays an important role on increasing the upconversion emission efficiency. A significant increase of the upconverted emission in the presence of Y^{3+} may be due to better miscibility of Er^{3+} in yttrium co-doped zirconia, which hinders the creation of large clusters [32] and circumvents the concentration quenching process for the same concentration of Er^{3+}. There is a possibility for the existence of erbium ions in either the yttrium or zirconium lattice site that prevents the aggregation resulting in the circumvention of the non-radiative process ($^2H_{11/2} \rightarrow {}^4I_{9/2}$) / ($^4I_{15/2} \rightarrow {}^4I_{13/2}$) due to cross relaxation. Therefore, $^4S_{3/2}$/ $^2H_{11/2}$ levels decay radiatively to the $^4I_{15/2}$ level. As a result, an overall increase in intensity is observed.

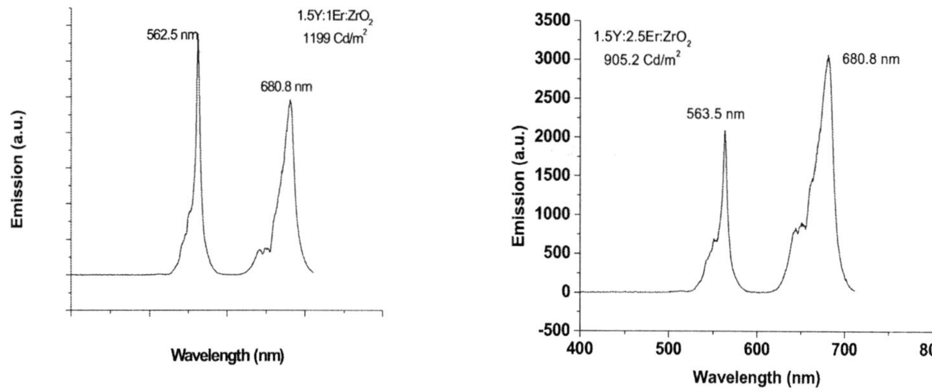

Figure 7. Upconverted emission spectra of different concentrations of Y^{3+}: Er^{3+} co-doped ZrO_2 nanoparticles heated at 1000 °C

Figure 8 presents the upconverted emission spectra of 1000°C heated ZrO_2 : Er^{3+} co-doped with Yb^{3+} -ions. The upconversion luminance values are 8850 and 5786 Cd/m^2 for 1000°C heated 0.25Er0.25Yb and 0.25Er1.25Yb doped ZrO_2 nanocrystals, respectively. However, without co-dopant (Yb^{3+}-ion), the value is 7132 Cd/m^2 for 0.25 mol% Er doped ZrO_2 samples. Here, we also observed increase of the upconverted emission in the presence of Yb^{3+} -ions. The concentration quenching process takes place in higher co-dopant concentration i.e 0.25 Er1.25Yb. A typical energy level diagram for the upconverted emission from a sample co-doped with Yb^{3+} and Er^{3+} ions under infrared excitation is shown in Figure 9 to explain the mechanism. In this system there are three possibilities for this upconversion process. The first possibility is simple excited state absorption (ESA), where Yb^{3+} ions do not play any role. A second possibility is energy transfer upconversion (ETU), where two excited ($^4I_{11/2}$) Er^{3+} ions interact with each other and one ion is de-excited to $^4I_{15/2}$ and the other is excited to $^4F_{7/2}$. This mechanism can be described as:

$$2\ ^4I_{11/2} \rightarrow\ ^4I_{15/2} +\ ^4F_{7/2} \tag{3}$$

A third mechanism for upconversion is energy transfer from Yb^{3+} ($^2F_{5/2}$) to Er^{3+} ($^4I_{11/2}$). The ET from Yb^{3+} ($^2F_{5/2}$) to Er^{3+} ($^4I_{11/2}$) depends on the Yb^{3+} concentration, i.e. the transfer process depends upon the donor and acceptor concentration [33]. During the lifetime of the $^4I_{11/2}$ level a second photon is absorbed by Yb^{3+} and energy transferred to Er^{3+} and finally the transition takes place from $^4I_{11/2}$ level to $^4F_{7/2}$ level. The possible channels for the green and red upconversion luminescence are given below:

$$^2F_{5/2}\,(Yb^{3+}) +\ ^4I_{11/2}\,(Er^{3+}) \rightarrow\ ^2F_{7/2}\,(Yb^{3+}) +\ ^4F_{7/2}\,(Er^{3+}) \tag{4}$$

The $^4F_{7/2}$ (Er^{3+}) state decays nonradiatively to the $^4S_{3/2}$/ $^2H_{11/2}$ and $^4F_{9/2}$ (Er^{3+}) levels. The green emission is observed from the $^4S_{3/2} \rightarrow\ ^4I_{15/2}$ transition while the $^4F_{9/2} \rightarrow\ ^4I_{15/2}$ transition produces red emission. It reveals that the presence of Yb^{3+} -ions in ZrO_2: Er^{3+} nanocrystals increases the overall upconverted emission intensity.

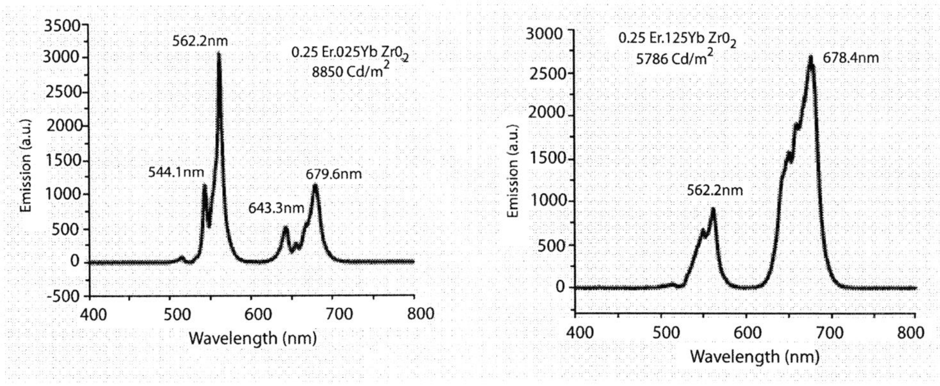

Figure 8. Upconverted emission spectra of different concentrations of Yb^{3+}: Er^{3+} co-doped ZrO_2 nanoparticles heated at 1000 °C

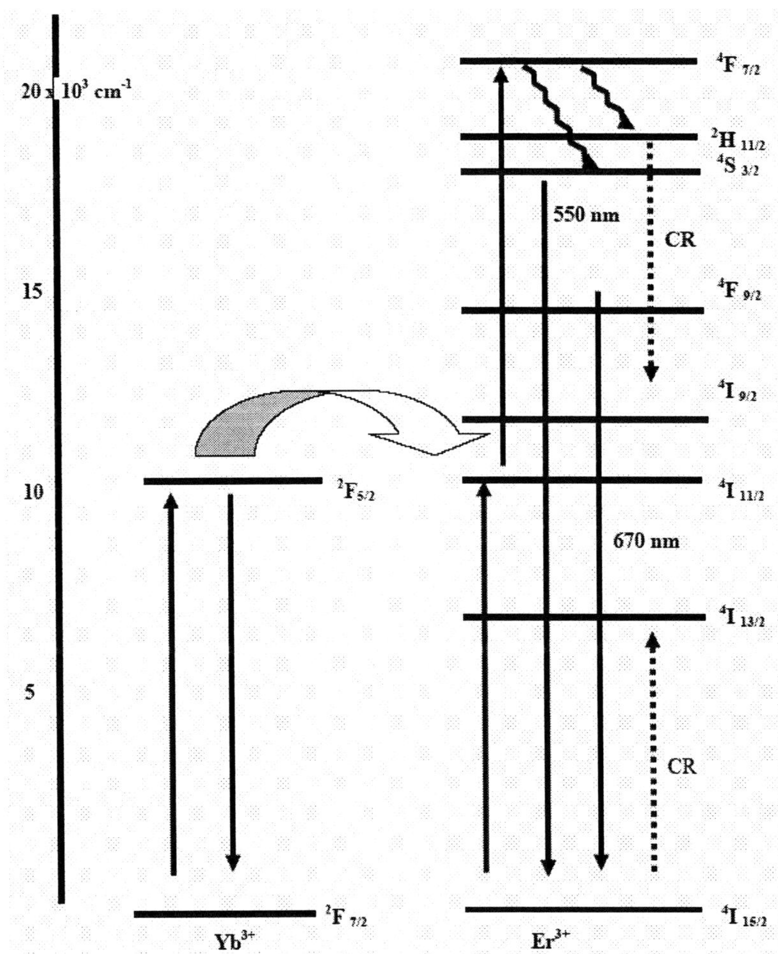

Figure 9. The energy level diagram for Yb^{3+} and Er^{3+} ions under infrared excitation

Figure 10 shows the upconverted emission spectra of Yb^{3+}: Tm^{3+} co-doped ZrO_2 nanocrystals prepared at 1000°C. It is interesting to note that the emission covers nearly the whole visible spectral region. Here, we demonstrate the white light generation by single wavelength excitation, which is significant for photonic application. A detailed investigation has been made in our previous work [19].

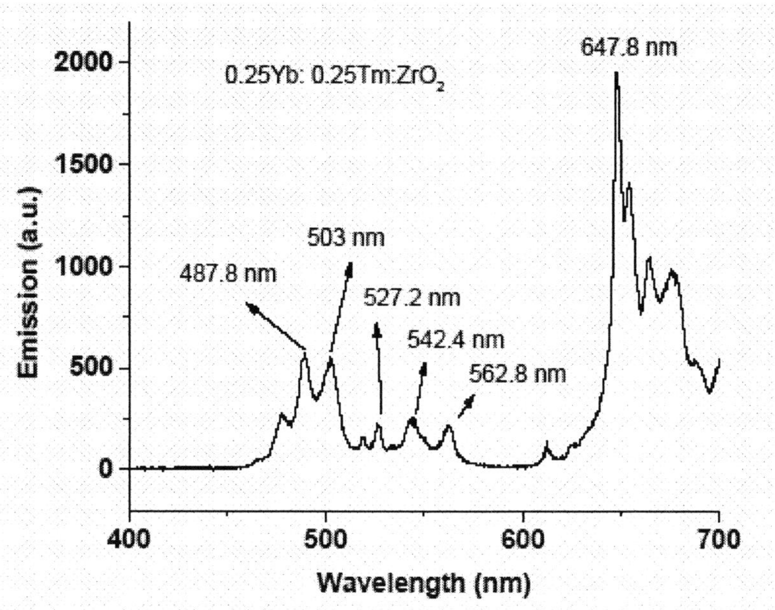

Figure 10. Upconverted emission spectra of Yb^{3+}: Tm^{3+} ions co doped ZrO_2 nanoparticles heated at 1000 °C

Effect of Crystal Size

Figure 11 shows the upconverted emission spectra of the 1.0 mol% Er^{3+} doped $BaTiO_3$ nanoparticles prepared at three different temperatures of heating, 700°C, 850°C and 1000^0C. The intensity of upconversion emission increases with increasing the temperature of heating. The upconversion luminance values are 343 Cd/m^2, 1448 Cd/m^2 and 9487 Cd/m^2 for the samples prepared at 700°C, 850°C and 1000^0C temperatures, respectively. The particles size increases with increasing the temperature of heating. The estimated crystallite sizes are 30, 50 and 80 nm for 700, 850 and 1000°C, respectively. In this case, same crystal phase (cubic) of $BaTiO_3$ is obtained even with different temperature of heating. Therefore, the enhancement of the green upconversion emission intensities related to the annealing temperature is ascribed by two main factors; one is the growth of the particles, which induces a decrease of the number of surface defects in the host; the second is the loss in the amount of organic residue and hydroxyl groups, which lowers the multiphonon relaxation rate. It is shown [34] that the probabilities for dopant pair-state formation depend on the particle size and the contributions of ions at the surface sites become increasing important as the size decreases. As a result, for the same concentration of dopant ions, the fraction of pairs will be different for different particles sizes. In normal crystal, the surface does not influence on pair formation because the number of nearest-neighbors is lower at the surface. However, the fraction of dopant ion-pairs for a given concentration is higher in small crystal because of their high surface area. Therefore, it could be the reason for increasing the upconversion luminance value with increasing the size for the same concentration. Therefore, it may be suggested that the resonance energy transfer is responsible for efficient upconversion, because resonance energy transfers extends the lifetimes of the excited states of the rare-earth ions.

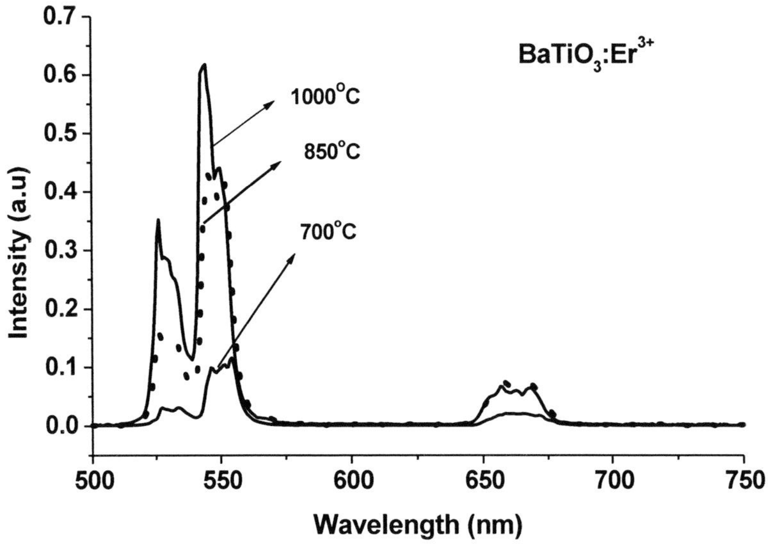

Figure 11. Upconverted emission spectra of 1.0 mol% Er^{3+} ions doped $BaTiO_3$ nanoparticles prepared at different temperatures, 700, 850 and 1000 °C

Power Dependence Study

The study of emission intensity versus pump intensity (pump power) is essential to understand the excitation mechanism in rare-earth doped upconverting materials. The two most common excitation processes that lead to emission from the higher energy states to the terminating state are energy transfer upconversion (ETU) and excited state absorption (ESA). It is well known for an unsaturated mechanism, the intensity of the upconverted luminescence, I_{upc}, is proportional to some power n of the excitation intensity I_i, i.e., $I_{upc} \alpha I_i^{\ n}$, where n = 2,3,... is the number of pump photons required to populate the emitting state and is determined from the slope of the line of the graph of intensity versus pump power in a log-log plot. The variations of the fluorescence intensity of the green and red emission versus the pump intensity in various compositions are plotted in Figures 12(a-d). It is seen that the emission intensity increases with increasing the pump power at low concentration of dopant ions. For 0.05 and 0.25 mol% Er doped samples, the green intensity increases but the red intensity increases for 1.0 and 2.5 mol% Er doped ZrO_2 samples. Most interesting result is that the unusual power-dependent upconversion luminescence is observed at higher concentration of dopant (Fig. 12c and 12d). As the power reaches to 0.7 W, the emission intensity approached a maximum, and then decreased or saturation as the power increased continuously. Figure 13 (a-b) shows the variations of the fluorescence intensity of the green and red emission versus the pump intensity in various compositions of Y-Er co-doped ZrO_2 nanocrystals.

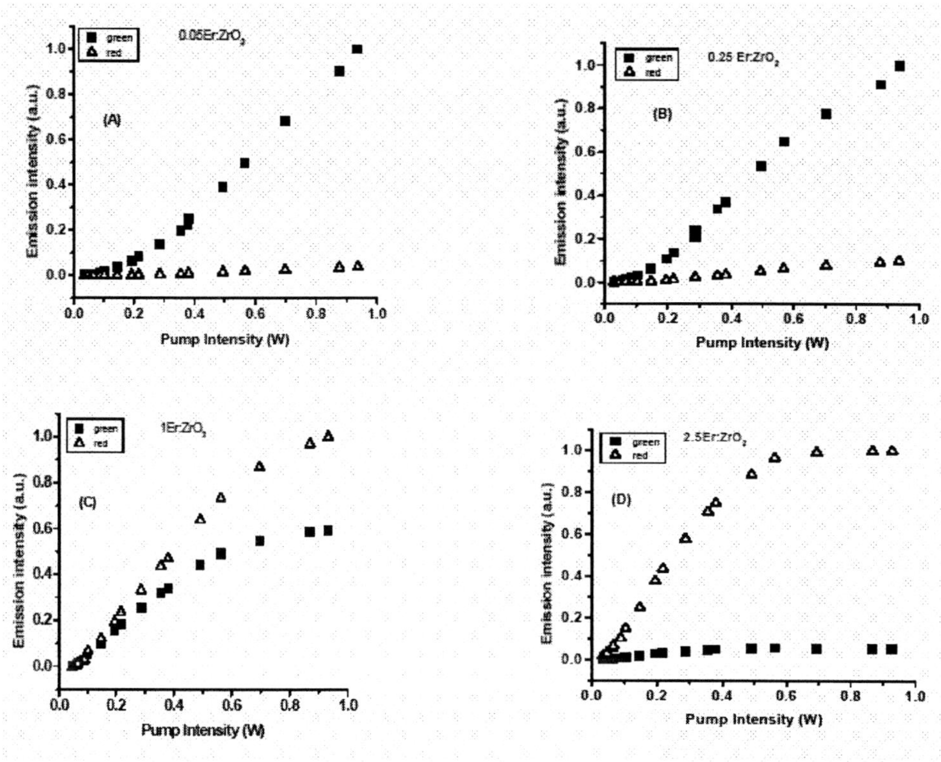

Figure 12. Power dependence upconverted emission intensity of different concentrations of Er^{3+} in ZrO_2 nanoparticles heated at 1000 °C

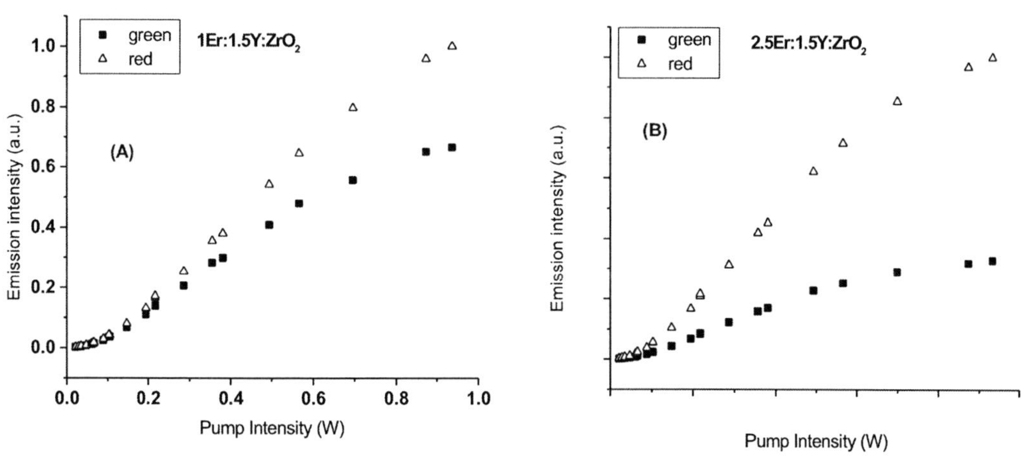

Figure 13. Power dependence upconverted emission intensity of Y^{3+}:Er^{3+} co-doped ZrO_2 nanoparticles heated at 1000 °C

The unusual power-dependent upconversion luminescence is also observed at high concentration in case of Y-Er co-doped samples. It is observed that the upconversion saturates at higher pump intensity or pump power. In case of parametric upconversion or higher order harmonic generations in quantum optics the cause of saturation is depletion of pump power. In such processes the upconverted intensity exhibits nth order dependence on the pump intensity, which levels off to linear behavior and conversion efficiency approaches unity. Unlike these systems the cause of saturation in ETU and ESA processes is the depletion of lower level populations. Therefore, in such cases at very high pump intensities the upconverted intensity should become constant or independent of pump intensity. Singh and Geusic [35] observed the "saturation" phenomenon of upconversion for higher pump power of $NdCl_3$ single crystal. Recently, Pollnau et.al [36] also studied the physical nature of the saturation phenomenon. There are several probabilities leading to the unusual power dependent upconversion luminescence behavior. As it is concentration dependence therefore this saturation behavior may be due to the thermal effect.

In order to analyze the UPC mechanism, the pump power dependence of the emission from $^4S_{3/2} \rightarrow {}^4I_{15/2}$ level at 550 nm was investigated on excitation at 980 nm for1000^{0}C heated 0.25 mol% Er^{3+} doped ZrO_2 nanocrystals and the results are shown in Fig. 14a.

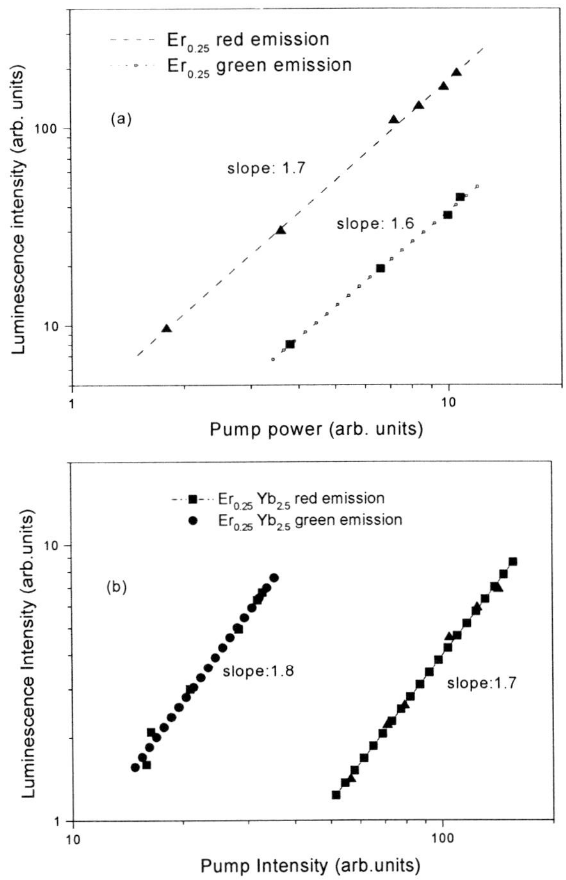

Figure 14. Pump power dependence upconversion emission (green and red) intensity plot for (a) 0.25 mol% Er^{3+} and (b) Er^{3+} (0.25) : Yb^{3+} (2.5 mol%) co-doped ZrO_2 nanocrystals prepared at 1000°C

The experimental data have been fit with a straight line with slope of ~2 (quadratic) which indicates that the green UPC emission from Er^{3+} is a two photon process for this emission band. The pump power dependence of the emission band at 670 nm was investigated on excitation intensity at 980 nm and the result is shown in Fig. 14a. Therefore, the experimental data for 550 and 670 nm emission bands have been fit with a straight line with slope of ~2 (quadratic), which confirms that both UPC emission lines are also a two-photon absorption process. Similarly, the pump power dependence of the emission from $^4S_{3/2} \rightarrow {}^4I_{15/2}$ level at 550 nm was investigated on excitation at 980 nm for 1000^0C heated 0.25 Er^{3+} 2.5 Yb^{3+} co-doped ZrO_2 nanocrystals and the results are shown in Fig. 14b.

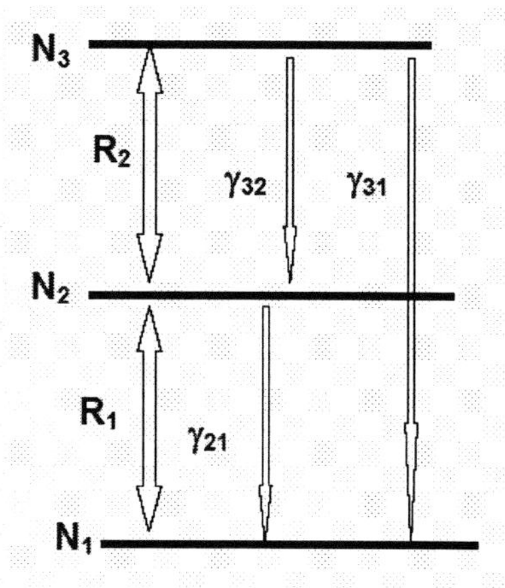

Figure 15. A three energy level diagram depicting two-photon upconversion with excited state absorption (ESA) process. N_i is the population density of the ith level, R_1 and R_2 are the first and second step excitation rate respectively and γ_{ij} is the transition rate from ith to jth level

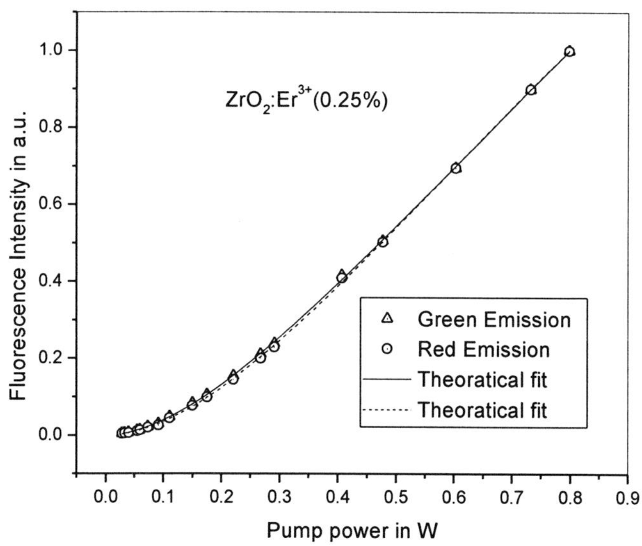

Figure 16. Variation of red and green upconverted emission intensities with pump power for 0.25 Er^{3+} ions doped ZrO_2 nanocrystals prepared at 1000°C

The experimental data for 550 and 670 nm emission bands have been fit with a straight line with slope of ~2 (quadratic), which confirms that the UPC emission lines are also a two-photon absorption process. We derived a relation between the fluorescence and the pump intensity by solving the rate equations based on the excitation and emission to understand the mechanism of the upconversion processes, i.e ESA or ETU processes. We take a simple two-photon stepwise excited-state absorption process as shown in Fig.15. The γ_{ij} is the decay rate or the transition probability from the ith to the jth level. $R_1 = \sigma_1 I_p/h\upsilon_p$ and $R_2 = \sigma_2 I_p/h\upsilon_p$ are the pump rates of the first and second steps, respectively. Here, σ_1 and σ_2 are the absorption cross sections of the first and second steps, respectively and υ_p is the frequency of the pump light beam. In this case only three levels are involved. The first excited state is further excited to the next excited stage and the fluorescence take place from the higher excited state as shown in Fig.15. If N_i is the population of the ith level and N is the total ion density, the rate equations for describing this case are given as

$$N_1 + N_2 + N_3 = N \qquad (5)$$

$$R_1 N_1 - (\gamma_{21} + R_1 + R_2)\, N_2 + (\gamma_{32} + R_2) N_3 = 0 \qquad (6)$$

$$R_2 N_2 - (\gamma_{31} + \gamma_{32} + R_2) N_3 = 0 \qquad (7)$$

By solving Eq. (5), (6) and (7), we obtained the steady state population N_3 of the upper excited state as

$$N_3 = N_s R_1^2 / \gamma_{21}(\gamma_{31}+\gamma_{32}) + (s\,\gamma_{21} + (2+s)\,\gamma_{31} + 2\,\gamma_{32})\, R_1 + 3sR_1^2 \qquad (8)$$

Where $s = \sigma_2/\sigma_1$. It is clear from Eq.(7) that the relation between the fluorescence intensity If and the pump intensity can be given by the following expression

$$I_f = \frac{KI_p^2}{(A + BI_p + CI_p^2)} \qquad (9)$$

where the coefficients K, A, B and C are given as

$$K = gNV\upsilon_f\sigma_1\sigma_2\gamma_{31}/h\upsilon_p^2$$
$$A = \gamma_{21}(\gamma_{31}+\gamma_{32})$$
$$B = [\sigma_2(\gamma_{21}+\gamma_{31}) + 2\sigma_1(\gamma_{31}+\gamma_{32})]/\, h\upsilon_p$$
$$C = 3\sigma_1\sigma_2/(h\upsilon_p)^2$$

Where g is the geometrical collection factor, N is the rare-earth ion density in the sample, I_f is the upconverted fluorescence intensity, I_P is the pumping intensity and V is the interaction volume. Therefore, for a two photon ESA upconversion process, the fluorescence intensity should be related to the pump intensity by the following Eq. 9. Variation of fluorescence intensity with pump power for 0.25 mol% Er^{3+} doped ZrO_2 nanocrystals prepared at 1000°C was recorded. It is seen that Eq. (9) provides an excellent fit to all sets of experimental data (Fig.16). This result confirms that an ESA process is predominately responsible for upconversion in these materials.

4. CONCLUSIONS

Results indicate that chemical synthesis through the sol-emulsion-gel method is an excellent route for the preparation of rare-earth doped oxide nanocrystals. In summary, the luminescence properties of rare-earth ions in nanocrystals are found to be dependent on the crystalline structure, co-dopant and dopant concentration. We have shown that the overall upconversion luminescence intensity decreases with an increasing concentration of the Er^{3+} ions in ZrO_2 matrix. We suggest that concentration quenching is responsible for the decrease in overall intensity. In the presence of Y^{3+} ions the overall upconversion emission intensity of Er^{3+} is found to increase. The Y^{3+} ions are presumed to prevent the cluster formation of erbium ions and therefore, the non-radiative path due to concentration quenching is reduced resulting in an increase in the upconverted emission intensity of Er^{3+}. The Yb^{3+} ions, which act as a sensitizer, also enhance the upconversion luminescence under infrared excitation. The analysis suggests that the modification of upconversion properties (tuning and efficiency) is due to the crystal structure of the host and concentration of ions. We have also confirmed that the upconversion process follows two-photon ESA process in low concentration dopant ions containing oxide nanocrystals.

ACKNOWLEDGEMENTS

The Department of Science and Technology (NSTI, No.SR/S5/NM-05/2003)) is acknowledged for financial support.

REFERENCES

[1] P.N. Prasad, *Nanophotonics*, John Wiley and Sons, NY, 2004.

[2] F. Auzel; *Chem. Rev*, 2004, 104,139-173.

[3] S. Heer, O. Lehmann, M. Haase, and H. U. Gudel, *Angew.Chem.Int.Ed.*, 2003, 42, 3179-3182

[4] A, R. S. Niedba, H. Feindt, K. Kordos, T. Vail, J. Burton, B, Bielska, S. Li, D. Milunic, P. Bourdelle and R. Vallejo, *Anal. Biochem.*, 2001, 293, 22-30.

[5] A. Patra, C. S. Friend, R. Kapoor, and P.N. Prasad, *Appl. Phys. Lett.* 2003, 83, 284-286

[6] D. Matsuura, *Appl. Phys. Lett.*, 2002, 81, 4526-4528.

[7] E. De La Rosa-Cruz, L. A. Diaz-Torres, R. A. Rodriguez-Rojas, M. A. Meneses-Nava, O. Barbosa-Garcia and P. Salas, *Appl. Phys. Lett.*, 2003, 83, 4903-4905.

[8] A. Patra, C. S. Friend, R. Kapoor and P. N. Prasad, *J. Phys. Chem. B.*, 2002, 106, 1909-1912.

[9] F. Vetrone, J. C.Boyer, J. A. Capobianco, A. Speghini and M. Bettinelli, *J. Phys. Chem. B,* 2003, 107, 10747-10752.

[10] F. Pandozzi, F. Vetrone, J. C.Boyer, R. Naccache, J. A. Capobianco, A. Speghini and M. Bettinelli, *J. Phys. Chem. B*, 2005, 109, 17400-17405.

[11] X. Wang, X. Kong, G. Shan, Yi Yu, Y. Sun, L. Feng, K. Chao, S. Lu and Y. Li, *J. Phys. Chem. B*, 2004, 108, 18408-18413.

[12] R. Kapoor, C. S. Friend, A. Biswas and P. N. Prasad, *Opt. Lett.*, 2000, 25, 338-340.

[13] A. Patra, C. S. Friend, R. Kapoor and P. N. Prasad, *Chem. Mater.*, 2003, 15, 3650-3655.

[14] H. Guo, N. Dong, M. Yin,W. Zhang, L. Lou and S. Xia, *J. Phys. Chem. B*, 2004, 108, 19205-19209.

[15] S. Srivakumar, F. C. J. M van Veggel and M. Raudsepp, *J. Am. Chem. Soc.* 2005, 127, 12464-12465.

[16] E. De la Rosa, P. Salas, H. Desirena, C. Angeles and R. A. Rodriguez, *Appl. Phys. Lett*, 2005, 87, 241912-1-3.

[17] M. A. R. C. Alencar, G. S. Maciel and Cid B. de Araújo and A. Patra, *Appl. Phys. Lett.*, 2004, 84, 4753-4755.

[18] P. Ghosh and A. Patra, *Pramana,* 2005, 65, 901-907.

[19] A. Patra, S. Saha, M. A. R. C. Alencar, N. Rakov and G. S. Maciel, *Chem. Phys. Lett*, 2005, 407 ,477-481.

[20] A. Patra, P. Ghosh, P. Saha Chowdhury, Márcio A. R. C. Alencar, W. Lozano B., N.Rakov, and G. S. Maciel, *J. Phys. Chem. B*, 2005, 109 ,10142 -10146.

[21] H. S. Yang, M. R. Geller and W. M. Dennis, *Phys. Rev. B*, 2000, 62, 9398-9401.

[22] R. S. Melzer and K.S. Hong, *Phys. Rev. B*, 2000, 61, 3396-3403.

[23] R.S. Meltzer, S. P. Feofilov, B. Tissue and H. B. Yuan, *Phys. Rev. B.,* 1999, 60, R14012-R14015.

[24] A. Patra, E. Sominski, R. Ramesh, Yu. Koltypin, Z. Zhong, A. Gedanken, H. Minti and R. Reisfeld, *J. Phys. Chem. B,* 1999, 103, 3361-3365.

[25] M. Chatterjee and A. Patra, *J. Am. Ceram. Soc.*2001, 87, 1439-44.

[26] T. Masui, K. Fujiwara, K. Machida, G. Adachi, T. Sakata and H. Mori, *Chem. Mater*, 1997, 9, 2197- 2204.

[27] I. Lisiecki, M. Bjorling, L. Motte, B. Ninham and M. P. Pileni, *Langmuir,* 1995, 11, 2385-2392.

[28] E. Dickinson, in *Controlled Particle, Droplet and Bubble Formation,* edited by D. J. Wedlock (Butterworth-Heinemann, Oxford, 1994), p. 191.

[29] E. F. Lopez, V. S. Escribano, M. Panizza, M. M. Carnasciali, G. Busca, *J. Mater. Chem.*, 2001, 11, 1891-1897.

[30] Y. W. Zhang, Y. Yang, S.J. Tian, C. S. Liao and C. H. Yan, *J. Mater. Chem*, 2002, 12, 219-224.

[31] R. S. Quimby, W. J. Miniscalco and B. Thompson, *J. Appl. Phys.*, 1994,76, 4472-4478.

[32] A. Gedanken, R. Reisfeld, E. Sominski, O. Palchik, Yu Koltypin, G. Panczer, M. Gaft and H. Minti, *J. Phys. Chem. B,* 2000, 104, 7057-7065.

[33] J. Silver, M. I. Martinez-Rubio, T. G. Ireland, G. R. Fern and R. Withnall, *J. Phys. Chem. B.* 2001, 105, 948-953 .

[34] J.F. Suyver, R. Meester, J.J. Kelly and A. Meijerink, Phys. Rev. B, 2001,64, 235408.

[35] S. Singh and J. E. Geusic, *Phys. Rev. Lett*, 1966, 17, 865-868.

[36] M. Pollnau, D. R. Gamelin, S. R. Luthi, H. U. Gudcl, M. P. Hchlcn, *Phys. Rev. B,* 2000, 61, 3337-3346.

In: Trends in Optical Materials Research
Editor: Owen G. Gardens, pp. 113-136

ISBN: 1-60021-290-5

Chapter 4

PHOTO-PHYSICS INVOLVED IN THE EXCITATION OF PYRYLIUM SALTS UNDER ULTRA-SHORT PULSED ILLUMINATION

I. Polyzos, G. Tsigaridas, M. Fakis, V. Giannetas and P. Persephonis

Physics Department, University of Patras, P.O. Box 1414, Patras 26500, Greece

ABSTRACT

Pyrylium salts constitute a new class of very promising materials for applications in the field of photonics. In the current article we present the photo-physics involved in the excitation of a group of representative pyrylium salts. This group consists of molecules with the same basic chemical structure differing only to a specific substituent. The two-photon absorption (TPA) properties of these molecules are investigated experimentally. Large TPA cross-section values are reported. Furthermore, it is shown that under ultra-short pulse irradiation these molecules are photobleached efficiently. The photobleaching process is realized via two, diverse, simultaneously present, mechanisms. It is proved experimentally that these mechanisms are step-wise multi-photon processes. In most of the cases more than two-photons are involved in the photobleaching process. Excited state absorption is found to play a major role in this process. Furthermore, the effect of substituent change on the photobleaching properties is investigated. It is shown that photobleaching depends strongly on the specific substituent. Finally, the decay dynamics of these molecules are studied through femtosecond fluorescence time-resolved spectroscopy.

1. INTRODUCTION

Laser-induced fluorescence of organic materials is widely used in the fields of physics, chemistry, biology and medicine. Over the past 50 years a wealth of information has been accumulated and the interest in this subject continuously increases. Molecules with diverse optical properties are required for special applications but in general the suitability of a molecule for every application depends on its absorption and emission coefficients as well as on its photostability.

A special case of absorption is Two-Photon Absorption (TPA) –an exotic subject until the invention of ultrafast lasers. With this term we mean the simultaneous absorption of two-photons to excite a molecule. The considerable attention given recently in TPA of organic molecules is motivated by the wide range of potential applications, such as upconverted lasing[1], optical power limiting[2,3], two-photon laser scanning microscopy[4-6], photodynamic therapy[7], microfabrication[8,10] and three dimensional optical data storage[11-13]. These applications demand organic chromophores with large two-photon absorption cross-sections [14-19] and high fluorescence quantum yields at desirable wavelengths. Also, the photostability of these molecules is crucial since its absence leads to the irreversible loss of emitted photons decreasing the sensitivity and the accuracy of measurements.

A common process that usually reduces the photostability of a molecule is photobleaching. This term includes all the photochemical reactions that permanent remove a molecule from absorption-emission cycle. It is the main limiting factor in single molecule spectroscopy [20-22]. More specifically, when the time scale of the process being studied is comparable with the survival time (photobleaching time) of the molecule no useful information for this process can be collected. Furthermore, photobleaching is a destructive parameter in laser scanning microscopy since the quality of obtained images depends on the amount of collected fluorescence (which is limited by photobleaching)[4,6].

Until recently, the fluorescence induced by a continuous wave laser and thereafter photobleaching was occurred in the whole illuminated area. Nowadays, infrared pulsed lasers are commonly used for this purpose. The high photon densities provided from these lasers sources allow molecules to absorb simultaneously two- or more photons [23]. The main advantage of this procedure is that the absorption is restricted to a small volume. Thus, photobleaching is confined to this small volume suppressing (but not eliminating) the above mentioned problems. Indeed, this property of two-photon excitation is used constructively in applications like fluorescence recovery after photobleaching (FRAP)[24] and 3-dimensional optical data storage[25-31]. The highly confined, in 3-dimensions, photobleaching is used for measuring diffusion processes of chromophore molecules in liquids and polymer films as well as for recording irreversibly photobleached patterns in polymer films. Therefore photobleaching is not always an undesirable effect.

From the above mentioned is obvious that besides one- and two-photon characterization of a molecule a better understanding of photobleaching mechanism is necessary. A large amount of work has been done for the better understanding of this process in the case of one-photon excitation [32-34]. Unlike, only a few data are available for photobleaching in the case of two-photon excitation [35-37]. The photophysics of photobleaching in the last case is not well known yet. As the intensity level increases the contribution of higher-order nonlinear phenomena becomes significant. Thus, at high intensities, the mechanism of photobleaching may change fundamentally.

Recently, in a novel class of organic compounds based on pyrylium salt [38-44] enhanced TPA properties were detected [45]. In the present work we systematically synthesised pyrylium-based molecules in order to optimize their TPA performance while keeping the values of quantum yield high. Through this procedure an increase of the TPA cross-section by a factor of 7 was observed and values as high as $1,8 \times 10^{-47}$ cm^4 s $photon^{-1}$ were measured

[46]. The molecules were further characterized by steady state and time resolved spectroscopy. In addition, the photophysics involved in this photobleaching process was investigated. It was proven that photobleaching occurs through two diverse mechanisms. One of the mechanisms is major and causes a fast bleaching whereas the other is minor and causes a slow bleaching. Finally, it is shown that both mechanisms are two-, three- or more-photon processes [47-49].

2. MATERIALS

The pyrylium salts that were investigated are 4-methoxyphenyl-2, 6-bis (4-methoxyphenyl) pyrylium tetrafluorobate (C_1), 4-methylphenyl-2, 6-bis (4-methoxyphenyl) pyrylium tetrafluorobate (C_2), 4-phenyl-2,6-bis (4-methoxyphenyl) pyrylium tetrafluorobate (C_3) and 4-(4-chlorophenyl)-2, 6-bis (4-methoxy- phenyl) pyrylium tetrafluorobate (C_4). In parenthesis are shown their abbreviated names (Figure 1). These molecules have the same basic chemical structure and differ only to a specific substituent, which is OCH_3, CH_3, H and Cl for C_1, C_2, C_3 and C_4 respectively (Table 1). The pyrylium salts were synthesized from the reactions of benzaldehyde or a substituted benzaldehyde with 4-methoxyacetophenone in the presence of borontrifluride etherate. They were characterized by elemental analyses as well as FT-IR and H^1-NMR spectroscopy. A detailed characterization of these pyrylium salts regarding one- and two-photon excitation as well as photobleaching will follow.

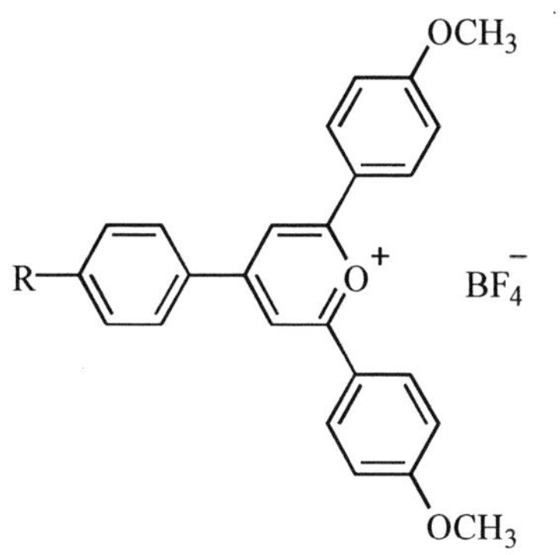

$C_1 : R = OCH_3$

$C_2 : R = CH_3$

$C_3 : R = H$

$C_4 : R = Cl$

Figure 1. Pyrylium salts studied in this work

Table 1. One-photon absorption and emission characteristics of pyrylium salts

Salt	R	λ_{abs_1}(nm)	λ_{abs_2}(nm)	$(\lambda_{abs_2} - \lambda_{abs_1})$(nm)	λ_{fl}(nm)	Stokes shift (nm)	φ_1	τ(nm)
C_1	OCH_3	388	overlapping	overlapping	531	143	0.68	0.8
C_2	CH_3	358	440	82	539	99	0.76	0.9
C_3	H	348	444	96	545	101	0.60	0.9
C_4	Cl	353	447	94	552	105	0.22	1.7

* λ_{abs_i} and λ_{fl} are the peak wavelengths of absorption and fluorescence spectra respectively

3. EXPERIMENTAL

3.1. Steady State Spectroscopy

The absorption spectra of the samples were taken with a Beckman DU-640 spectrophotometer. The emission spectra were recorded with a Perkin-Elmer LS55B spectrophotometer. The dyes were dissolved in acetone (concentration 10^{-4}M) and were disposed in square quartz cells.

3.2. Time Resolved-Spectroscopy

The fluorescence dynamics of the pyrylium chromophores were examined by a femtosecond time resolved upconversion spectroscopy [50,51]. The experimental set-up consists of a mode-locked Ti:Sapphire laser (Tsunami, Spectra Physics) pumped by a frequency doubled Nd:YVO_4 continuous wave laser (Millennia V_s, Spectra Physics). The laser emits pulses at 800 nm with 80 fs duration and 82 MHz repetition rate. The laser beam is focused onto a BBO crystal (1 mm thickness, Fujian Castech) generating a second harmonic beam at 400nm (6 mW average power) which excites the polymeric sample. The remaining fundamental laser beam (gate beam), after passing through an optical delay line, (driven by a Newport ESP300 DC motor) is focused together with the fluorescence of the sample onto a second BBO crystal (1.5 mm thickness, Fujian Castech) where an upconversion beam is generated (type I phase matching). This upconversion beam is separated from the fundamental and fluorescence beams by an iris and filters (UG-11 glass filters), and is directed into a 0.25 m monochromator (ORIEL 77200). Finally, it is detected by a photomultiplier (Becker-Hickl PMH-100-3) connected with a photon counter (Becker-Hickl PMS-400). The fluorescence decays were determined by measuring the intensity of the upconversion beam as a function of the delay between the fluorescence and gate beams. The dynamics of the molecules was detected under magic angle conditions (54.7^0). The magic angle between the polarization of the excitation and gate beams was used to avoid dependence of the dynamics on rotational motions of the molecules. For determining the temporal resolution of our system, an upconversion signal (cross correlation signal) between the gate and the scattered from the sample excitation pulses was obtained. Phase matching in this case was achieved by rotating the polarization of the excitation pulse so that the excitation and gate pulses are parallel polarized. The full width at half maximum of this upconversion signal is ~250 fs and corresponds to the temporal resolution of the system. The spectral resolution of our system is 2.5 nm.

3.3. Two-Photon Excitation

The two-photon excitation (TPE) cross-section spectra of the Ci (i=1,2,3,4) compounds were determined by measuring the fluorescence signal generated by the TPA process. The mode-locked Ti:Sapphire laser described above was used as the excitation source. The laser beam was shaped to have a nearly top-hat spatial distribution using a 5-fold beam expander.

After passing through a long-wave dichroic mirror the excitation beam was focused onto the sample with the use of a 16X achromatic microscope objective lens (Melles Griot). The samples investigated were acetone solutions (concentration 10^{-4}M) of Ci in 1cm pathlenth quartz cuvettes. The fluorescence of each sample was collected by the same lens and separated from the input beam by the dichroic mirror (Melles Griot) and 3 short pass filters. The excitation power was measured at the exit of the sample by a pyroelectric detector and a power meter while a photomultiplier tube (PMH-100-3) and a single photon counting system were used to measure the fluorescence signal.

3.4. Photobleaching

In order to investigate the photobleaching properties of pyrylium salts the samples were continuously irradiated with ultrashort laser pulses while the resulted fluorescence was recorded at the same time (Figure 2). The laser system used in our experiments was again the mode-locked Ti:Sapphire laser described above. A 5 fold expanding telescope was used to expand the laser beam in order to uniformly illuminate the back-aperture of an oil immersion objective lens (100X, N.A.: 1,2). This tight focusing geometry made available the high intensities required for the nonlinear excitation of the samples. The irradiation power was controlled with a combination of a motorized λ/2 plate and an ultrafast polarizer (Melles Griot, 16PPB200) and measured at the back-aperture of the lens. The fluorescence signal was collected with the same lens used for excitation and separated from the input beam by a cold mirror reflecting from 400 to 600nm (Thor-Labs, FM03). The remaining infrared irradiation was rejected with a sequence of three short pass filters. A photomultiplier tube (PMH-100-4) and a single photon counting system (BeckerandHickl, PMS300) were used for the measurement of fluorescence.

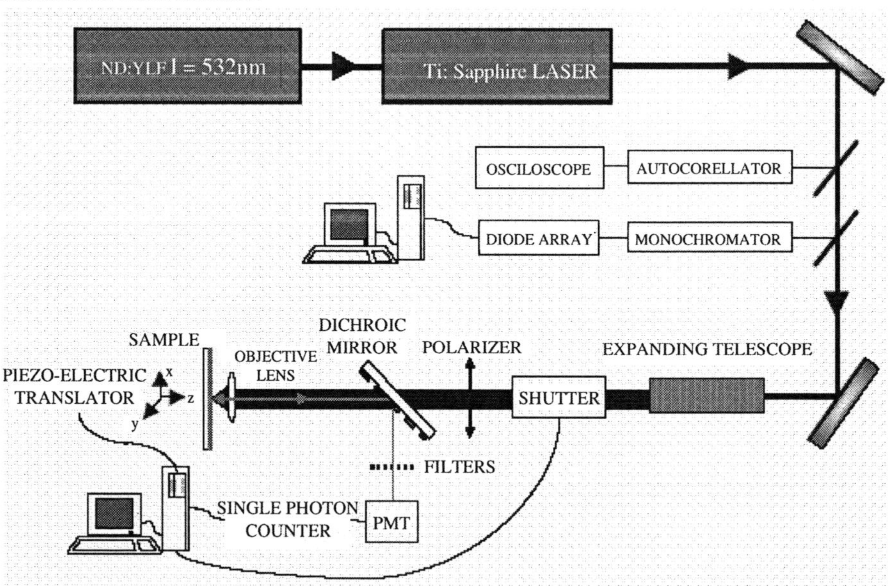

Figure 2. Experimental setup used for the photobleaching of pyrylium salts

In order to record the sharp decay in the fluorescence caused by photobleaching a fast shutter system (rise time 3 ms, Newport 845HP) was synchronized with the single-photon counting electronics. The time resolution of the experimental setup was 250nsec which is much smaller than the resolution needed in our measurements. Each decay curve consists of 1500 to 12000 points depending on the time resolution selected. The choice of time-resolution was made with respect to the initial fluorescence signal and in every case was as small as needed for recording correctly this signal.

The photobleaching experiments were conducted on polymer films of PMMA doped with pyrylium salts. To prepare the films, salt (purified 99%) and PMMA (Aldrich) were mixed together with weight ratio 1%. The mixture was diluted in Tetrahydrofuran (THF) and casted over a microscope slide glass to form the film. The evaporation of the solvent was made in the ambient air and at room–temperature (25 ^{0}C). The same procedure was followed for all pyrylium salts.

4. RESULTS AND DISCUSSION

4.1. Steady State Spectroscopy

The one-photon absorption and fluorescence spectra are shown in Figure 3. The spectra do not have any special structure. It is interesting to note that there are two peaks in the visible absorption spectrum for compounds C_2, C_3 and C_4. The positions of the main and secondary absorption peaks vary from 348 to 358 nm and 440 to 447 nm, respectively. In contrast compound C_4 has a single-absorption peak at 388 nm with a very broad line width. This should be attributed to the overlapping of the two expected peaks. The fluorescence spectra are quite similar for all compounds and the peaks ranging from 531 to 552 nm.
The absorption and fluorescence spectra were used for the determination of the fluorescence quantum yields (φ_1) [52,53]. For the estimation of φ_1 Rhodamine B was used as a reference. The results are depicted in Table 1.

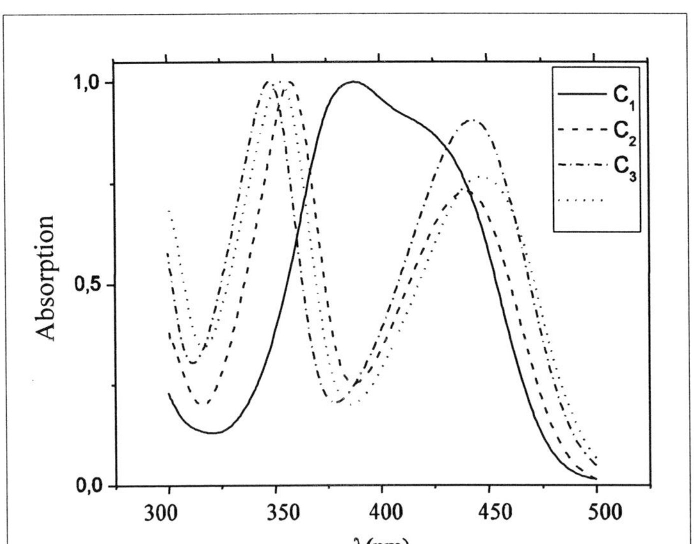

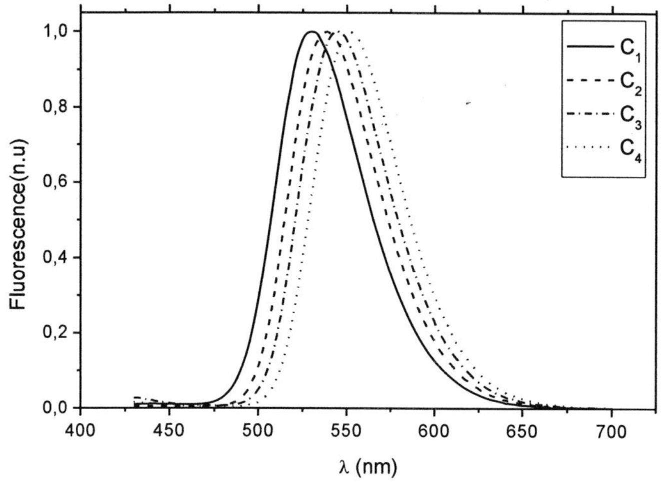

Figure 3. One-photon absorption and fluorescence spectra of pyrylium salts

4.2. Time Resolved-Spectroscopy

The fluorescence decay dynamics of the four pyrylium chromophores are shown in Figure 4.

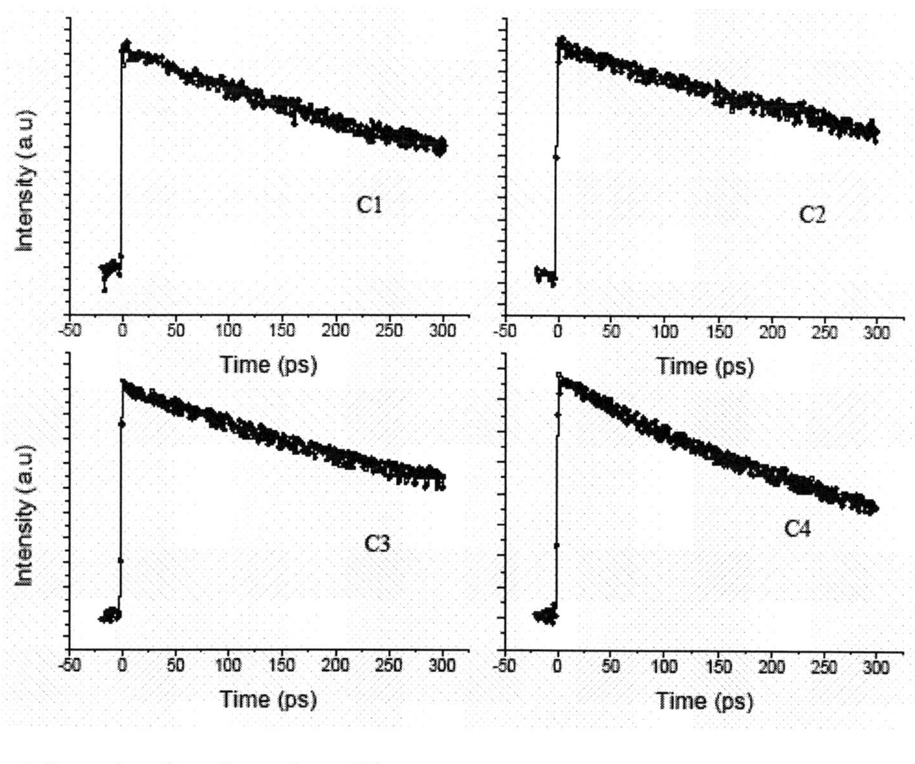

Figure 4. Dynamics of pyrylium salts at 400nm

All compounds were studied in acetone solutions with concentration 5 x 10^{-3}M. The excitation and detection wavelengths were 400 and 510nm respectively. The chromophores decay monoexponentially and the results were fitted by the following equation

$$I(t) = I_0 + A_1 \exp(-t/\tau)$$

where I_0 is a constant parameter representing the background noise, A_1 is a pre-exponential function and τ is the fluorescence decay time. Pyrylium C_2 showed the slower decay with τ=700ps. This is consistent with its highest fluorescence quantum yield. On the other hand, the shortest decay time, equal to 370ps, was observed from C_4, which also has the lowest quantum yield. Finally, the decay times of chromophores C1 and C_3 were found equal to 530 and 560 ps respectively. Using the decay times obtained from the experimental results shown in Figure 4 and the quantum yields, the radiative decay times τ_{SP} were also determined and they were equal to 0.8, 0.9, 0.9 and 1.7 ns for C_1, C_2, C_3 and C_4 respectively. All the decay parameters of the pyrylium chromophores are summarized in Table 1.

4.3. Two-Photon Excitation

Theoretically, the two-photon excited fluorescence should be proportional to the square of the excitation power. In Figure 5 the fluorescence signals of the chromophore molecules versus incident power at various excitation wavelengths are shown. In all cases no significant deviation from the square law was found verifying that the observed fluorescence results only from two-photon excitation. Effects like saturation, photobleaching, excited state absorption and intersystem crossing are negligible.

The fluorescence signal of each compound was calibrated in reference to a Rhodamine B solution (concentration 10^{-4}M) in methanol having a well-known TPE cross-section spectrum [13]. The relative fluorescence is measured instead of the absolute one because the latter requires the accurate knowledge of pulse duration, bandwidth, shape and temporal coherence of the beam. The determined TPE spectra (δ_{TPE}) for all molecules studied are shown in Fig. 6.

However, the TPE cross-section δ_{TPE} is related to the TPA cross-section δ with the following equation

$$\delta_{TPE} = \varphi_2 \delta$$

where φ_2 is the two-photon fluorescence quantum yield of the molecule.

The corresponding δ values were calculated through the above equation by using the previously determined one-photon fluorescence quantum yield φ_1 and assuming that one-photon and two-photon quantum yields are equal [54]. All the results are given in Table 2.

Initially, the TPE spectrum of the compound C_1(R = OCH_3) was recorded and high TPE cross-section values were measured. For improving the TPA properties of C_1 three other pyrylium salts C_2, C_3 and C_4 were synthesized by changing the chemical structure of the substituent R. Compound C_2 (R = CH_3) showed significantly larger δ and somewhat higher φ value than C_1 ones. Compound C_3 (R = H), as compared to C_1 and C_2, displayed a larger δ but a little smaller quantum yield. More particularly, the TPA cross-section of C_3 was found to be

2,5 and 4,8 times greater than those of C_1 and C_2 respectively. Finally, compound C_4 ($R = Cl$) exhibited larger δ (7 times greater than C_1) but considerably lower φ_1. Obviously, these divergences of the TPA properties are due to the different charge distribution of the substituents. Specifically when R is a donor (CH_3, OCH_3) the molecule's TPA cross-section is large and the quantum yield is high.

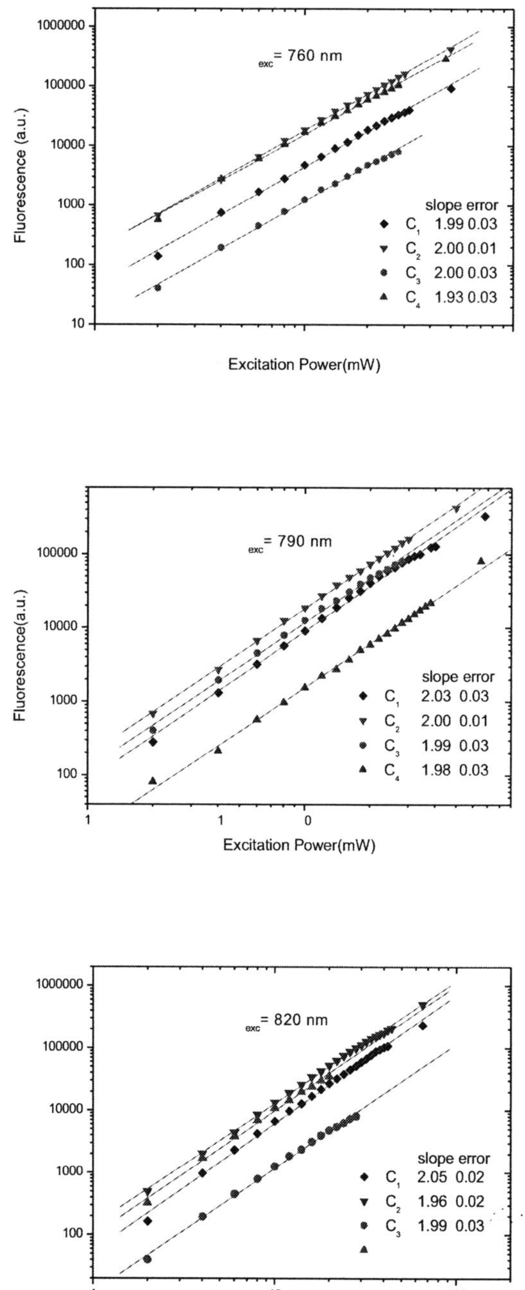

Figure 5. Log-to-log plot of Fluorescence versus excitation power for pyrylium salts C1, C2, C3 and C4 at three different wavelengths

When R is neither donor nor acceptor (H) the quantum yield decreases slightly but the cross-section value increases significantly. Finally, when R is an acceptor (Cl) a great enhancement of δ is observed in contrast to the considerable decrease of the quantum yield. The conclusion based on the above remarkable observations is that the TPA properties of the synthesized molecules are closely related to the substituent's ability of donating or accepting electrons.

Table 2. TP excitation and TP absorption cross-section values

Salt	R	λ_{max} (nm)	$\eta_2\sigma^{(2)}$(GM)*	$\eta_2=\eta_1$	$\sigma^{(2)}$(GM)
C_1	OCH_3	810	182	0,68	267
C_2	CH_3	760	513	0,76	675
C_3	H	740	767	0,60	1278
C_4	Cl	740	390	0,22	1772

*1 GM=10^{-50} cm^4 s photon^{-1}

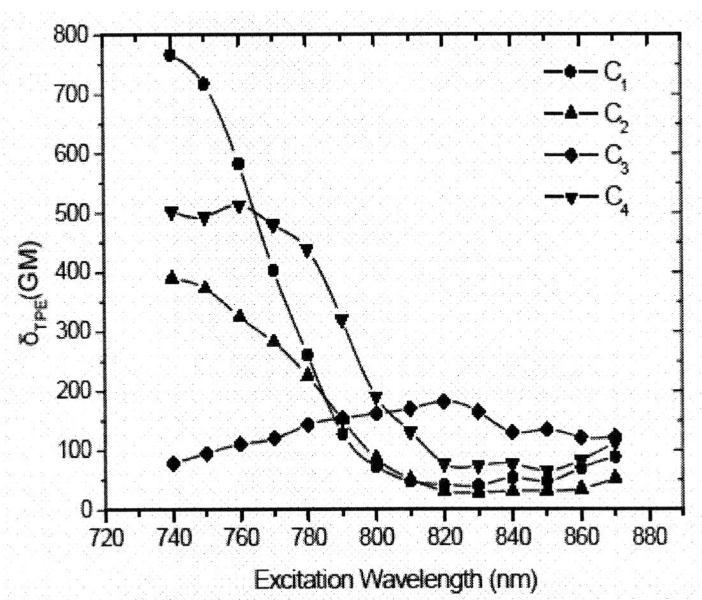

Figure 6. Two-photon excitation cross-section spectra for pyrylium salts C1, C2, C3 and C4

4.4. Photobleaching

Initially, the photobleaching of pyrylium salt C_3 under ultra-short pulsed illumination was examined. The salt C_3 was chosen among the other salts because of its large TPA cross-section and high quantum yield.

At very low power of irradiation photobleaching is negligible and the pyrylium salt C_3 should be excited by TPA (because of its large TPA cross-section reported previously). This prospect was confirmed experimentally. Figure 7a shows that the fluorescence of the optical material, under ultrashort pulsed illumination, depends on the illumination power quadratic for the entire spectral range off our measurements.

For excitation power values greater than 1.5 mW the process of photobleaching becomes significant and a decay of the fluorescence is observed (Figure 7b). Specifically, as irradiation time increases more molecules are removed from the normal absorption-emission cycle. The large amount of molecules stops to emit photons at the very beginning of irradiation leading to the steep decrease in fluorescence shown in the figure. The photobleaching process continues as irradiation time increases but with a much slower rate in comparison with the beginning. The only function that fits to our experimental data for fluorescence signal versus irradiation time is a double exponential decay.

$$F(t) = F_0 + A_1 e^{-k_1 t} + A_2 e^{-k_2 t}$$

In this expression F is the fluorescence photon flux (photons/sec), k_1 and k_2 are the decrease rates of fluorescence, called here photobleaching rates (1/sec), and F_0 is a noise constant (photons/sec). A_1 and A_2 are the initial contributions to fluorescence signal from each exponential decay (photons/sec) since at the begging of irradiation (t=0) the fluorescence photon flux is $F(0) = F_0 + A_1 + A_2$.The existence of two independent exponential decays denotes that the photobleaching process is occurred through two different mechanisms.

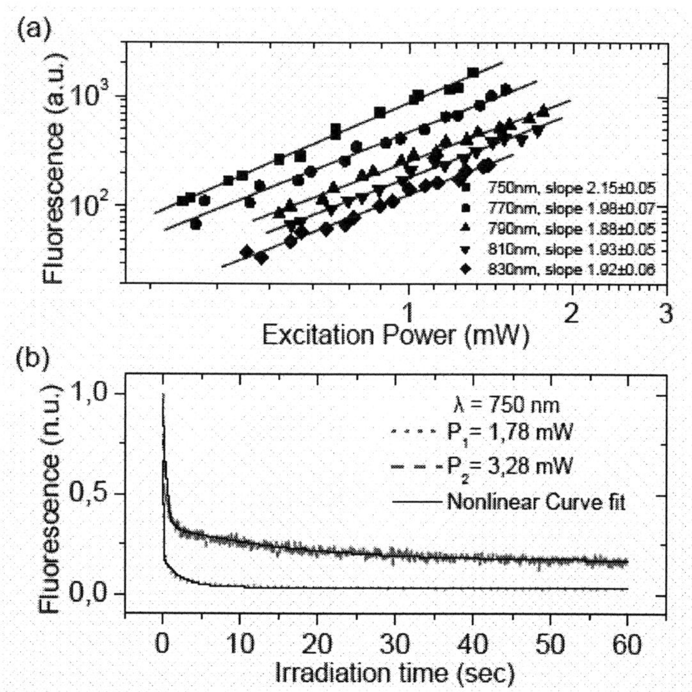

Figure 7. (a) Two-photon induced fluorescence intensity from the photobleaching polymer as a function of the incident power for different excitation wavelengths (log-tolog plots). Curves are fitted linearly. The slopes, shown in the graph, are ~ 2 denoting two-photon excitation in all wavelengths. Pulse duration τ=80fs (b) Fluorescence versus time for two irradiation power values. The best-fitting function is a double-exponential decay

It must be underlined the simultaneous existence of both mechanisms at the first few milliseconds of irradiation. This is critical since a delay in the appearance of one of the mechanisms could lead to the conclusion that it originates from a photoproduct. This does not happen in our measurements and photobleaching occurs from the pyrylium salt itself. A further study of these mechanisms will follow.

The ratio A between the parameters A_1 and A_2 as well as the photobleaching rates k_1 and k_2 as functions of irradiation power are shown in Figure 8. It is evident that the impact of each photobleaching mechanism to the total fluorescence decay is different. The ratio A increases significantly as the irradiation increases and ranges from 2 to 23 for the power values of our measurements (Figure 8a). Thus, photobleaching mainly occurs via the mechanism with higher probability than via the lower one, especially for high values of irradiation power.

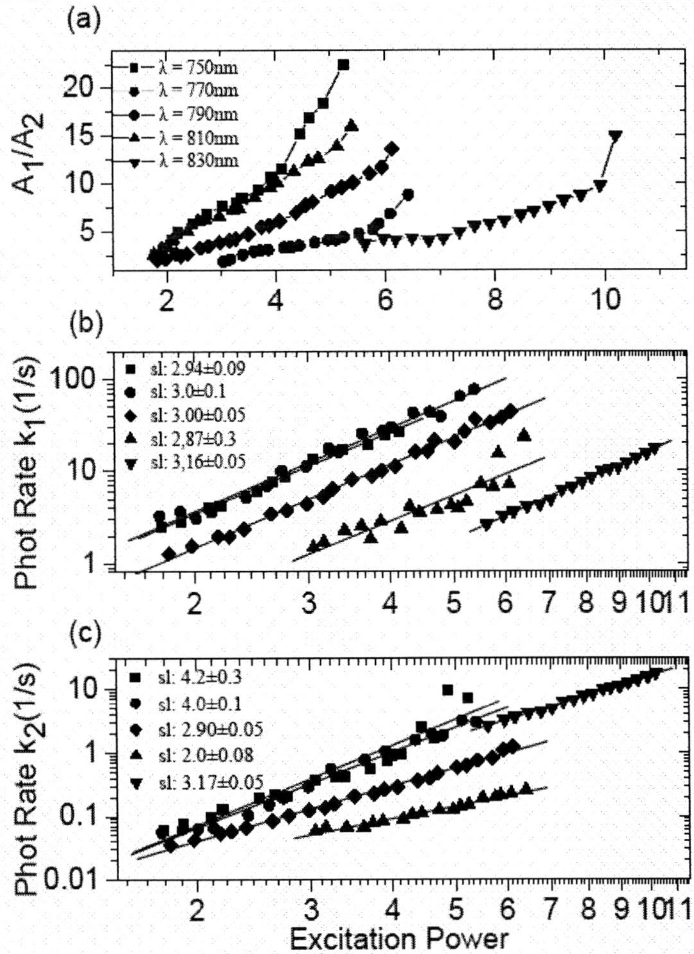

Figure 8. (a) Ratio A between A1 and A2 versus irradiation power. (b),(c) Log-log plots of the photobleaching rates k1,2 versus irradiation power. The excitation wavelength and the fitted slopes are shown in the graphs. The power range of each dataset selected with respect to the starting point of bleaching at each wavelength

In analogy with fluorescence we expected the photobleaching rates to depend on the square of the incident power. However, our results showed that the photobleaching rate k_1 depends on irradiation power cubic, in all wavelengths studied (Figure 8b). Hence, unlike fluorescence, photobleaching is not a two- but a three-photon process. It is worth mentioning that for high power values an increase in the slope of k_1 versus irradiation power was observed. The data are not included in the graph because of its high uncertainties but may proclaim the absorption of more than three photons. The photobleaching rate k_2 shows similar behaviour with k_1 with the exception of a wavelength dependence. Analytically, it was found that photobleaching through the slow mechanism is a four photon process at 750 and 770 nm, a three-photon process at 790 nm, a two-photon process at 810 nm and finally a three-photon process at 830 nm (Figure 8c). The main difference between k_1 and k_2 according to our measurements is that k_1 is almost 100 times greater than k_2. This result, leads to the conclusion that the first and fast mechanism is responsible for the steep decrease of the fluorescence at the first milliseconds of sample irradiation (Figure 7b).

The photons that interact with the memory material during photobleaching may be simultaneously or sequentially absorbed. One way to distinguish these procedures is to study the photobleaching dependence on excitation pulse duration (τ). In the case of simultaneous absorption of three photons the photobleaching coefficients are expected to be inversely proportional to τ^2[48] Our results in Figure 9a are shown that the photobleaching rate k_1 of the first mechanism at 770 nm depends inversely on τ .This is the case of simultaneous two-photon absorption, although in our case there is a three photon process. The explanation of this observation is that photobleaching through this mechanism is a combination of a simultaneous TPA and a successive absorption of one more photon from the excited state (the latter process is independent on pulse duration). As mentioned before the second mechanism at 770nm is a four-photon process. The rate k_2 depends inversely on τ^2. Hence, the most possible scenario that describes this behaviour is the simultaneous absorption of two photons (proportional to $1/\tau$) followed by the simultaneous absorption of two additional photons from the excited state (also proportional to $1/\tau$).

An additional evidence for the above conclusions is shown in Figure 9. This figure depicts the dependence of A_1, k_1 (figure 9b) and A_2, k_2 (figure 9c) on the wavelength in comparison with the TPA spectrum of P. The spectrum of the photobleaching rate k_1 deviates significantly from the TPA spectrum (maximum deviation~50%). Hence, more or other electronic states are involved in the photobleaching process in comparison with TPA. The spectrum of A_1 has the same form with the TPA one since their maximum deviation is only 13%. So, we can write that $A_1(\lambda) \propto \delta(\lambda)$, where δ is the TPA cross-section. This indicates that the molecule absorbs simultaneously two photons to reach the first excited state. The above discussion supports the previous conclusion that photobleaching is caused by the simultaneous absorption of two photons and the sequential absorption of one more photon to reach the bleaching state. Analogous observations stand for the slow mechanism (figure 9c). Thus, photobleaching occurs through an excited state absorption and not through the simultaneous absorption of three or four photons from the fundamental state.

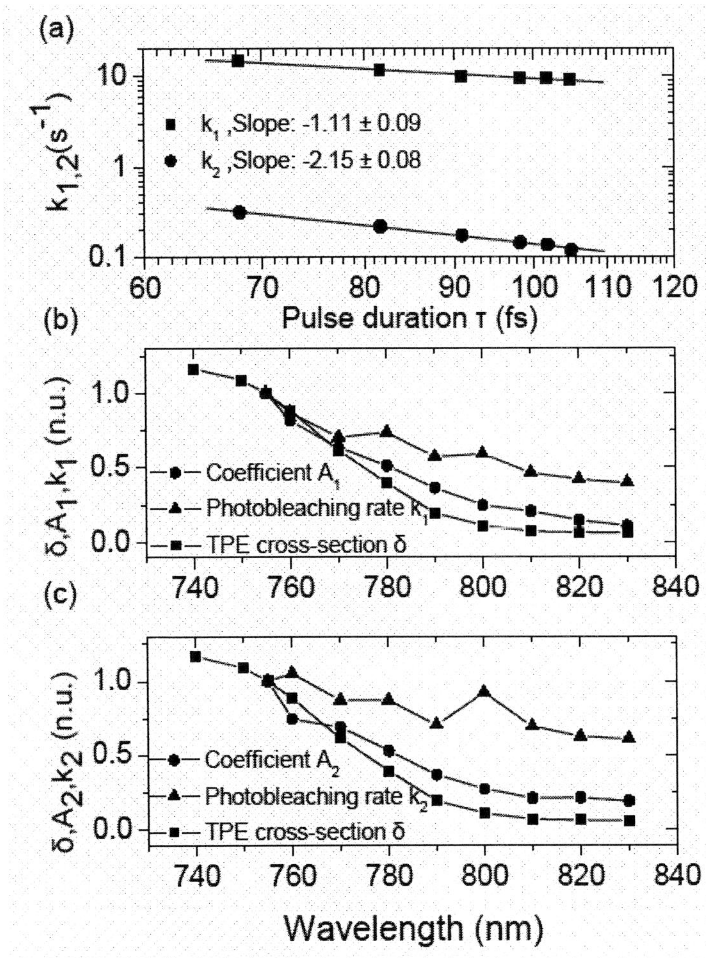

Figure 9. (a) Logarithmic plots of photobleaching rates k1,2 versus pulse width. The fitted slopes are included in the graph (sl).(b),(c) Comparison of TP excitation cross-section spectrum (δ) with weight constants *A1,2* and photobleaching rates *k1,2* spectra of pyrylium salt (concentration=1% wt. Irradiation power P=7.5 mW, τ= 80fs)

Figure 10 shows a suggested simplified Jablonski diagram for the photobleaching pathways [34]. According to this model, photobleaching may occur from an excited state absorption in a singlet or a triplet manifold. The excited singlet state is reached with the simultaneous absorption of two-photons whereas the triplet state is reached with the intervention of intersystem crossing. The small contribution to fluorescence signal from the slow mechanism is an indication that it is a triplet state absorption (fluorescence is mainly related to the first excited state S_1). This assumption is in agreement with its slow nature in comparison with the other mechanism. Thus, we suggest that the fast and dominant photobleaching mechanism is a singlet to singlet process whereas the minor and slow one is a triplet to triplet process.

The high order photobleaching of C_3 explains in a very persuasive way the efficiency in recording observed in our previous work [46]. The fast photobleaching mechanism, being a step-wise, three-photon process, is responsible for the fast and sharp recording. The already small volume excited by simultaneous two-photon absorption is restricted more. In particular, the excited state absorption occurs inside this volume but its linear dependence on intensity results in sharper photobleaching in the central area of the volume (where the light intensity is higher). This stepwise photobleaching-recording mechanism avoids problems that could be appeared in simultaneous three-photon photobleaching such as the need for large intensities, the beam homogeneity, the dissociation of the molecule etc.

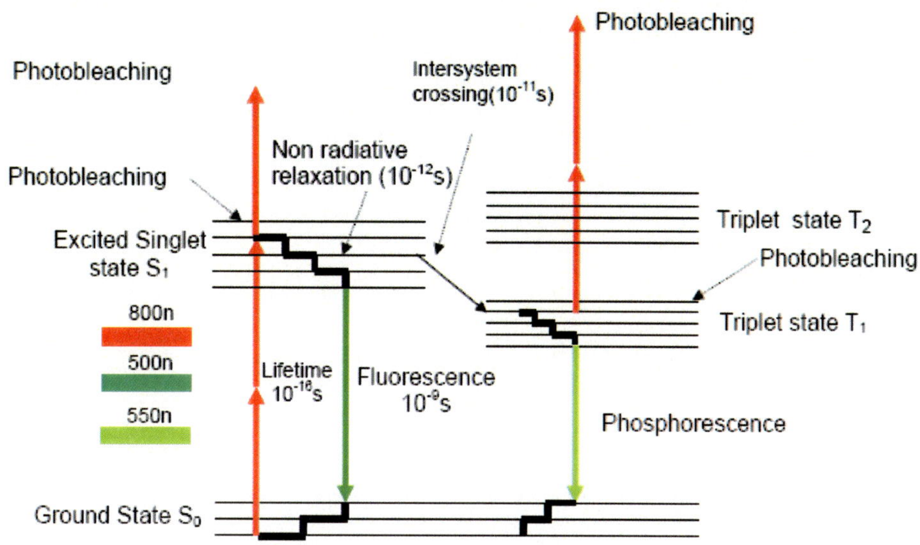

Figure 10. Proposed simplified Jablonski diagram describing the high order photobleaching of pyrylium salts

In the following of this chapter, the investigation of the photobleaching behavior is extended to the other pyrylium salts. As mentioned above, these salts have the same basic chemical structure and differ only to a specific substituent. Thereafter, the effect of substitution on the photobleaching properties of these pyrylium salts will be presented.

Figure 11 shows the continuous decrease in the fluorescence signal caused by the continuous irradiation of the pyrylium salts samples studied in this work. The function that fits to the experimental data is, in every case, the double exponential decay described in the photobleaching of C_3. Thus, the *two* photobleaching mechanisms are present to all pyrylium salts studied, independently of the specific chemical structure The influence of substituent change is therefore restricted only to the specific parameters of photobleaching. These parameters will be investigated afterwards.

Figure 12 shows the initial contribution to fluorescence signal from each photobleaching mechanism. Regarding the first mechanism, it is clear that C_3 is the most fluorescent molecule. This may be attributed partially to its large two-photon excitation cross-section (Table 3). The fluorescence signal resulting from C_2 is higher than the corresponding of C_4 and lower than that of C_3 in agreement with their cross section values. Finally, C_1 is the less fluorescent molecule since it has the smallest two-photon excitation (TPE) cross-section. Similar observations stand for the second mechanism.

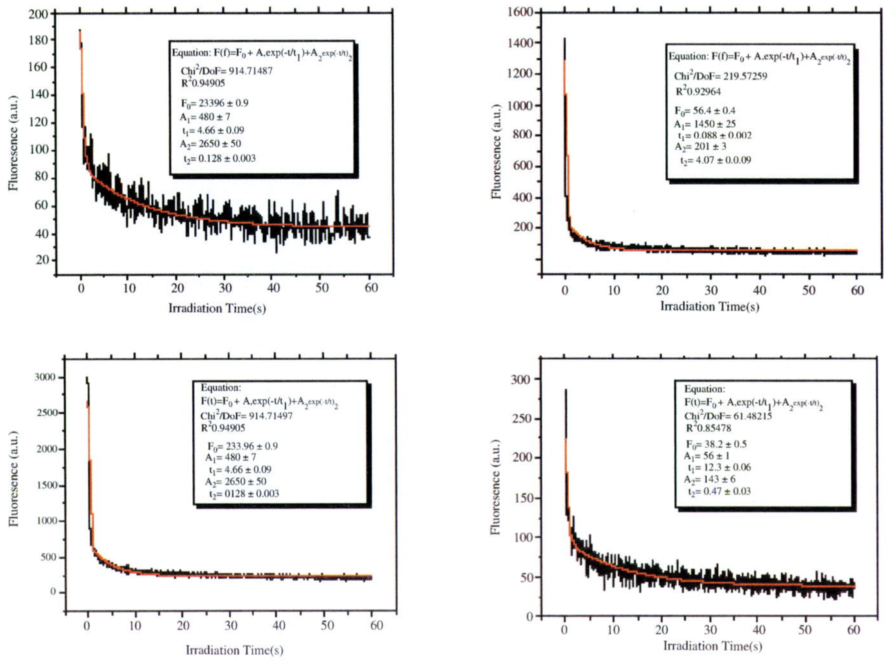

Figure 11. Decrease of fluorescence signal caused by photobleaching for pyrylium salts C1, C2, C3 and C4 The best fitting function is a double exponential decay. (excitation wavelength λexc=750nm, irradiation power P~3.7mW, pulse duration τ≈100fs)

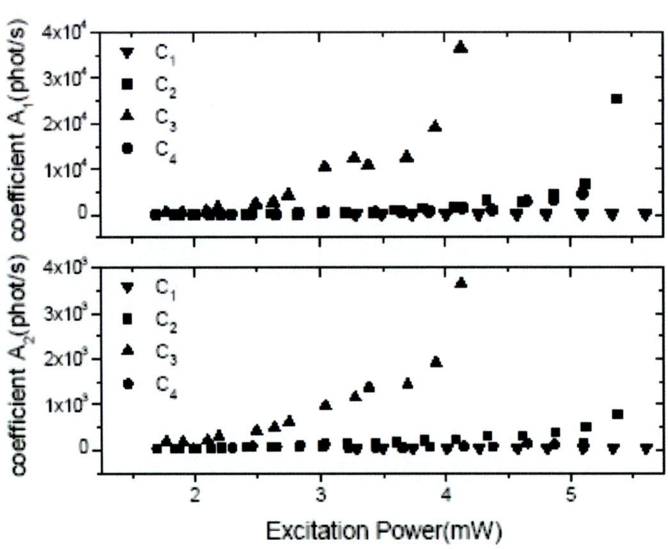

Figure 12. Contribution from each photobleaching mechanism to total fluorescence, A1,2, for C1,C2,C3 and C4 (λexc=750 nm ,τ≈100fs)

The strong dependence of the initial fluorescence on the excitation power, observed in all pyrylium salts, is an indication that higher-order processes take place during photobleaching.

The above rather qualitative analysis reveals in a preliminary way how the photobleaching parameters, A_1 and A_2, are affected by the changes in the basic chemical structure.

Figure 13 shows the relative initial contribution to fluorescence signal from the two mechanisms. It can be seen that for all pyrylium salts the ratio A_1/A_2 is much greater than one. This means that the first mechanism contributes much more to fluorescence signal in comparison with the second one. Furthermore, this ratio is not constant but increases with irradiation power. Hence, at large power values the molecules show a preferred excitation via the first mechanism. This characteristic is more intense for C_4 in comparison with the other salts. The ratio A_1/A_2 of C_4 increases more quickly for irradiation powers above 3mW than the corresponding ratio of the other molecules. This indicates a further confinement of the second mechanism, relatively to the first one, caused by the specific substituent (Cl). As depicted in figure 4, at a power value of about 5 mW the ratio between A_1 and A_2 becomes 6, 13, 22 and 33 for C_1, C_2, C_3 and C_4 respectively. The domination of the first mechanism is more than obvious.

The photobleaching rates of C_3 found to depend on the third or even the forth power of irradiation intensity. The behavior of photobleaching rates $k_{1,2}$ for the group of pyrylium salts we worked on will be exhibited below. The dependence of k_1 versus irradiation power for C_1, C_2, C_3 and C_4 is shown in Figure 14. The slopes of the logarithmic plots, which are shown in the graph, are 2 for C_1 and about 3 for the other three salts. Hence, our measurements show a square dependence on excitation power for C_1 and a cubical one, common for the other three pyrylium salts. However, there is a quantitative divergence in the values of photobleaching rates. More specifically, C_3 has the higher photobleaching rate in comparison with the other pyrylium salts. C_4 and C_2 are following respectively.

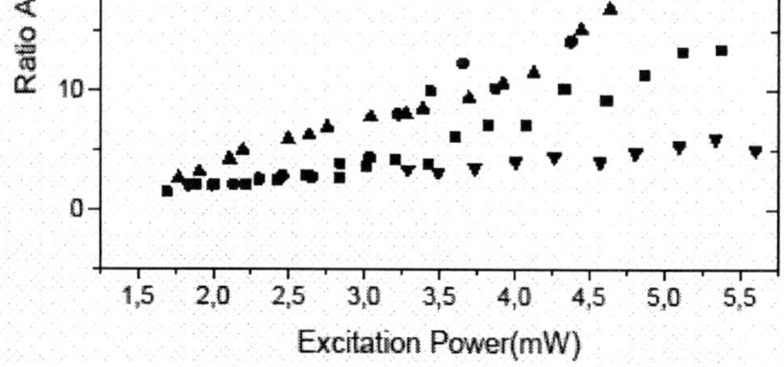

Figure 13. Relative contribution to initial fluorescence from the two-mechanisms versus irradiation power for C1, C2, C3 and C4

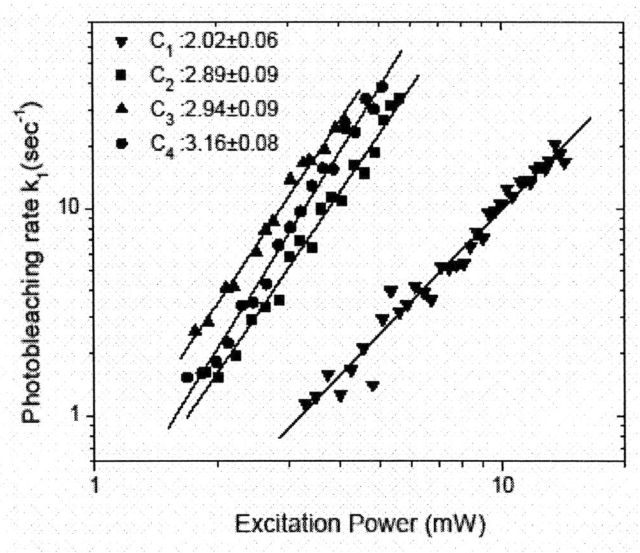

Figure 14. Log-log plots of the photobleaching rates k1 versus irradiation power for each pyrylium salt. The fitted slopes are shown in the graph. The power range of each dataset was selected with respect to the starting point of bleaching

It would be expected C_2 to be photobleached more efficiently than C_4 according to two-photon excitation cross-sections (Table 3). As it is mentioned above the initial contributions to fluorescence signal from each molecule follow this "rule". Nevertheless, this is not resulted from our measurements for k_1. This discrepancy is important since reveals the fact that photobleaching and fluorescence are occurred either from different electronic states of the molecule either from the same electronic states but with different efficiencies.

The photobleaching rate k_2 versus irradiation power is depicted in Figure 15. It can be seen easily that there is a significant difference in the behaviour of C_3 and the other pyrylium salts. k_2 depends on the square of irradiation power for C_1, C_2 and C_4 whereas on the forth power for C_3. The small change in the chemical structure caused a significant change in the nature of the second photobleaching mechanism. Furthermore, C_3 has the higher photobleaching rate k_2 (because of its higher-order power dependence), C_2 and C_4 have lower ones but comparable with each other and finally C_1 has the lowest one.

Table 3. Comparison between Two-photon excitation and Photobleaching parameters

	Substituent	$\varphi\delta$(GM)*	A_1(photons/s)‡	$k_1(s^{-1})$‡	A_2(photons/s)‡	$k_2(s^{-1})$‡
C_1	OCH_3	182	$2*10^2$	2,9	37	0,06
C_2	CH_3	513	$6.8*10^3$	26,8	$5*10^2$	0,5
C_3	H	767	$1.5*10^6$	130	$3.3*10^4$	7
C_4	Cl	390	$4.7*10^3$	38	$1.4*10^2$	0,5

* φ: quantum yield, δ: TPA cross section, 1 GM=10^{-50} $cm^4 s$ /(photon molecule)

‡ Irradiation power ~5 mW

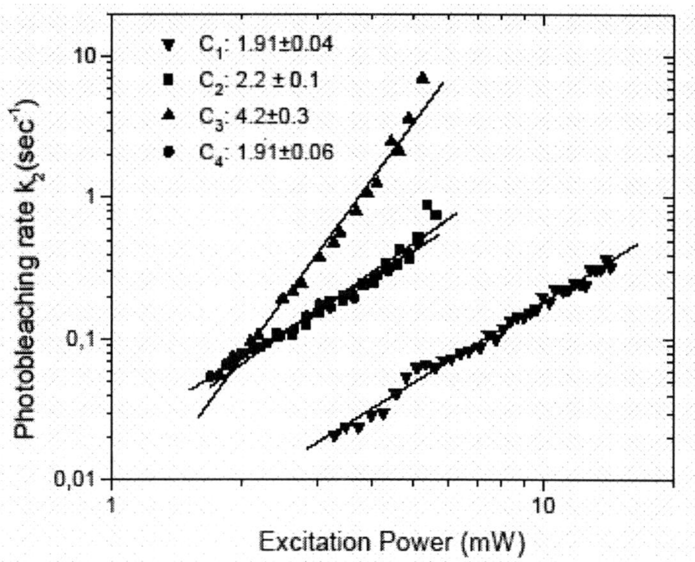

Figure 15. Log-log plots of the photobleaching rates k2 versus irradiation power for each pyrylium salt. The fitted slopes are shown in the graph. The power range of each dataset was selected with respect to the starting point of bleaching

The group of investigated pyrylium salts found to have large two-photon absorption cross-sections, ranging from 267 to 1722 GM [46]. Hence, it was expected at very low power values (where photobleaching is negligible) fluorescence to depend on the square of the incident power. This was confirmed experimentally for all pyrylium salts. The meaning of this result is that the molecule absorbs two photons to reach the first excited state and then relaxes by emitting one photon. Unwanted effects like self-quenching, saturation etc are insignificant. At higher power levels photobleaching makes the situation more complicated. The cubical dependence of photobleaching rate k_1 on irradiation power for the three pyrylium salts shows that the bleaching state is reached with the absorption of three-photons. The analysis followed for pyrylium salt C_3 proved that this process is not a simultaneous three photon absorption but a step-wise one. In more detail, two photons are absorbed simultaneously to reach the first excited state and one more photon is absorbed from this state to reach the bleaching state. Thus, photobleaching through this mechanism is an excited state absorption process. In the case of C_1 the square dependence of k_1 on irradiation power indicates that only two photons are involved in the bleaching procedure. However, at the wavelength of our measurements C_1 can be excited only with a two-photon excitation process leading to the conclusion that photobleaching is occurred from the two-photon excited state. As we noticed above the substituent change affects fluorescence and photobleaching in a different way. For example, the CH_3 (C_2) as a substituent raises fluorescence in comparison with Cl (C_4). In the case of photobleaching rate k_1 the situation is reverse, namely the substituent change lowers photobleaching efficiency. This observation shows that the substituent change causes considerable modifications not only on the two-photon absorption properties but also on the excited state ones. Indeed, these modifications are realized in a different manner.

The substitution effect on the other mechanism is more decisive. When the substituent is H (C_3) the mechanism is a step-wise four photon process [48]. Two photons are absorbed simultaneously to reach the excited state and from this state with the simultaneous absorption of two more photons the molecule is bleached. Photobleaching is therefore again an excited state absorption. For the other three pyrylium salts, where the substituents are OCH_3 (C_1), CH_3 (C_2), and Cl (C_4), photobleaching occurs with the simultaneous absorption of only two-photons. In this case photobleaching is not caused by an excited state absorption.

Many times the irradiation of a molecule with ultra-short laser pulses causes the generation of photoproducts affecting significantly the photobleaching mechanisms. The slow nature of the second mechanism may lead someone to assume that it occurs from a fluorescent photoproduct and not from the pyrylium salt. A discussion on this assumption is following. As shown in figure 13 for pyrylium salt C_3 at the low power value P=1.88mW the ratio of the initial fluorescence signals, A_1/A_2, is 2.6. This means that the contribution to fluorescence signal from the two mechanisms is comparable and thus the first mechanism does not overlap the second one. The photobleaching rates k_1 and k_2 are $2.53s^{-1}$ and $0.055s^{-1}$ respectively (Figure 14 and Figure 15). These values correspond to a decrease of the fluorescence signal at the 1/e of the maximum fluorescence at ~400ms ($1/k_1$) and ~16.5 sec ($1/k_2$) for mechanism 1 and 2 respectively. The slow mechanism is therefore more than 40 times slower than the fast one. If we assume that the second mechanism occurs from a photoproduct at the first 400 ms of irradiation an accumulation of fluorescent photoproducts should be generated (due to the very slow photobleaching of these photoproducts). This accumulation would lead to an increase of fluorescence through this mechanism. Therefore at the begging of irradiation (0-400ms) a significant deviation from the double exponential decay should be observed. The time scale of the phenomenon described above is large enough to be recorded accurately from our experimental setup (time resolution of only a few ms is sufficient). However, such a deviation was not observed in our measurements. In every case a double exponential decay was recorded. Thereafter, we conclude that the second mechanism is not generated from some fluorescent photoproduct but from the pyrylium salt itself. We have no indication for a divergence from this behaviour at the whole power range of the photobleaching measurements. Similar results stand for the other pyrylium salts C_1, C_2, and C_4. The above discussion makes the reliability of our measurements and results more evident.

According to J.Mertz and C.Eggeling et al photobleaching may occur from an excited state absorption in a singlet or a triplet manifold [55,34]. As we showed for molecule C_3 the excited singlet state is reached with the simultaneous absorption of two-photons whereas the triplet state is reached with the intervention of intersystem crossing. The fast and dominant photobleaching mechanism is a singlet to singlet process whereas the minor and slow one is a triplet to triplet process. The same observations stand for the other three pyrylium salts. Furthermore, in the cases where the slow mechanism is not an excited state absorption (as mentioned above) the molecule is bleached in the triplet state without the extra absorption of one photon. This proposal is enforced from the very low measured photobleaching rates k_2 for these salts (Table 3). Consequently, the general Jablonski diagram, proposed for the photobleaching of C_3, stands for all pyrylium salts studied.

5. Conclusion

In conclusion, the one- and two-photon absorption (TPA) properties of a novel group of pyrylium-based compounds were studied (steady state spectroscopy as well as time resolved spectroscopy). The molecules of this group were synthesized by systematically changing the chemical structure of a specific substituent in an initial chromophore. TPA cross-sections as large as 1,8 x 10^{-47} cm^4 s $photon^{-1}$ and high quantum yields were obtained. Furthermore, an experimental investigation of the photobleaching properties of the pyrylium salts under ultra-short pulsed illumination was realized. It was confirmed that two photobleaching mechanisms are simultaneously present to all salts studied. In addition, it was proved that these mechanisms are multiphoton processes. The involvement of two, three or even four photons in these processes was observed, depending on the specific chemical structure. Furthermore, the contribution of each mechanism to the total photobleaching process as well as the efficiency of photobleaching was found to depend essentially on the substituent changes in the basic chemical structure. It was shown that photobleaching and fluorescence are mainly not occurred from the same electronic states. Finally, a simplified Jablonski diagram, describing the photobleaching process, is given.

Acknowledgement

I.Polyzos was supported by the Research Committee of the University of Patras in the frame of the program "K. Karatheodoris 2004, cod. B.392". Dr. M. Fakis acknowledges financial support from the State Scholarships Foundation (IKY).

References

[1] G. S. He, J. D. Bhawalkar, C. F. Zhao, C. K. Park and P. N. *Prasad, Appl. Phys. Lett.* 68, 3549 (1996)
[2] J. E. Ehrlich, X. L. Wu, I.-Y.S Lee, Z.-H. Hu, H. Rockel, S. R.Murder, J. W. Perry, *Opt.Lett* 22, 1843 (1997)
[3] O'Flaherty, S.M., Murphy, R., Hold, S.V., Cadek, M., Coleman, J.N., Blau, W.J. *Journal of Physical Chemistry* B 107 (4), pp. 958-964,2003.
[4] Denk, J. H. Strickler, W. W. Webb, *Science* 248, 73 (1990)
[5] Ohulchanskyy, T.Y., Pudavar, H.E., Yarmoluk, S.M., Yashchuk, V.M., Bergey, E.J., Prasad, P.N. *Photochemistry and Photobiology* 77 (2), pp. 138-145 2003.
[6] Diaspro, M. Robello, J. *Photochem. Photobiol. B:Biol* 55, 1 (2000)
[7] W. G. Fisher, W. P. Jr Partridge, C. Dees, E. A. Wachter, *Photchem. Photobiol.* 66, 141 (1997)
[8] H. B. Sun, S. Matsuo and H. Misawa, *Appl. Phys. Lett.* 74, 786 (1999).
[9] S. Maruo, O. Nakamura and S. Kawata, *Opt.Lett.* 22, 132 (1997)
[10] Zhou, H.-P., Li, D.-M., Zhang, J.-Z., Zhu, Y.-M., Wu, J.-Y., Hu, Z.-J., Yang, J.-X., Wang, C.-K. *Chemical Physics* 322 (3), pp. 459-470 2006.
[11] D. A. Parthenopoulos and P. Retzepis, *Science* 45, 843 (1989)

[12] R. Sivarman, S.J. Clarson, B. K. Lee, A. J. Steckl, B. A. Reinhardt, *Appl. Phys. Lett.* 77 328 (2000)

[13] D. Day and M. Gu, *Appl. Phys. Lett.* 80, 2404 (2002)

[14] M. Albota, D. Beljonne, J.-L. Brédas, J. E. Ehrlich, J.-Y. Fu, A. A. Heikal, S. E. Hess, T. Kogej, M. D. Levin, S. R. Marder, D. McCord-Maughon, J. W. Perry, H. Röckel, M. Rumi, G. Subramaniam, W. W. Webb, X.-L. Wu, and C. Xu, *Science* 281 1653 (1998)

[15] Zhou, X., Ren, A.-M., Feng, J.-K., Liu, X.-J. *Chemical Physics Letters* 362 (5-6), pp. 541-548,2002

[16] Liu, Z.-Q., Fang, Q., Wang, D., Cao, D.-X., Xue, G., Yu, W.-T., Lei, H. *Chemistry - A European Journal* 9 (20), pp. 5074-5084 2003.

[17] Andrade, A.A., Yamaki, S.B., Misoguti, L., Zilio, S.C., Atvars, T.D.Z., Oliveira Jr., O.N., Mendonça, C.R. *Optical Materials* 27 (3), pp. 441-444 2004.

[18] Yi, W., Feng, W., Xu, Y., Wu, H. *Japanese Journal of Applied Physics*, Part 1: Regular Papers and Short Notes and Review Papers 44 (5 A), pp. 3022-3027 2005

[19] Kiran, P.P., Reddy, D.R., Dharmadhikari, A.K., Maiya, B.G., Kumar, G.R., Rao, D.N. *Chemical Physics Letters* 418 (4-6), pp. 442-447 2006.

[20] Sanchez, E.J.; Novotny, L.; Holtom, G.R.; and Xie, X.Sunney. *J.Phys.chem A* 1997,101, 7019

[21] Tamarat, Ph.; Maali, A.; Lounis, B. and Orrit, M. *J.Phys.Chem.A*. 2000,104

[22] Dijk, M.A.V; Kapitein, L.C.; Mameren, J.v.; Schmidt, C.F. and Peterman, E.J.G. *J.Phys.Chem. B* 2004, 108, 6479

[23] Brousmiche, D.W.; Serin, J.M.; Frechet, J.M.J.; He, G.S.; Lin, T.C.; Chung, S.J.; Prasad, P.N.; Kannan, R. and Tan, L.S. *J. Phys. Chem. B* 2004 108, 8592.

[24] Keuren, E.V. and Schrof, W. *Macromol*. 2003 36 5002.

[25] Nakano, M.; Kooriya, T.; Kuragaito, T.; Egami, C.; Kawata, Y.; Tsuchimori, M.; Watanabe, O. *Appl.Phys.Lett.* 2004, 85, 176.

[26] Dyer, Daniel J.; Cumpston, Brian H.; McCord-Maughton, Dianne; Thayumanavan, S.; Barlow, Stephen; Perry, Joseph W.; Marder, Seth R., *Nonlin.Opt.:Quant. Opt*. 2004, 31(1-4), 175.

[27] Gu, M. and Day, D. *Opt. Lett.* 1999 24, 288.

[28] Olson, C. E.; Previte, M. J. R.; Fourkas, J. T. *Nature. Materials*. 2002, 1(4), 225.

[29] Kubitscheck, U.; Heinrich, O.; Peters, R. *Bioimaging* 1996 4, 158.

[30] Tang, H., Xing, H., Jiang, B., Cai, J., Huang, W., Xia, A. *Proceedings of SPIE* - 6150 II, art. no. 61504U 2006.

[31] Jiang, B., Shen, Z., Cai, J., Tang, H., Xing, H., Huang, W. *Proceedings of SPIE* - 6150 II, art. no. 61503Q 2006.

[32] Bonnett, R. and Martinez G. *Tetrahedron*. 2001, 57, 9513.

[33] Gohde, W.; JR., Fischer, U.C.; Fuchs, H.; Tittel, J.; Basche, Th.; Brauchle, Ch.; Hermann, A. and Mullen, K. *J.Phys.Chem A*. 1998, 102, 9109.

[34] Eggeling, C.; Widengren, J.; Rigler, R. and Seidel, C.A.M. Anal.Chem. 1998, 70, 2651.

[35] Chirico, G.; Cannone, F. and Diaspro A. *J.Phys.D:Appl.Phys.* 2003, 36,1682.

[36] Kao, F.-J.; Wang, Y.-M.; Chen J.-Ch.; Cheng P.-Ch.; Chen R.-W.; Lin B.-L. *Opt.Comm*. 2002 201, 85.

[37] Dittrich, P.S. and Schwille, P *Appl.Phys B* 2001, 73, 829.

[38] Parreira, R.L.T., Galembeck, S.E. *Journal of Molecular Structure: THEOCHEM* 760 (1-3), pp. 59-73 (2006)

[39] Gong, W.-T., Li, X.-C., Ning, G.-L., Zhu, L., Wang, L., Lin, Y. *Spectrochimica Acta - Part A: Molecular and Biomolecular Spectroscopy 62* (4-5), pp. 835-839 (2005)

[40] Zhou, Y.-F., Feng, S.-Y. *ChemPhysChem* 3 (11), pp. 969-972 2002.

[41] Balaban T.S. et al. *Science of Synthesis*, 14 (2003) 138.

[42] Moghimi, A., Rastegar, M.F., Ghandi, M. *Canadian Journal of Chemistry* 82 (12), pp. 1716-1724 (2004)

[43] Gong, W.-T., Li, X.-C., Ning, G.-L., Lin, Y. *Journal of Chemical Research* (6), pp. 444 2004.

[44] M. Fakis, G. Tsigaridas, I. Polyzos, V. Giannetas, P. Persephonis, I. Spiliopoulos, J. Mikroyannidis, *Chemical Physics Letters*, 342 (2001), 155-161.

[45] M. Fakis, J. Polyzos, G. Tsigaridas, J. Parthenios, A. Fragos, V. Giannetas, P. Persephonis' J. Mikroyannidis, *Chem. Phys. Lett*, 323, 111 (2000)

[46] I Polyzos, G. Tsigaridas, M. Fakis, V. Giannetas, P. Persephonis , J. Mikroyannidis, *Chemical Physics Letters*, 369 (2003), 264-268.

[47] Polyzos, G. Tsigaridas, M. Fakis, V. Giannetas and P. Persephonis, *Optics Letters* 30 (2005) 2654-2656.

[48] Polyzos, G. Tsigaridas, M. Fakis, V. Giannetas, P. Persephonis, and J. Mikroyannidis, *Journal of Physical Chemistry B* 110 (2006) 2593-2597.

[49] I Polyzos, G Tsigaridas, M Fakis, V Giannetas, P Persephonis and J Mikroyannidis, *Journal of Physics: Conference Series* 10 (2005) 230-233.

[50] M. Fakis, I. Polyzos, G. Tsigaridas, V. Giannetas , P. Persephonis, *Chemical Physics Letters* 394 (2004) 372-396.

[51] D. Anestopoulos, G. Tsigaridas, M. Fakis, I. Polyzos, P. Persephonis and V. Giannetas, *Journal of Physical Chemistry B* 109 (2005) 9476-9481.

[52] N. Derras, G. A. Crosby, J. Phys. Chem. 75, 991 (1971)

[53] Lakowicz J. R. (1999), Principles of Fluorescence Spectroscopy, Kluwer Academic, Plenum Publishers, New York C. Xu and W.W. Webb, J.*Opt.Soc.Am.B* 13, 481 (1996)

[54] Mertz J.,*Eur.Phys.J. D* 1998, 3, 53.

In: Trends in Optical Materials Research
Editor: Owen G. Gardens, pp. 137-183

ISBN: 1-60021-290-5

Chapter 5

STUDY OF THE OPTICAL AND THERMAL PROPERTIES OF PHOSPHATE GLASSES

Eric Tong Yih Lee and Elizabeth M. Taylor
Optoelectronics Research Centre,
University of Southampton, Highfield, Southampton SO17 1BJ, UK

ABSTRACT

Lightwave devices such as gratings and filters have a drawback where the central wavelength of the devices is sensitive to changes in temperature, which compromises their performance. This problem is particularly acute in systems that employ Wavelength Division Multiplexing (WDM), where the stability of wavelength with regards to temperature is essential. An athermal waveguide, which is inherently insensitive to temperature, is an ideal choice to solve this problem instead of using a passive temperature control unit, which could limit the optimum performance of a WDM system.

Glasses are considered ideal host materials for athermalisation as they are versatile and can be physically and compositionally tailored to meet various optical and thermal requirements. Generally, athermal glasses must exhibit negative refractive index change with temperature ($\mathrm{d}n/\mathrm{d}T$) to counter the effect of thermal expansion (α). Hence, phosphate-based glasses are ideal candidates for athermalisation as they exhibit negative $\mathrm{d}n/\mathrm{d}T$. Various compositions of ternary potassium aluminophosphate glasses, binary calcium phosphate and barium phosphate glasses and ternary barium borophosphate and sodium borophosphate glasses were examined in this research work to assess their potential for athermalisation. The optical and thermal properties of the phosphate glasses were studied, particularly α and $\mathrm{d}n/\mathrm{d}T$, and the compositional effects of the glasses on these properties were also investigated. Results showed that potassium aluminophosphate glasses were strongly influenced by the strengthening effect of Al^{3+} and the trend observed for the binary phosphate glasses depended on the availability of terminal oxygens to coordinate the cations, while the change in the properties of borophosphate glasses depended on the effect and coordination number of boron.

Glasses with negative $\mathrm{d}n/\mathrm{d}T$ were achieved from the phosphate glasses studied. The results showed that α and $\mathrm{d}n/\mathrm{d}T$ of the glasses were inter-dependent and semi-empirical equations relating these properties could be defined, enabling the estimation of $\mathrm{d}n/\mathrm{d}T$ of any phosphate glasses within the same system. In addition, the results enabled the identification of the specific glass composition of a particular phosphate glass system with the minimum α value required in order for the glass to exhibit negative d*n*/d*T*. This research work represents the first comprehensive measurement of d*n*/d*T* of phosphate glasses.

1. INTRODUCTION

The advent of Dense Wavelength Division Multiplexing (DWDM) technology has efficiently increased the transmission bandwidth of the optical fibre networks and enabled all optical routing functions such as defining channel frequencies, filtering noise and add/drop multiplexing to be implemented. Wavelength discriminating devices such as resonators, filters and gratings are sensitive to temperature due to the temperature dependence of the optical path length, *S*, of the devices. This can change the central resonance response of the devices significantly, which can be as large as 0.01 nm/K for silica-based waveguide filters and 0.08 nm/K for semiconductor-based waveguide filters (at the wavelength, λ, of 1.55 μm) [1-5]. Hence the use of such devices in DWDM systems can adversely affect the efficiency of these systems where channels are compactly spaced to increase transmission bandwidth. Actively temperature-stabilised components would therefore be required to maintain stable spectral responses for the DWDM systems over a large temperature range in order to minimise crosstalk between the closely-spaced channels. However, this will be an obstacle to the broad use of DWDM systems from the viewpoint of cost, device size and reliability. Therefore an alternative solution is to develop an athermal waveguide in which the optical path length change with temperature is zero.

Glasses are suitable for athermal waveguide development due to their versatility, where their chemical and physical properties can be exploited to develop more efficient, low-cost and compact planar and fibre devices for the information industry. Chemically, glass compositions can be tailored to meet optical and thermal requirements such as selecting the spectroscopic features of dopant ions, photosensitivity, refractive index (*n*), glass transition temperature (T_g), thermal expansion coefficient (α) and thermo-optic coefficient (d*n*/d*T*), among others. Also, glasses are suitable optical material for processing as they can be physically moulded, pressed, extruded and drawn into fibres. Athermal glasses are currently used in high power laser to minimise thermally induced distortion in lenses [6-7].

In order to achieve athermalisation, glasses must exhibit intrinsic negative d*n*/d*T* to counter the effect of positive thermal expansion. In this respect, the family of phosphate, borate and borophosphate glasses are ideal candidates for athermalisation as they inherently exhibit negative d*n*/d*T*. Therefore phosphate-based glasses are potential candidates for athermalisation and can be compositionally tailored via the addition of suitable glass modifying cations. Table 1 lists the properties of the pure glass network formers [8].

In this chapter, potassium aluminophosphate, calcium phosphate, barium phosphate, barium borophosphate and sodium borophosphate glasses were studied and presented in order to assess the viability of such glasses for athermalisation applications. The main objective was to study the optical and thermal properties of these different phosphate glass families and examine the compositional effects of the glasses on these properties. These glasses were chosen because of their inherent negative d*n*/d*T* and high chemical durability. Table 2 shows the properties of some metaphosphate glasses.

Table 1. Properties of pure network formers.

Glasses	n_e (at λ = 546.1 nm)	α (×10^{-6}/°C)	dn/dT (×10^{-6}/°C)
GeO_2	1.6056	7.1	+19.4
SiO_2	1.4601	0.6	+10.4
B_2O_3	1.4650	14.4	-35.0
P_2O_5	1.4930	13.7	-92.2

Table 2. Properties of metaphosphate glasses

Glasses	n_e [9]	T_g (°C) [9]	α [9] (×10^{-6}/°C)	dn/dT [8] (×10^{-6}/°C)
$Zn(PO_3)_2$	1.5260	310	5.8	+5.1
$Al(PO_3)_3$	1.5310	-	5.7	+5.0
$Y(PO_3)_3$	-	-	-	+4.9
$Mg(PO_3)_2$	1.4974	510	8.3	+2.9
$La(PO_3)_3$	1.6000	570	7.8	+0.9
$Ca(PO_3)_2$	1.5469	500	11.0	-2.5
$Sr(PO_3)_2$	1.5614	495	12.4	-5.9
$LiPO_3$	-	-	-	-6.1
$Pb(PO_3)_2$	1.7720	350	14.6	-10.4
$Ba(PO_3)_2$	1.5899	455	14.2	-10.6
$NaPO_3$	1.4840	275	25.0	-15.0
KPO_3	1.4740 [10]	228 [10]	-	-23.3

2. Background

This section presents a topical review examining the properties and structures of phosphate glasses, which form the basis of the investigations in this chapter. The basic phosphate tetrahedral sites are presented and used to explain the different phosphate networks, as set by the oxygen-to-phosphorus (O/P) ratio.

The concept of athermalisation is also examined and the athermal condition is derived for a bulk material. Finally starting from the Lorentz-Lorenz equation, the concept of dn/dT and the various parameters influencing the dn/dT of glasses are also presented.

2.1. Phosphate Glasses

The glass former component in phosphate glasses is P_2O_5. Pure P_2O_5 has a melting point of 560°C and a boiling point of 605°C. Some of the properties of pure P_2O_5 glass include n_e = 1.4930 (at λ = 546.1 nm), α = 13.7×10^{-6}/°C and dn/dT = -92.2×10^{-6}/°C [8]. Hypothetically, P_2O_5 has a UV edge of less than 150 nm (~ 8.27 eV). However, because of its hygroscopic and volatile nature [11], P_2O_5 generally contains water and is more realistically represented by the HPO_3 glass structure with OH bonds, thus shifting the UV edge towards longer wavelengths. Various alkali, alkaline earth, transition metal and rare earth oxides can be added to modify the properties of phosphate glasses. However, phosphate glasses have poor chemical durability which often limits their practical applications. This can be overcome by adding oxides such as PbO [12], Al_2O_3 and Fe_2O_3 [13,14] to the phosphate glasses.

The basic building blocks of crystalline and amorphous phosphates are the P-tetrahedra, which link via covalent bridging oxygens to form various phosphate anions. These tetrahedral units can be classified using the Q-site model proposed by Lippma *et al.* [15], originally used to describe the structures of binary silicate glasses. Based on this model, the phosphate anions can be described using the Q^i terminology, where '*i*' represents the number of bridging oxygens (BO) per tetrahedron, as shown schematically in Figure 1.

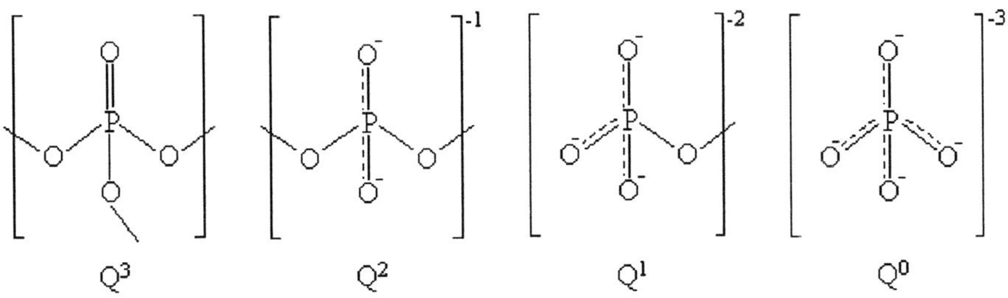

Figure 1. Phosphate tetrahedral structures

P_2O_5 is different from the other single component glass formers due to the existence of a terminal oxygen (TO) on each network forming cation. This terminal oxygen is a short double-bonded (P=O) bond, with a significant π-bond character that accounts for the additional valence electron [16]. The presence of such terminating oxygens reduces the cross-link density of the phosphate glass structure, resulting in a higher thermal expansion coefficient (α), and a lower transition temperature (T_g), compared to silicate and borate glasses [17].

The phosphate glass networks can be classified by the O/P ratio, which describes the tetrahedral linkages via bridging oxygens between neighbouring P-tetrahedra [18], as summarised in Table 3. Vitreous P_2O_5 with O/P = 2.5 has a three-dimensional structural network of cross-linked Q^3 tetrahedra with three covalent bridging oxygens (P-O-P bonds) linked to neighbouring tetrahedra and one terminal oxygen (P=O bond). The addition of a modifying oxide (MeO) depolymerises the structural network by converting BOs to non-bridging oxygens (NBOs), resulting in (P-O-Me) links, and in the process forming an ultraphosphate network ($2.5 \leq O/P < 3.0$) dominated by Q^3 tetrahedra. As more MeO is added, the three-dimensional network is converted into metaphosphate chains and rings based on Q^2 tetrahedra (O/P = 3.0) [12,19], which are interconnected through ionic cross-links with modifying cations between NBOs of two different chains. Further increases in the modifier content results in polyphosphate compositions (O/P > 3.0) based on Q^2 chains terminated by Q^1 tetrahedra where the average chain lengths are shorter than that of the metaphosphate compositions. At the pyrophosphate stoichiometry (O/P = 3.5), the network is dominated by phosphate dimers (Q^1-Q^1), while at the orthophosphate stoichiometry (O/P = 4.0), the glasses contain isolated Q^0 tetrahedra. Figure 2 shows a three-dimensional network based on Q^3 tetrahedra (left side) and metaphosphate chains of Q^2 tetrahedra (right side).

Table 3. Phosphate glass classifications and their O/P ratios

Phosphate network	*Ultra-phosphate*	*Meta-phosphate*	*Polyphosphate*	
			Pyro-phosphate	*Ortho-phosphate*
O/P	< 3.0	3.0	3.5	4.0

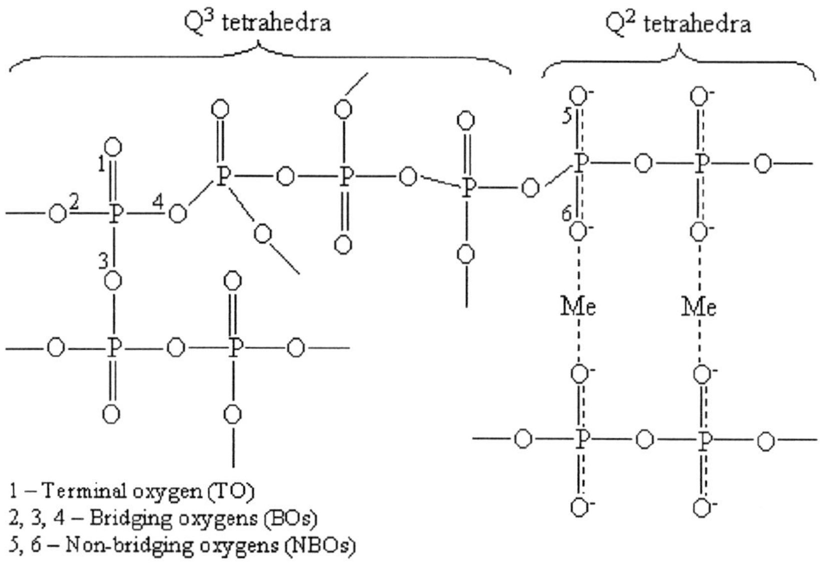

Figure 2. Phosphate structural network

2.2. Concept of Athermalisation

The optical path length, *S*, of an optical material is sensitive to changes in temperature, due to the combined effects of thermal expansion and thermal refractive index change. Athermalisation is a process which eliminates the temperature dependence of the optical path length.

The optical path length of a homogeneous bulk material can be defined as [1]

$$S = nL \tag{1}$$

where n is the refractive index of the medium and L is the length of the medium. The temperature dependence of S can then be defined as

$$\frac{1}{L}\left(\frac{dS}{dT}\right) = n\alpha + \frac{dn}{dT} \tag{2}$$

where α is the thermal expansion coefficient ($\alpha = (1/L)(\mathrm{d}L/\mathrm{d}T)$), $\mathrm{d}n/\mathrm{d}T$ is the thermo-optic coefficient and T is temperature. Athermalisation is achieved by equating Equation (2) to zero ($\mathrm{d}S/\mathrm{d}T = 0$), giving the athermal condition as

$$\frac{dn}{dT} = -n\alpha \tag{3}$$

As glasses generally exhibit positive n and α, dn/dT must be negative to satisfy Eq. (3).

2.3. d*n*/d*T* of Glasses

The thermo-optic coefficient of glasses can be understood via the Lorentz-Lorenz equation for refractive index [20].

$$\frac{n^2 - 1}{n^2 + 2} = \frac{4}{3}\frac{\pi N_A P_e}{V_m} \tag{4}$$

where n is the refractive index, N_A is the Avogadro's number (6.022×10^{23}), P_e is the electronic polarisability and V_m is the molar volume. The equation for the refractive index change with temperature, dn/dT, can be derived from Equation (4) and is defined by Equation (5), with the approximation of (d(lnV_m)/dT ~ 3α) [21].

$$\frac{dn}{dT} = A[\varphi - 3\alpha] \tag{5}$$

where $A = [(n^2\text{-}1)(n^2\text{+}2)/6n]$, φ = d(lnP_e)/dT and T is temperature. Equation (5) shows that dn/dT is affected by n, P_e and α of the glass, where α must be sufficiently large such that $3\alpha > \varphi$ in order to exhibit negative dn/dT.

By assuming that polarisability is increased by an increase in the inter-ionic distance due to a change in temperature, Equation (5) can be rewritten as Equation (6), with $\varphi = [\partial(\ln P_e)/\partial(\ln r)]_T [d(\ln r)/dT]$ and $\alpha = [d(\ln r)/dT]$ [22].

$$\frac{dn}{dT} = A(\gamma - 3)\alpha \tag{6}$$

where γ is the change in polarisability due to a change in inter-ionic distance ($\gamma = [\partial(\ln P_e)/\partial(\ln r)]_T$), and r is the inter-ionic distance. As A and α are positive for glasses, the sign of dn/dT depends on γ.

3. Glass Preparation and Characterisation

The general principles of glass melting procedures employed and the various measurement techniques used to investigate the optical and thermal properties of the different phosphate glasses studied within this chapter are described in this section.

3.1. Glass Synthesis Process

Glass synthesis generally consists of three steps : preparing the appropriate mixture of raw materials, melting the reaction mixture and finally annealing the glass.

Batching of raw materials was carried out in a dry nitrogen-purged glovebox, where the materials were weighed and mixed in a glass beaker, homogenised with a porcelain spoon and then transferred into a silica crucible. Melting was carried out in an oxygen atmosphere for 60 - 90 minutes within a melting furnace, which was preheated to temperatures between 1000°C and 1500°C (melting temperature, T_m), depending on the glass compositions. The crucible was covered with a platinum lid to minimise evaporation and the loss of materials during the melting process. Once, the melting process was completed, the melt was cast into a graphite casting mould placed within an annealing furnace, which had been pre-heated approximately to the glass T_g. In order to produce well-annealed, stress-free and striae-free glasses, the glasses were held at a temperature near their T_g for 60 minutes before being slowly cooled to room temperature at a rate of 0.5°C/min. Figure 3 shows the glass synthesis route employed. As silica crucibles were used for the melting of the reaction mixtures, the actual compositions of the glasses typically contained approximately 1 mol% silica as the crucibles can dissolve during melting at high temperatures and traces of silica mixed in with the glass melts.

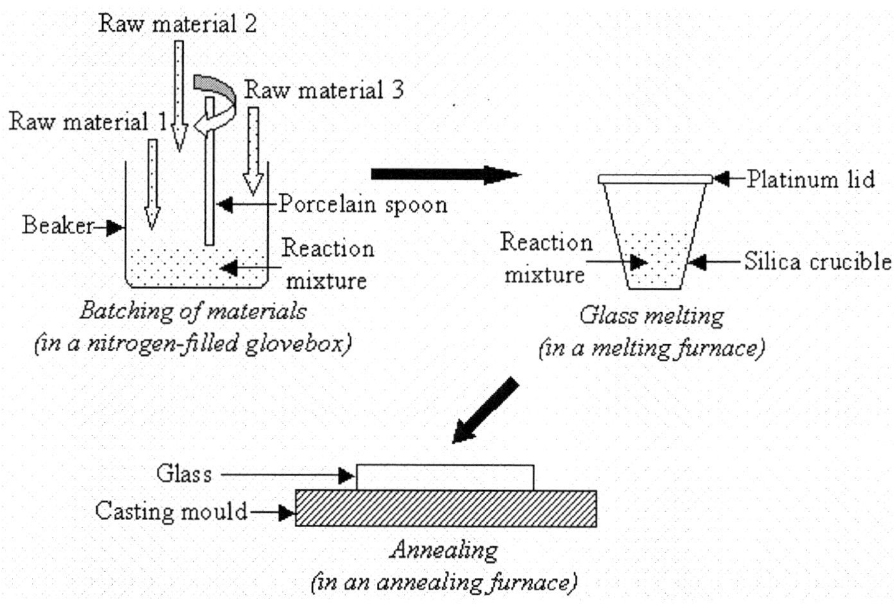

Figure 3. Glass synthesis process

In a homogeneous glass, stresses can be caused by the rapid cooling of the glass through the glass transition region (approximated by the glass T_g), leaving inadequate time for the constituent atoms to occupy the lowest energy sites. Within the transition region, microstructure changes can occur. These changes require an increasing amount of time to occur as the temperature is reduced. Hence, most glass properties, such as T_g and α, depend on time and temperature (thermal history). This means that a change in the heating and cooling rate can induce variations in glass properties. For a glass cooled at a low rate, the glass is able to shrink to a metastable equilibrium volume and become dense. However, if the glass undergoes rapid cooling, there is insufficient time for contraction and the glass structure freezes at a greater volume. Upon heating to the glass transition region, the structure elements of the glass are rearranged into denser and lower energy sites, leading to a reduction of the glass volume. Such structural changes can affect the glass property measurements.

Once the glasses were removed from the annealing furnace and before any glass properties were measured, the glasses were polished to produce parallel, smooth, flat and optically transparent surfaces. A three-step procedure using three different solutions containing 9 μm calcined alumina powder, 3 μm calcined alumina powder and 0.125 μm silica particles, respectively, were used. The glasses were first lapped using the coarse grit size of 9 μm on a cast iron plate to shape, flatten as well as to thin the glasses to the required dimensions. This was then followed by the solution containing grit size of 3 μm to reduce scratches and edge damage caused by the previous step. The final step involving the 0.125 μm silica particles solution on a polyurethane plate was to improve the quality of the surface by removing surface roughness, scratches and edge damage.

3.2. Characterisation of Thermal Properties

3.2.1. Glass Transition Temperature, T_g

A Differential Thermal Analyser (Perkin Elmer DTA7) equipped with a high temperature furnace was used to measure the glass transition temperature in a nitrogen atmosphere. There was a matched pair of thermocouples located within the furnace, one monitoring an alumina cup containing alumina powder as reference while the other an alumina cup with a piece of glass sample, about 20 - 30 mg, in alumina powder. As the furnace ramped up, the thermocouples measured the differential temperature, ΔT, between the two alumina cups, due to the endothermic and exothermic reactions in the glass.

For T_g measurement, the glass samples were heated at a rate of 40°C/min from room temperature to about 150°C below their respective T_g and then heated at a rate of 10°C/min to about 900°C. Repeated measurements of samples with similar nominal composition showed an uncertainty of ±5°C. An example of a typical DTA trace is shown in Figure 4, recorded as ΔT between the alumina cups against sample temperature, T. The glass transition temperature, T_g, can be estimated from the intercept of the tangents to the curves. In addition, the onset of crystallisation, T_x, peak crystallisation, T_p, and melting, T_m, temperatures can also be determined.

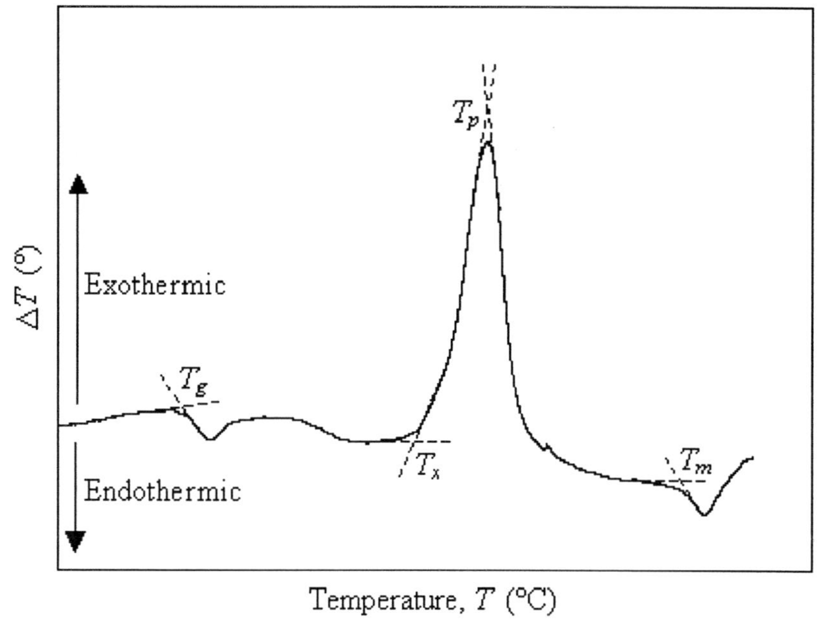

Figure 4. An example of T_g measurement data

As the temperature increases, the disordered rigid glassy structure changes to a 'liquid' state with increased molecular mobility and flow. This is denoted by the first endothermic event (T_g). Further increase in temperature results in nucleation and crystal growth, characterised by the onset of crystallisation (T_x) and an exothermic peak (T_p). When the DTA curve returns to the baseline after the melting event (T_m), the liquidus stage is reached. The thermal stability of the glass is generally determined from (T_x-T_g) and (T_p-T_x). (T_x-T_g) should be large (ideally ≥ 150°C) to provide the maximum working temperature range for the glass before crystallisation sets in. (T_p-T_x) should also be large as this generally means that there are a number of crystal events happening and competing against each other such that crystallisation will not progress rapidly once started.

3.2.2. Thermal Expansion Coefficient, α

The thermal expansion coefficient was measured using a Thermo-Mechanical Analyser (Perkin Elmer TMA7), which allows the measurement of dimensional changes in materials as a function of temperature or time. The glass sample was held in place by a probe in the sample holder, within a furnace. As the furnace ramped up, the sample expanded and caused a change in the probe position, which was recorded.

α was measured in the temperature range of 50 - 300°C with a force of 110 mN. The glass samples were heated to 300°C at 5°C/min. Repeated measurements of samples showed an experimental error within $\pm 0.2\times10^{-6}$/°C. Figure 5 shows an example of the expected TMA curve, presented as sample height, L, against temperature, T. The expansion coefficient, α, is defined as $(1/L)(\mathrm{d}L/\mathrm{d}T)$ where $\mathrm{d}L/\mathrm{d}T$ can be determined from the slope. If the measurement was carried out to higher temperatures, the glass transition temperature, T_g, of the glass can also be measured, as indicated in Figure 5 by a change in the slope representing a change in the glass viscosity.

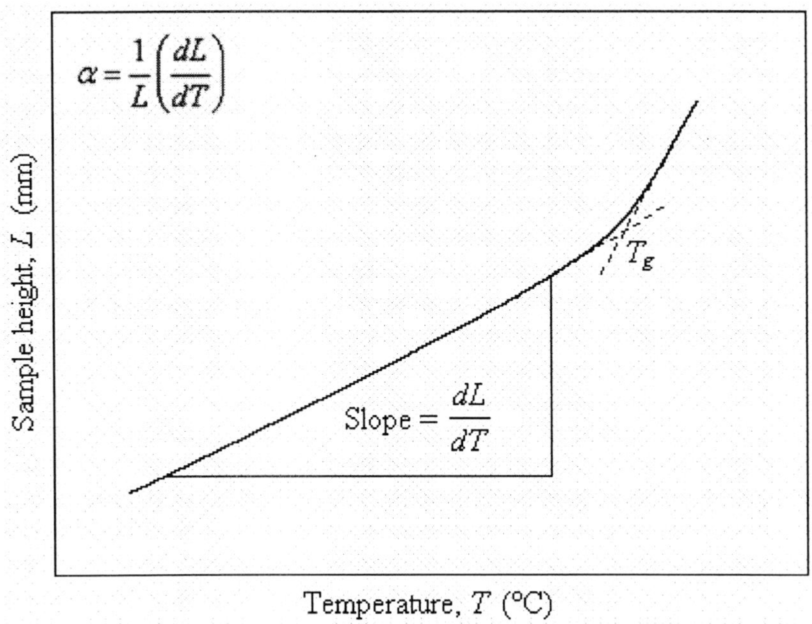

Figure 5. An example of α measurement data

3.3. Characterisation of Optical Properties

3.3.1. Refractive Index, n

An Abbe '60' refractometer (Bellingham and Stanley Ltd.) was used to measure the refractive index of the glasses at room temperature, by means of the total internal reflection phenomenon, at the sodium D-line (λ = 589.3 nm). Measurement error was estimated to be within ±0.0005, due to instrument limitation.

There was a prism within the refractometer, with a high refractive index ~ 1.7. The source radiation was incident on the first face of the prism and diffusely scattered into the prism and directed towards the second face of the prism, which was in contact with the glass sample. Depending on the contact angle with the sample, the diffused light either partly propagates into the sample or was total internally reflected and emerged through the third face of the prism. Hence the field of view observing the third face will be divided into a dark region (for contact angles less than the critical angle) and a bright region (for contact angles greater than the critical angle). Knowing the properties of the prism, the instrument was calibrated such that the observed boundary of these two regions directly corresponds to the sample refractive index.

3.3.2. OH Content, r_{OH}

Infrared measurements of relative OH concentrations were collected using a Fourier transform infrared (FTIR) spectrophotometer (Perkin Elmer System 2000), which records the interaction of IR radiation with experimental samples, measuring the frequencies at which the sample absorbs the radiation and the intensities of the absorptions. The source radiation was directed onto the sample, which then passed through a beamsplitter and split into two perpendicular directions towards a stationary mirror and a moving mirror, respectively, creating a difference in the path lengths when the light paths were reflected back to the beamsplitter and recombined. An interferogram was created from this interference effect. The recombined beam was then passed through the sample, which absorbed the wavelengths specific to the sample, and corrected for the interferogram to produce the absorption data.

Measurements were taken in the transmission mode in the range of 2000 - 5000 cm^{-1} (5000 – 2000 nm) with a resolution of 4 cm^{-1}, averaged over 30 scans. Background subtraction was performed for each sample analysis to eliminate the effects of IR-active atmospheric components such as carbon dioxide (CO_2) and water (H_2O). Figure 6 shows an example of an FTIR spectrum in transmission mode. For phosphate glasses, there is a broad OH absorption band at around 3000 cm^{-1} (~ 3.3 μm) [23]. From the FTIR scan, the relative OH concentration can be expressed in terms of the OH absorption coefficient, r_{OH} [24]:

$$r_{OH} = \frac{\left(-ln\left(\frac{T_{3000}}{T_{5000}}\right)\right)}{l} \qquad (7)$$

Equation (7) determines the relative concentration of OH from the transmission % at 3000 cm^{-1} (T_{3000}), normalised to the transmission % at 5000 cm^{-1} (T_{5000}) and the sample thickness, l. T_{5000} serves as a background transmission and includes the Fresnel reflection losses.

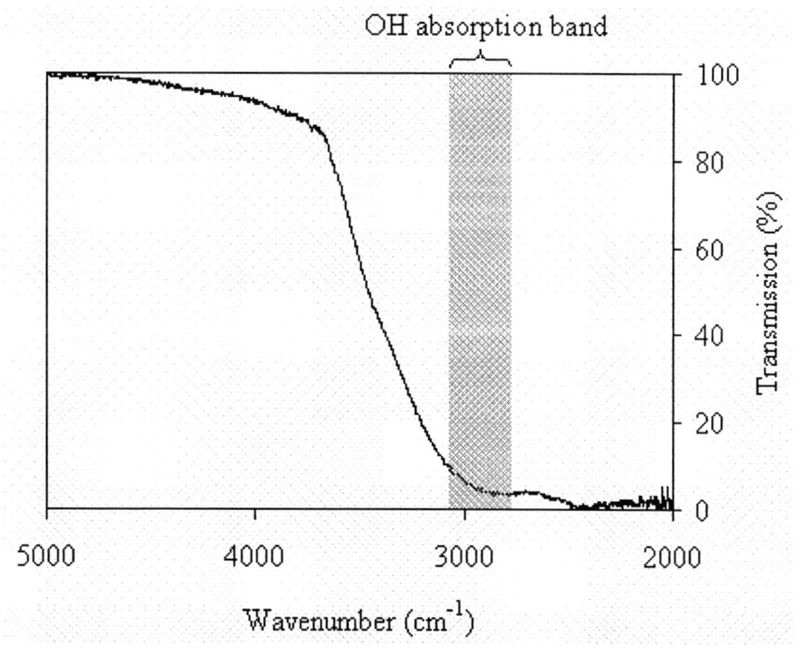

Figure 6. An example of FTIR spectrum for OH measurement, corrected for Fresnel reflection

3.3.3. Thermo-Optic Coefficient, Dn/Dt

3.3.3.1. Interferometer Set-up

An interferometric technique was set up to measure the thermo-optic coefficient, dn/dT, as shown schematically in Figure 7. It was based on the work described by Jewell *et al.* [25], but with modifications to the apparatus set-up and sample design.

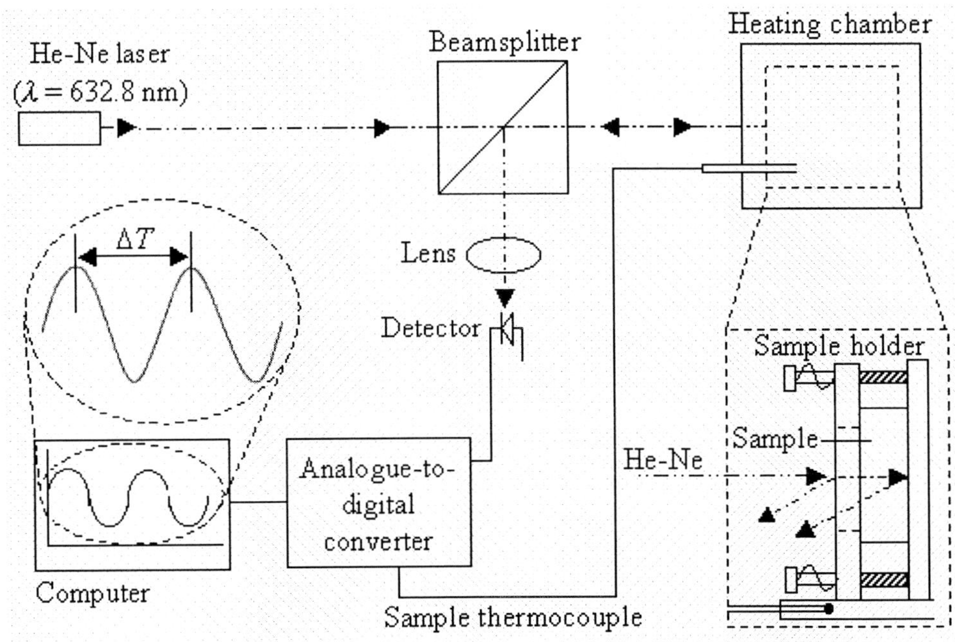

Figure 7. Interferometer set-up

The sample was held in place by a sample holder, which consisted of two aluminium plates, within a heating chamber. The front plate was graphite-coated to prevent any reflection from the plate and provide good thermal contact with the sample and also was spring loaded to provide good sample stability and prevent strain-induced distortion from thermal cycling during measurements, while the back plate was fixed in place.

Light from a He-Ne laser was used to illuminate the sample. The light was then reflected from the front and back faces of the sample and formed an interference pattern, which could be detected and collected by a silicon photodetector, via the beamsplitter. The photodetector was connected to an analogue-to-digital converter together with the thermocouple of the temperature controller in order to feed the required data into a computer for analysis. This thermocouple was attached to the base of the sample holder where this position offered the most stable temperature ramping. Heating was carried out at a slow rate of 40°C/hour to allow sufficient time for the sample to reach the desired temperature.

3.3.3.2. Analysis of Fringe Pattern

The interference pattern can then be analysed to calculate dn/dT. The photodetector was able to detect the intensity of the interference pattern. The detector was first positioned on either a bright fringe or a dark fringe. As the sample was heated, the optical path length between the front and back faces of the sample changed due to the combined effects of α and dn/dT of the sample. When the net path length increased, the fringes moved and were recorded as oscillations in the intensity of the light upon the detector. Hence the results as observed on a computer showed an approximate sinusoidal wave.

The general relative phase of two interfering beams, $\Delta\phi$, as a function of temperature, T, is given as [26]

$$\Delta\phi = \left(\frac{4\pi}{\lambda}\right) l \left[n\alpha + \left(\frac{\partial n}{\partial T}\right)\right] T \tag{8}$$

where λ is the laser wavelength, l is the glass thickness, n is the refractive index, α is the thermal expansion coefficient, $\partial n/\partial T$ is the thermo-optic coefficient and T is temperature.

By measuring ΔT, corresponding to the temperature difference between successive maxima or minima of the fringes, or equivalently to a phase difference of $\Delta\phi = 2\pi$, Equation (8) can be rearranged to Equation (9) below to calculate dn/dT [25,27],

$$\frac{dn}{dT} = \left(\frac{\lambda}{2l\Delta T}\right) - n\alpha \tag{9}$$

The thermo-optic coefficient was measured in the temperature range of 20 - 100°C at a rate of 40°C/hour. dn/dT values were reproducible to ±5%, due to the uncertainty of ±0.1°C for ΔT, extracted from the measurements to calculate dn/dT, as well as the errors in the measured values of n and α.

4. Aluminophosphate

This section presents the preparation and compositions of the ternary potassium aluminophosphate glasses examined in this study and the results and discussion of the optical and thermal properties of these glasses.

4.1. Introduction

Potassium phosphate (KPO_3) exhibits the most negative dn/dT at -23.3×10^{-6}/°C amongst the various metaphosphate glasses (Table 2) and hence is a potential candidate for athermalisation applications, where a negative dn/dT value is required to satisfy the athermal Equation (3). However, alkali earth phosphate glasses suffer from hydrolysis, which limits their practical applications. This can be overcome by adding aluminium to improve the chemical durability of the glasses [28,29], despite aluminophosphate ($Al(PO_3)_3$) exhibiting positive dn/dT of $+5.0\times10^{-6}$/°C.

The presence of aluminium generally increases the cross-linking density of the phosphate glass structures due to the conversion of Q^3 tetrahedral units to Q^2 tetrahedral units, through the formation of strong covalent bridging aluminophosphate (P-O-Al) bonds between phosphate chains, as shown in Figure 8. This structural depiction is based on similar structures of sodium aluminophosphate glasses and evidence that octahedral aluminium, Al(6), is the preferred moiety in aluminophosphate glass networks with metaphosphate stoichiometries [30,31].

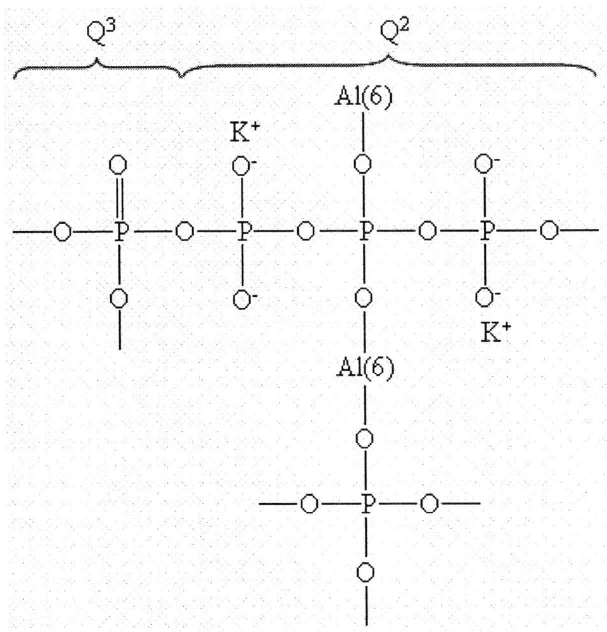

Figure 8. Potassium aluminophosphate structure

The main aim in this section is to assess the optical and thermal properties of potassium aluminophosphate glasses (designated as KAP) and examine the compositional effects (ultraphosphate, metaphosphate, and polyphosphate compositions) on these properties.

4.2. Glass Preparation and Characterisation

Ternary KAP glasses of the K_2O-Al_2O_3-P_2O_5 system were prepared, encompassing stoichiometries of ultraphosphates, metaphosphates and polyphosphates, with oxygen-to-phosphorus (O/P) ratio ranging approximately between 2.8 and 3.1. Four different series of KAP glasses were synthesised, corresponding to (mol%):

- Series 1 : xK_2O-yAl_2O_3-$(x+3y)P_2O_5$ metaphosphate composition with $10 \leq x \leq 40$, $5 \leq y \leq 20$.
- Series 2 : xK_2O-$(35-x)Al_2O_3$-$65P_2O_5$ with $15 \leq x \leq 30$ (P_2O_5 was fixed at 65 mol%).
- Series 3 : $20K_2O$-yAl_2O_3-$(80-y)P_2O_5$ with $10 \leq y \leq 17.5$ (K_2O was fixed at 20 mol%).
- Series 4 : xK_2O-$15Al_2O_3$-$(85-x)P_2O_5$ with $10 \leq x \leq 25$ (Al_2O_3 was fixed at 15 mol%).

Table 4 shows the nominal batch compositions of the KAP glasses prepared while Figure 9 shows the glass melting region for the KAP glasses.

Table 4. Batch compositions of potassium aluminophosphate (KAP) glasses

Sample	Batch composition (mol%)			O/P	Phosphate
	K_2O	Al_2O_3	P_2O_5		
KAP-1a	40	5	55	3.00	Meta-
KAP-1b	30	10	60	3.00	Meta-
KAP-1c	20	15	65	3.00	Meta-
KAP-1d	16.5	16.75	66.75	3.00	Meta-
KAP-1e	10	20	70	3.00	Meta-
KAP-2a	30	5	65	2.85	Ultra-
KAP-2b	27.5	7.5	65	2.88	Ultra-
KAP-2c	25	10	65	2.92	Ultra-
KAP-2d	15	20	65	3.08	Poly-
KAP-3a	20	10	70	2.86	Ultra-
KAP-3b	20	12.5	67.5	2.93	Ultra-
KAP-3c	20	17.5	62.5	3.08	Poly-
KAP-4a	10	15	75	2.87	Ultra-
KAP-4b	15	15	70	2.98	Ultra-
KAP-4c	25	15	60	3.08	Poly-

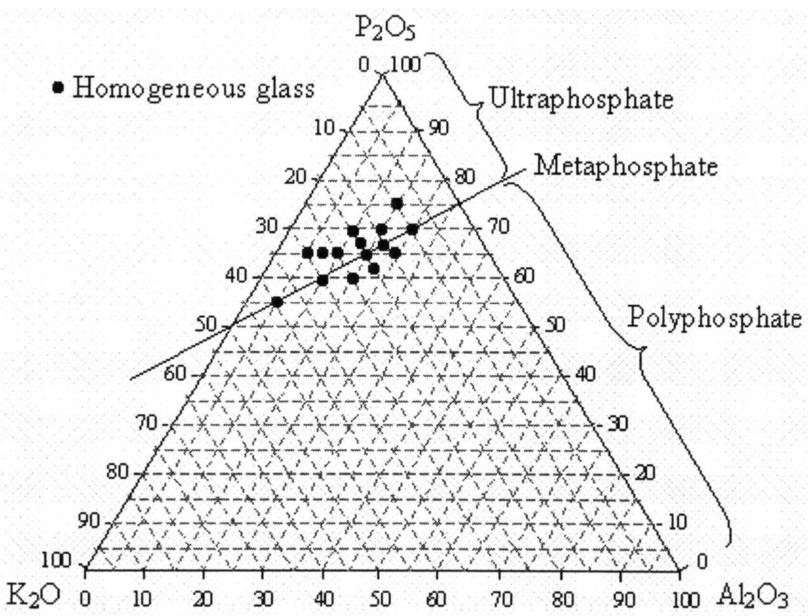

Figure 9. Potassium aluminophosphate (KAP) glass melting

Batches were prepared from the appropriate mixtures of high purity powder K_2CO_3 (99.9%), $Al(OH)_3$ (99.9%) and P_2O_5 (99.99%), which were then synthesised according to the procedures described in Section 3.1. The reaction mixtures were melted under an oxygen atmosphere for 60 - 90 minutes at temperatures between 1050°C and 1500°C and then annealed. Finally, the as-prepared glasses were polished to produce flat and optically transparent surfaces for property characterisation.

4.3. Results and Discussion

The optical and thermal properties of the four series of KAP glasses were measured and are shown in Tables 5 and 6, respectively. Sample *KAP-1c*, corresponding to the metaphosphate composition of $20K_2O-15Al_2O_3-65P_2O_5$, with its associated properties were included within the results for KAP series 2–4 glasses to show the systematic changes in the glass properties as the glass network changed from ultraphosphate to metaphosphate and finally to polyphosphate.

Table 5. Optical properties of KAP glasses

Sample	n (λ = 589.3 nm)	dn/dT ($\times 10^{-6}$/°C) (λ = 632.8 nm, 20 - 100°C)	r_{OH} (cm^{-1})
KAP-1a	1.484	-16.61	35.88
KAP-1b	1.493	-8.54	30.61
KAP-1c	1.505	-2.76	27.87
KAP-1d	1.511	+1.65	34.32
KAP-1e	1.519	+3.41	30.20
KAP-2a	1.469	-14.99	33.17
KAP-2b	1.488	-11.39	31.26
KAP-2c	1.495	-5.48	28.51
KAP-1c	1.505	-2.76	27.87
KAP-2d	1.515	+3.30	12.68
KAP-3a	1.497	-4.83	28.96
KAP-3b	1.502	-4.45	26.46
KAP-1c	1.505	-2.76	27.87
KAP-3c	1.508	-0.99	18.52
KAP-4a	1.515	+6.63	28.12
KAP-4b	1.509	-0.47	24.37
KAP-1c	1.505	-2.76	27.87
KAP-4c	1.502	-4.66	16.84

Table 6. Thermal properties of KAP glasses

Sample	T_g (°C)	T_x (°C)	T_p (°C)	α (×10^{-6}/°C) (50 - 120°C)
KAP-1a	326	452	552	21.25
KAP-1b	418	633	697	15.20
KAP-1c	507	759	849	11.02
KAP-1d	543	771	880	8.71
KAP-1e	621	773	823	7.67
KAP-2a	272	558	670	19.70
KAP-2b	329	636	696	17.40
KAP-2c	400	618	782	13.34
KAP-1c	507	759	849	11.02
KAP-2d	567	814	882	7.23
KAP-3a	415	755	800	12.91
KAP-3b	475	736	793	12.15
KAP-1c	507	759	849	11.02
KAP-3c	526	702	798	10.15
KAP-4a	580	>900	>900	6.84
KAP-4b	539	749	>850	9.80
KAP-1c	507	759	849	11.02
KAP-4c	479	704	748	13.13

4.3.1. OH Content

The presence of high field strength cation (defined as the ratio of the ion valence to the square of the bond distance between the ion and oxygen, z/a^2 [32]), such as Al^{3+}, generally reduces the OH content of a glass, due to the formation of hydration-resistant bonds with the phosphate non-bridging oxygens [19]. However, the relative OH contents of the KAP glasses showed little changes with respect to the glass compositions, as shown in Table 5. This could be due to the synthesis process employed, notably the long melting time and sealed crucible, which created an equilibrium within the crucible, preventing further evaporation of water. There were, however, exceptions such as *KAP-2d* ($15K_2O$-$20Al_2O_3$-$65P_2O_5$), *KAP-3c* ($20K_2O$-$17.5Al_2O_3$-$62.5P_2O_5$) and *KAP-4c* ($25K_2O$-$15Al_2O_3$-$60P_2O_5$) glasses, which contained relatively lower OH contents. This was due to the high alumina contents for *KAP-2d* and *KAP-3c* and the high K_2O and low P_2O_5 contents for *KAP-4c*. *KAP-1e* ($10K_2O$-$20Al_2O_3$-$70P_2O_5$), despite containing 20 mol% alumina, had higher OH content due to its high P_2O_5, which made it hygroscopic. The lack of correlation between the OH content and alumina content suggested that the optical and thermal properties of the KAP glasses measured in this study were not significantly influenced by the presence of (P-OH) bonds. Figure 10 shows the infrared transmission spectra for some KAP series 2 glasses. As the alumina content in the glass was increased, the absorption band around 3000 cm^{-1} attributed to OH was reduced.

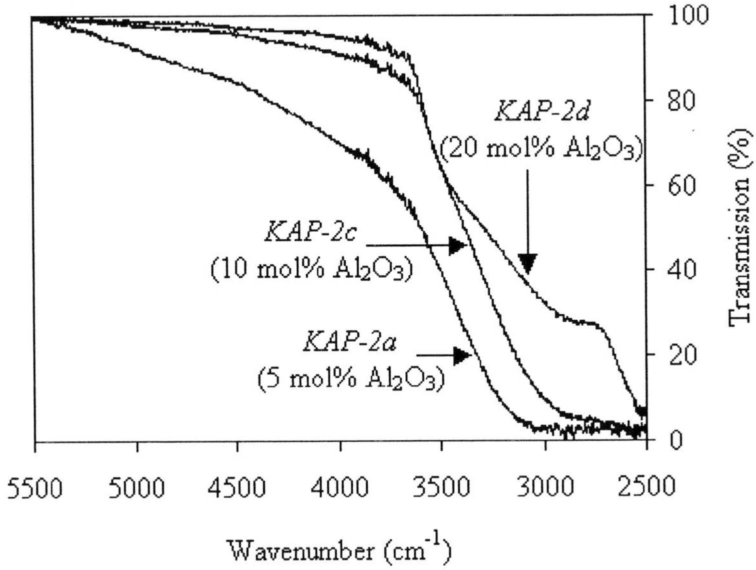

Figure 10. Infrared transmission spectra through xK_2O-$(35-x)Al_2O_3$-$65P_2O_5$ (series 2) glasses. The spectra are normalised for a 1 mm thick sample and corrected for Fresnel loss

4.3.2. Refractive Index

Table 5 shows that the refractive index, n, of KAP series 1-3 glasses increased monotonically with the alumina content while that of KAP series 4 glasses decreased with an increase in K_2O content, with alumina fixed at 15 mol%. Figures 11a and 11b show the variations of n for KAP series 1-3 glasses with Al_2O_3 and KAP series 4 glasses with K_2O, respectively.

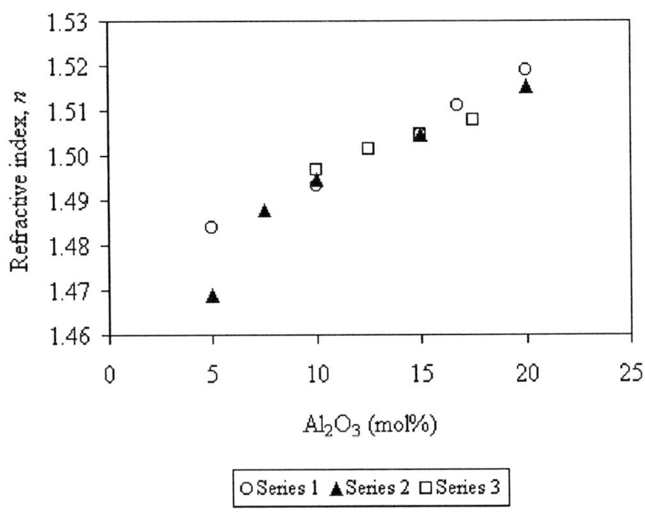

Figure 11a. Refractive index (n) of KAP series 1 (metaphosphate), 2 (65 mol% P_2O_5) and 3 (20 mol% K_2O) glasses, measured at $\lambda = 589.3$ nm

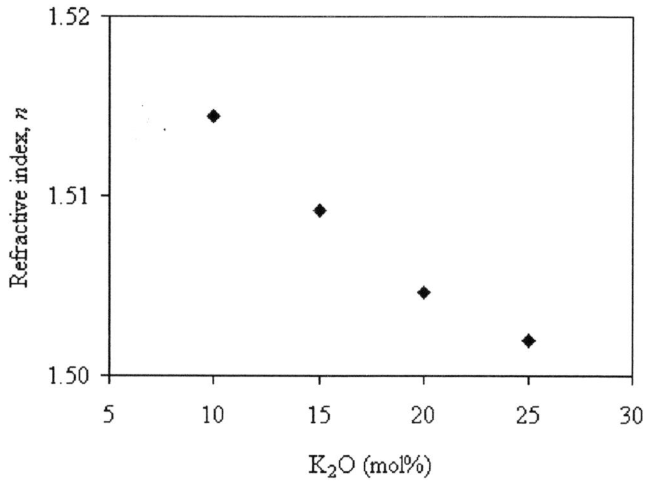

Figure 11b. Refractive index (n) of KAP series 4 (15 mol% Al_2O_3) glasses, measured at $\lambda = 589.3$ nm

The Lorentz-Lorenz equation (Equation (4)) shows that the refractive index of a glass depends on its molar volume and to a smaller extent, its polarisability. Hence, a change in the K^+, Al^{3+} and P^{5+} contents would be expected to cause a change in n, due to the differences in the molar volume and the ionic refractivity of the cations.

The addition of cations to phosphate glasses causes the glass structure to expand to accommodate the cations at interstitial sites and also depolymerises the glass network by converting bridging oxygens to non-bridging oxygens to create a more open structure. This increase in molar volume decreases the refractive index of the glasses, as observed in Figure 11b when K_2O was substituted for P_2O_5, in addition to the change in stoichiometry from ultraphosphates to polyphosphates. However, the addition of a high field strength cation such as Al^{3+} polymerises the glass structural network by forming strong (P-O-Al) bridging bonds between phosphate chains, creating a more rigid network and thus reducing the molar volume, subsequently leading to higher refractive index. Therefore, the substitution of Al_2O_3 for K_2O and P_2O_5 (KAP series 1), Al_2O_3 for K_2O (KAP series 2) and Al_2O_3 for P_2O_5 (KAP series 3) increased the refractive index while the substitution of K_2O for P_2O_5 decreased the refractive index (KAP series 4). Figure 8 shows the depolymerising effect of K^+ and the polymerising effect of Al^{3+} on the potassium aluminophosphate glass structure.

4.3.3. Glass Transition Temperature and Thermal Expansion Coefficient

Figures 12a and 13a show the variations of T_g and α, respectively, of KAP series 1-3 glasses as a function of Al_2O_3 mol% while Figures 12b and 13b show the variations of T_g and α, respectively, of KAP series 4 glasses as a function of K_2O mol%. The incorporation of alumina into the KAP series 1-3 glasses increased the glass T_g and decreased the glass α, consistent with the strengthening and polymerising effects of Al^{3+} on the glass structural network. The formation of bridging (P-O-Al) bonds increased the cross-link density of the network, creating a stronger structure thereby resulting in increased T_g and decreased α. Similar observations had also been made for other aluminophosphate glasses [10,28,33,34]. However, KAP series 4 glasses experienced reversed trends compared to the other three series with the incorporation of K_2O (with the Al_2O_3 content fixed). This was consistent with the depolymerising effect of the low field strength K^+, which bonded with non-bridging oxygens and weakened the glass structure, resulting in decreased T_g and increased α.

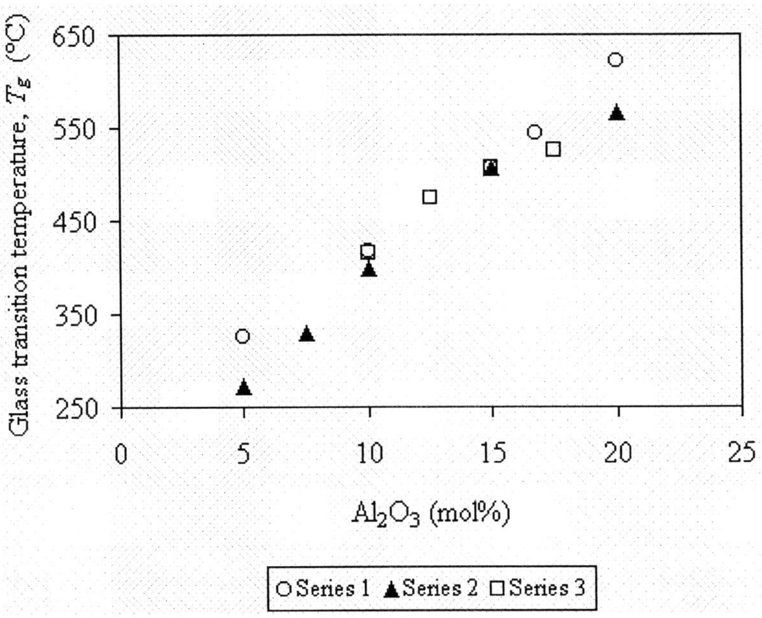

Figure 12a. Glass transition temperature (T_g) of KAP series 1 (metaphosphate), 2 (65 mol% P_2O_5) and 3 (20 mol% K_2O) glasses

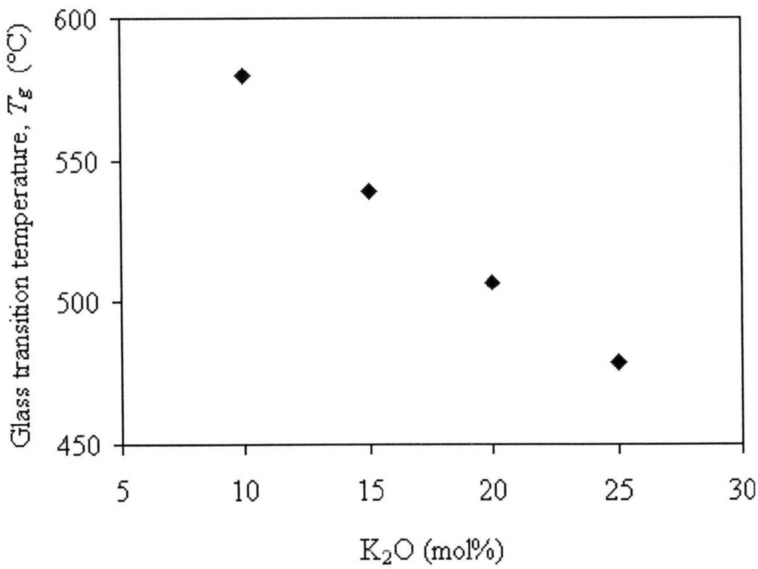

Figure 12b. Glass transition temperature (T_g) of KAP series 4 (15 mol% Al_2O_3) glasses

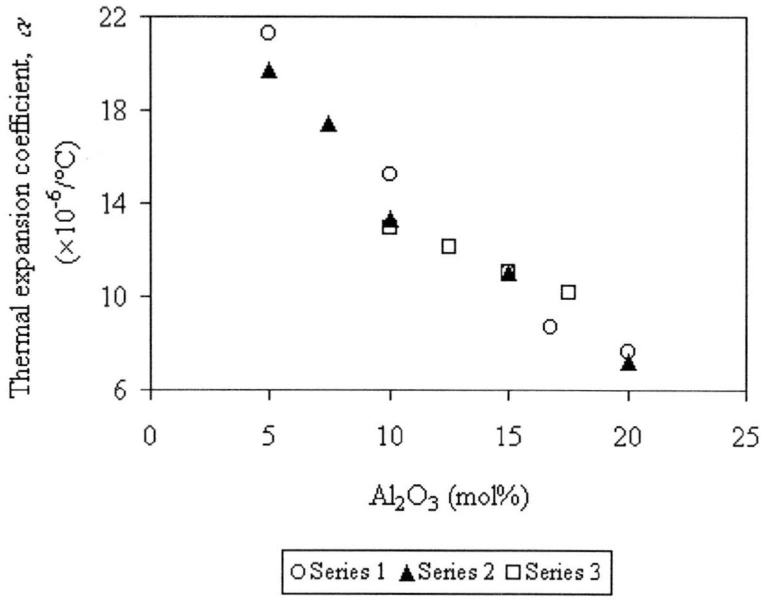

Figure 13a. Thermal expansion coefficient (α) of KAP series 1 (metaphosphate), 2 (65 mol% P_2O_5) and 3 (20 mol% K_2O) glasses, measured over 50 - 120°C

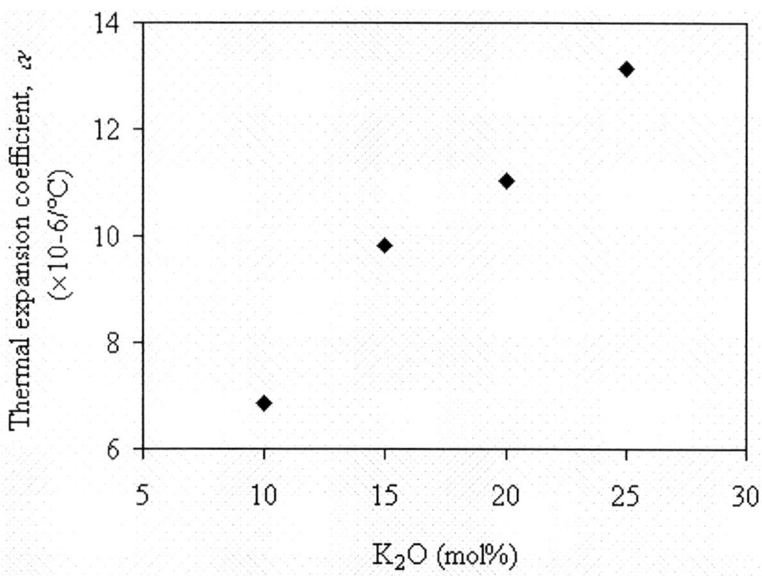

Figure 13b. Thermal expansion coefficient (α) of KAP series 4 (15 mol% Al_2O_3) glasses, measured over 50 - 120°C

4.3.4. Thermo-Optic Coefficient

The compositional effects on d*n*/d*T* of KAP series 1-3 glasses as a function of alumina content are shown in Figure 14a, where the addition of alumina changed the negative values of d*n*/d*T* towards positive values. Figure 14b shows the change in d*n*/d*T* of KAP series 4 glasses towards negative values as the K_2O content increased.

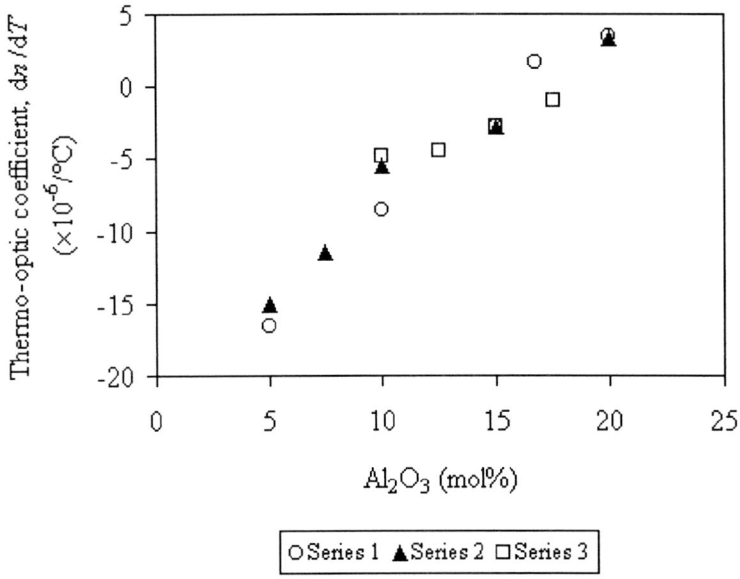

Figure 14a. Thermo-optic coefficient (d*n*/d*T*) of KAP series 1 (metaphosphate), 2 (65 mol% P_2O_5) and 3 (20 mol% K_2O) glasses, measured at $\lambda = 632.8$ nm over 20 – 100 °C

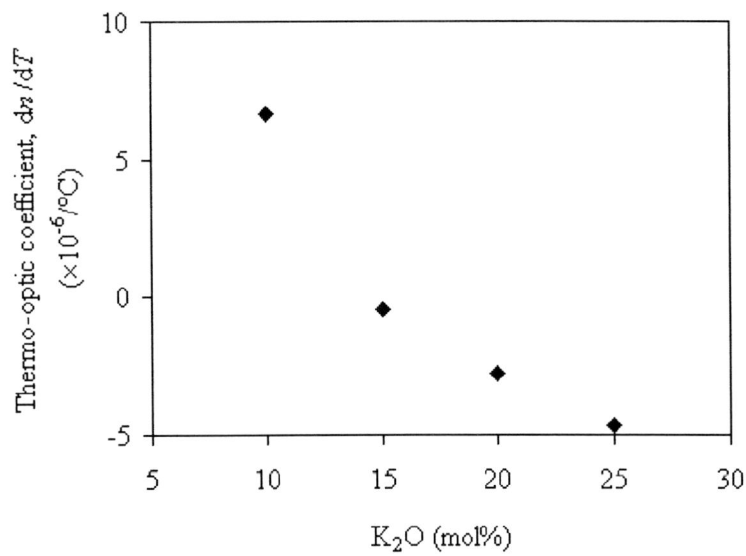

Figure 14b. Thermo-optic coefficient (d*n*/d*T*) of KAP series 4 (15 mol% Al_2O_3) glasses, measured at λ = 632.8 nm over 20 – 100 °C

The behaviour of dn/dT is highly dependent on the glass α, as described by Equations (5) and (6) and shown in Figures 15 and 16. In Figure 15, the glass α and dn/dT are plotted against $Al(PO_3)_3$ mole fraction for the KAP glasses. As the $Al(PO_3)_3$ mole fraction increased, its corresponding α decreased while the magnitude of dn/dT decreased, so much so that at approximately 0.47 $Al(PO_3)_3$ mole fraction and beyond, the glasses exhibited positive dn/dT because polarisability overcame the effect of thermal expansion ($3\alpha < \varphi$). The minimum α value required for the KAP glasses to exhibit negative dn/dT was ~ 9.5×10^{-6}/°C.

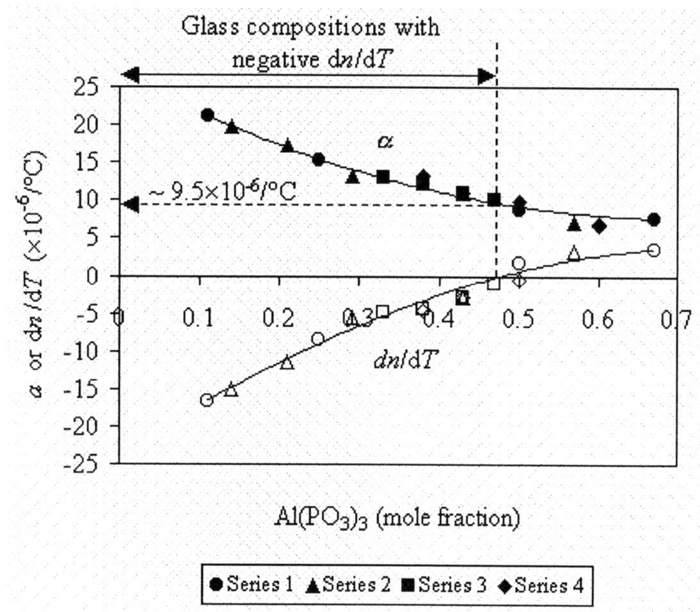

Figure 15. α and dn/dT of KAP glasses as a function of $Al(PO_3)_3$ mole fraction

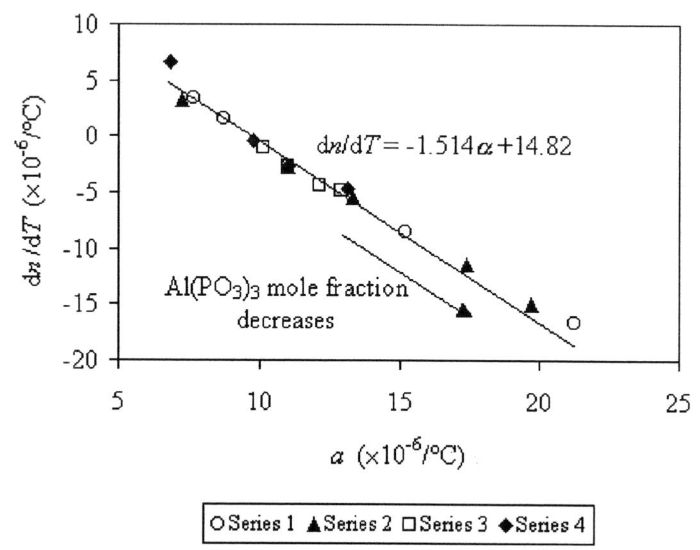

Figure 16. dn/dT of KAP glasses as a function of α

Figure 15 also shows that α and dn/dT lie on the same, but separate curves (lines are drawn as guides to the eyes) for the different KAP glass series. This suggested that there was a gradual and systematic change from ultraphosphates to metaphosphates and finally to polyphosphates. In Figure 16, this inter-dependence of α and dn/dT is better visualised and a semi-empirical equation of $dn/dT = -1.514\alpha + 14.82$ can be defined for KAP glass compositions for compositions within the ranges of $1.469 \leq n \leq 1.519$ and $6.84\times10^{-6}/°C \leq \alpha \leq 21.25\times10^{-6}/°C$.

Figure 17 shows a plot of γ against $Al(PO_3)_3$ mole fraction. The γ values had been calculated based on Equation (6) using the glasses' respective n, α, and dn/dT values and found to be approximately $1.61 \leq \gamma \leq 4.59$. In order to achieve negative dn/dT, the condition must be $\gamma < 3$ and this corresponded to KAP glass compositions with $Al(PO_3)_3$ mole fraction < 0.5.

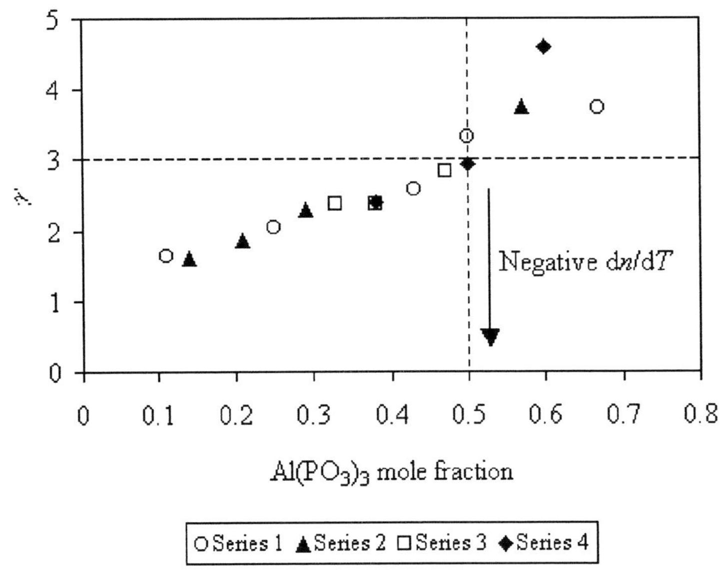

Figure 17. γ of KAP glasses as a function of $Al(PO_3)_3$ mole fraction

4.4. Summary

Various compositions of ternary potassium aluminophosphate glasses encompassing ultraphosphate, metaphosphate and polyphosphate compositions were synthesised. Characterisation of the glasses showed that their optical and thermal properties were strongly influenced by the strengthening effect of Al^{3+} through the formation of (P-O-Al) bridging bonds between phosphate chains. This cross-linking effect strengthened the glass structural network and created a compact network, leading to an increase in refractive indices and glass transition temperatures, a decrease in thermal expansion coefficients and consequently positive thermo-optic coefficients. However, these observed trends with increasing alumina content were reversed when a low field strength modifier such as K^+ was incorporated.

dn/dT and α were linked and $\alpha > 9.5\times10^{-6}$/°C was required to obtain negative dn/dT for the potassium aluminophosphate glasses. A semi-empirical fit of dn/dT = -1.514α + 14.82 was obtained, allowing the deduction of dn/dT of any potassium aluminophosphate glasses within the compositional ranges of $1.469 \le n \le 1.519$ and $6.84\times10^{-6}/°C \le \alpha \le 21.25\times10^{-6}/°C$. Glasses with negative dn/dT up to -16.61×10^{-6}/°C were achieved, confirming the potential athermalisation applications offered by these glasses.

5. Binary Phosphate

This section presents the preparation and compositions of the binary calcium phosphate and barium phosphate glasses examined in this study and the results and discussion of the optical and thermal properties of these glasses.

5.1. Introduction

Binary alkaline earth metal phosphate glasses, such as calcium phosphate and barium phosphate glasses, look promising for athermalisation because they exhibit negative dn/dT. Binary alkali metal phosphates could be promising too, if not for their poor chemical durability that limits their practical uses.

Binary phosphate glasses generally consist of Q^3, Q^2 and Q^1 tetrahedral units, depending on the compositions. However, due to the hygroscopicity and volatility of P_2O_5 [11], water plays an important role in ultraphosphate glasses as it can depolymerise the phosphate network by forming (P-OH) bonds. This reduces the fraction of Q^3 units present in the glass network to a lower number than that expected from the nominal alkali-to-phosphorus ratio [35] and hence (P-OH) bonds must be considered when describing the structures and properties of such glasses. However, as more modifying cation (MeO) is added, glass durability is improved due to the formation of (P-O-Me) bonds.

Hoppe [36,37] presented a structural model describing the effect of modifier coordination environments (having coordination number CN_{Me}) on the properties and structures of ultraphosphate glasses, based on the availability of terminal oxygens (TOs) to coordinate the modifier ions (Me^{z+}, where z is the ion valence). The number of terminal oxygens, M_{TO}, per modifying ion for glasses with the stoichiometry $x(Me_{2/z}O)$-$(100-x)P_2O_5$ is given by Equation (10).

$$M_{TO} = z(100/x) \tag{10}$$

Two structurally distinct regions can be observed for the cases of $M_{TO} > CN_{Me}$ (region I) and $M_{TO} < CN_{Me}$ (region II). In region I, there are sufficient TOs to coordinate with each Me^{z+} ion so that these ions can exist at isolated sites within the network (Figure 18(a)). However, in region II, there are insufficient TOs and the Me^{z+} ions must share the available TOs by forming (Me-O-Me) bonds which cross-link neighbouring Q^2 polyhedra (Figure 18(b)) and increase the network cross-link density. In this work, various compositions of binary calcium phosphate glasses (designated as CaP) and barium phosphate glasses (designated as BaP) were synthesised and their optical and thermal properties characterised.

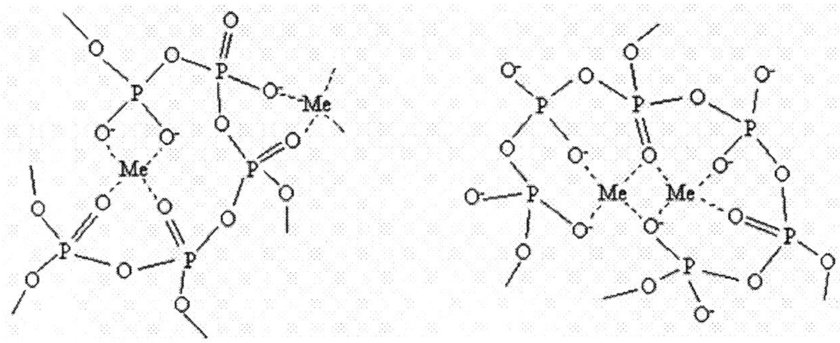

Figure 18. Binary phosphate structure when (a) $M_{TO} > CN_{Me}$ and (b) $M_{TO} < CN_{Me}$

5.2. Glass Preparation and Characterisation

Two sets of binary phosphate glasses were synthesised, corresponding to (mol%) :

- xCaO-(100-x)P_2O_5 in the range $20 \leq x \leq 50$
- xBaO-(100-x)P_2O_5 in the range $20 \leq x \leq 50$

Different CaP and BaP glass compositions were prepared from high purity powder of $CaCO_3$ (99.95%), $BaCO_3$ (99.95%) and P_2O_5 (99.99%), according to the melting procedures in Section 3.1. The batches were melted in an oxygen atmosphere at temperatures between 1000°C and 1400°C and then annealed and finally polished for characterisation. Table 7 shows the nominal batch compositions of the CaP and BaP glasses that were prepared.

Table 7. Batch compositions of calcium phosphate (CaP) and barium phosphate (BaP) glasses

Sample	Batch composition (mol%)	
	CaO	P_2O_5
CaP-1a	50	50
CaP-1b	40	60
CaP-1c	30	70
CaP-1d	20	80
Sample	BaO	P_2O_5
BaP-1a	60	40
BaP-1b	55	45
BaP-1c	50	50
BaP-1d	45	55
BaP-1e	40	60
BaP-1f	30	70
BaP-1g	20	80

5.3. Results and Discussion

The binary phosphate glasses were characterised using the techniques described in Section 3.2 and the optical and thermal properties of the glasses are given in Tables 8-9 and Figures 19-24 (lines are drawn as guides to the eyes). Attempts to produce homogeneous glasses for *BaP-1a* ($60BaO-40P_2O_5$) and *BaP-1b* ($55BaO-45P_2O_5$), corresponding to polyphosphate stoichiometries, were not successful as these samples developed phase-separation.

Table 8. Optical properties of CaP and BaP glasses

Sample	n (λ = 589.3 nm)	dn/dT ($\times10^{-6}$/°C) (λ = 632.8 nm, 20 - 100°C)	r_{OH} (cm^{-1})
CaP-1a	1.551	-2.98	8.08
CaP-1b	1.535	-0.92	12.86
CaP-1c	1.526	-0.39	15.27
CaP-1d	1.527	+0.36	14.28
BaP-1a	Phase-separated sample		
BaP-1b	Phase-separated sample		
BaP-1c	1.585	-10.24	14.04
BaP-1d	1.575	-9.14	13.24
BaP-1e	1.569	-8.19	14.73
BaP-1f	1.556	-7.06	21.92
BaP-1g	1.547	-5.76	22.26

Table 9. Thermal properties of CaP and BaP glasses

Sample	T_g (°C)	T_x (°C)	T_p (°C)	α ($\times10^{-6}$/°C) (50 - 120°C)
CaP-1a	547	719	799	10.22
CaP-1b	510	734	788	9.39
CaP-1c	504	781	841	9.10
CaP-1d	533	811	875	8.60
BaP-1a	Phase-separated sample			
BaP-1b	Phase-separated sample			
BaP-1c	478	651	699	13.67
BaP-1d	458	637	673	13.35
BaP-1e	452	645	694	13.07
BaP-1f	421	674	766	12.38
BaP-1g	419	704	>780	11.62

5.3.1. OH Content

As discussed in Section 5.1, water plays a major role in the structures and properties of ultraphosphates by depolymerising the glass structural network, due to the hygroscopic nature of P_2O_5. Generally, the addition of modifying cations to phosphate glasses reduces the OH content due to the conversion of phosphate bridging oxygens (BOs) to non-bridging oxygens (NBOs) to form ionic bonds with the cations, thus minimising the available phosphate BOs to form (P-OH) bonds. From the results in Table 8, the CaP and BaP glasses have fairly similar OH content relatively, despite the difference in the amount and the cation of the modifying oxides, most probably due to the synthesis method employed. However, samples *CaP-1a*, *BaP-1f* and *BaP-1g* exhibited slightly different relative OH contents. *CaP-1a* (50CaO-50P_2O_5) contained lower OH content due to its metaphosphate composition while *BaP-1f* (30BaO-70 P_2O_5) and *BaP-1g* (20BaO-80 P_2O_5) contained slightly higher OH contents due to their ultraphosphate compositions, despite efforts to minimise water contamination during the melting process and this would significantly affect the properties of these two BaP glasses. CaP glasses are generally more resistant to hydrolysis than BaP glasses because of the higher field strength of Ca^{2+} [8]. This is evident from Figure 19, which shows that the transmission of the *BaP-1c* sample with 50 mol% BaO at around 3000 cm^{-1} (coinciding with the OH absorption band) is lower compared to the transmission of the *CaP-1a*, having 50 mol% CaO.

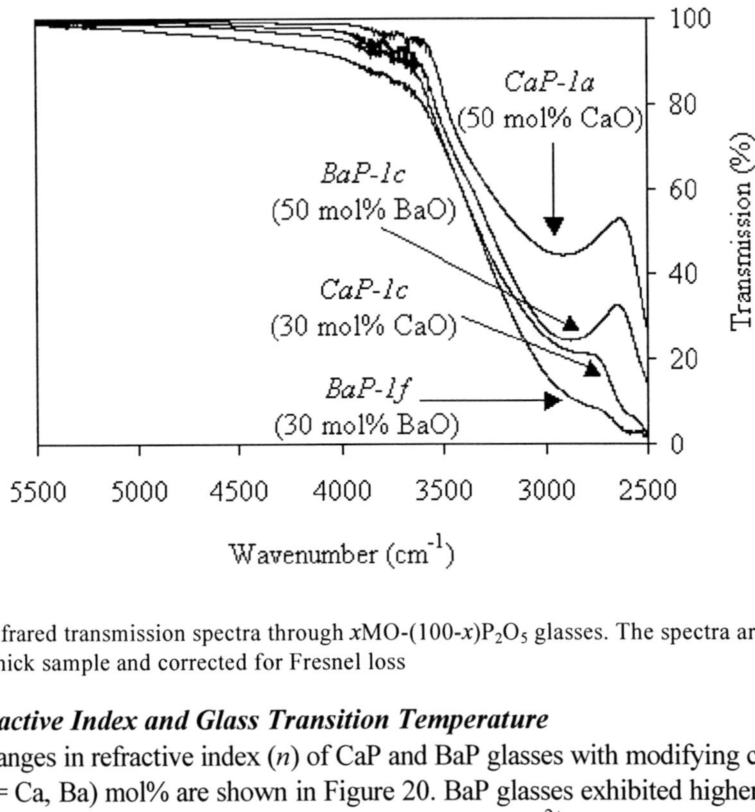

Figure 19. Infrared transmission spectra through xMO-(100-x)P_2O_5 glasses. The spectra are normalised for a 1 mm thick sample and corrected for Fresnel loss

5.3.2. Refractive Index and Glass Transition Temperature

The changes in refractive index (n) of CaP and BaP glasses with modifying cations (MeO where Me = Ca, Ba) mol% are shown in Figure 20. BaP glasses exhibited higher n compared to CaP glasses duc to thc larger and more polarisable Ba^{2+} ion, in accordance with the Lorentz-Lorenz equation [20]. The initial addition of CaO in CaP glasses decreased n to a minimum point, before increasing for CaO contents beyond 25 mol%, while for BaP glasses, n increased monotonically with BaO mol%.

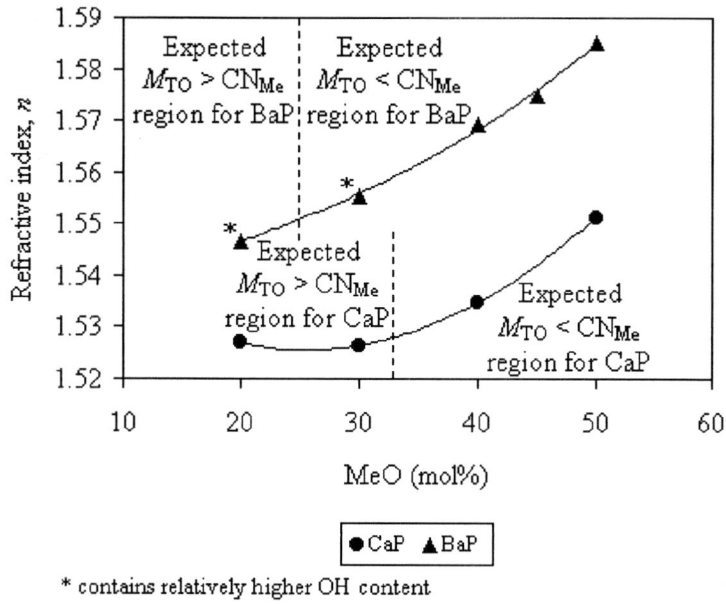

Figure 20. Refractive index (n) of CaP and BaP glasses, measured at $\lambda = 589.3$ nm

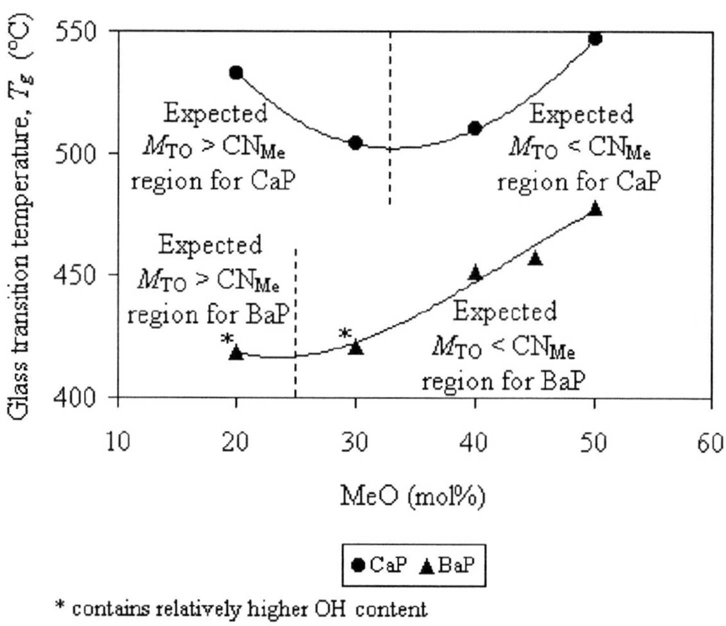

Figure 21. Glass transition temperature (T_g) of CaP and BaP glasses

The results for glass transition temperature (T_g) followed fairly similar trends as that for n, except that the minimum point observed for the CaP glasses was shifted to a higher CaO mol% and more pronounced while a minimum point can be observed at around 25 mol% BaO for the BaP glasses. These results are substantiated by results reported in literature where minimum points were also observed for Li and Na ultraphosphate density [38] and Na and Cs ultraphosphate T_g [39].

Such behaviours can be explained based on the model proposed by Hoppe [36,37], described in Section 5.1. In the $M_{TO} > CN_{Me}$ region, the addition of cations expands the phosphate network to accommodate these cations as well as depolymerises the network and reduces the structural cross-link density. This leads to a decrease in density, n and T_g. In the $M_{TO} < CN_{Me}$ region, the formation of modifier sub-structures creates a more compact network by cross-linking phosphate polyhedra, thereby strengthening the glass network. This leads to an increase in density, n and T_g.

The structural transition from a network with isolated modifier sites to a network with modifier sub-structures can be predicted based on the coordination number (CN_{Me}) of the cation, where $CN_{Ca} \sim 6$ and $CN_{Ba} \sim 8$ [37]. Using Equation (10) and setting $M_{TO} = CN_{Me}$, the transition is expected to occur at 33 mol% CaO for CaP glasses and 25 mol% BaO for BaP glasses, as denoted in Figures 20 and 21. These predictions are in agreement with the positions of the minima points observed for T_g for the CaP and BaP glasses. The non-conformity of the n results to the above prediction suggested that this can be attributed to the relatively higher OH contents in the ultraphosphate glasses. The presence of water reduced M_{TO} through the formation of (P-OH) bonds, which effectively extended the $M_{TO} < CN_{Me}$ region to lower MeO mol% than expected theoretically or even eliminated the existence of the $M_{TO} > CN_{Me}$ region altogether. Hence, the cations immediately formed sub-structures the moment they were added, creating a compact network. This assessment is supported by the observation of a minimum point in the density of anhydrous sodium ultraphosphates [38], which was absent in the sodium ultraphosphate glasses prepared in open crucibles [40]. These results also suggested that T_g was less sensitive to the OH contents in glasses compared to n.

5.3.3. Thermal Expansion Coefficient and Thermo-optic Coefficient

Figure 22 shows the changes in α and dn/dT of CaP and BaP glasses as a function of MeO mol%. As the MeO contents increased, α increased and dn/dT progressed towards negative values. This was due to the depolymerising effect of modifying cations, which converted BOs to NBOs, reducing the structural cross-linking and strength of the glass network and leading to an increase in α. The increase in α subsequently decreased dn/dT, as expected from Equation (5). The results also showed that in order to exhibit negative dn/dT, CaP glasses must contain a minimum of ~ 28 mol% CaO with α exceeding ~ 9.0×10^{-6}/°C, while BaP glasses exhibited negative dn/dT throughout the entire compositional range investigated. There was no evidence of a maximum or minimum point for α and dn/dT throughout the whole range of glasses examined.

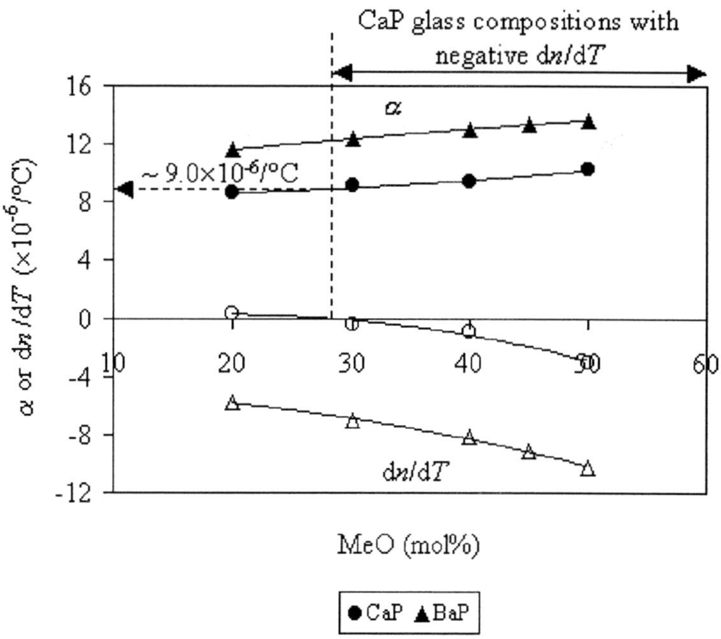

Figure 22. α and dn/dT of CaP and BaP glasses as a function of MeO mol%

Figure 23 shows the linear inter-dependence between α and dn/dT for CaP and BaP glasses. A semi-empirical fit of dn/dT = -2.088α + 18.50 can be defined for CaP glass compositions within $1.526 \leq n \leq 1.551$ and 8.60×10^{-6}/°C $\leq \alpha \leq$ 10.22×10^{-6}/°C and a fit of dn/dT = -2.104α + 18.90 for BaP glass compositions within $1.547 \leq n \leq 1.585$ and 11.62×10^{-6}/°C $\leq \alpha \leq 13.67\times10^{-6}$/°C.

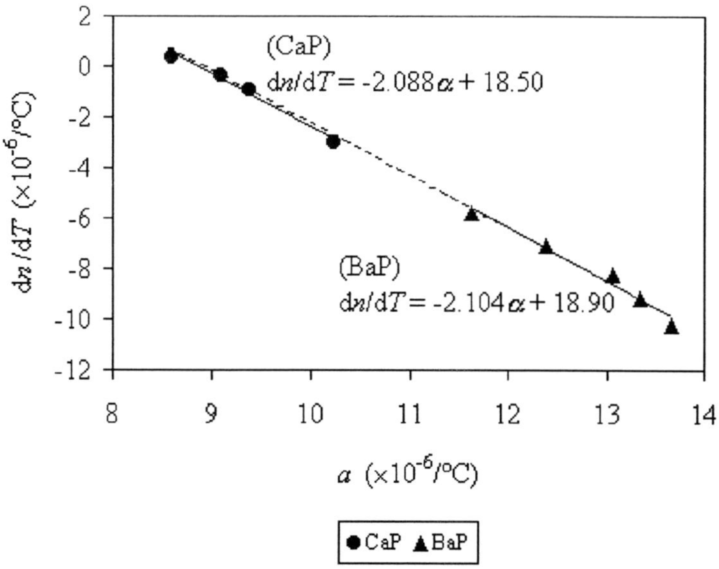

Figure 23. dn/dT of CaP and BaP glasses as a function of α.

By extrapolating the BaP results in Figure 23, both sets of CaP and BaP data practically followed a similar line, implying that $d(\ln P_e)/dT$ is predominantly due to the phosphate matrix, irrespective of the modifying cations. Based on this assumption, the semi-empirical equation can be generalised for any calcium phosphate, barium phosphate and possibly any binary alkaline earth metal phosphate glass compositions within $8.60\times10^{-6}/°C \leq \alpha \leq 13.67\times10^{-6}/°C$.

Figure 24 shows a plot of γ against MeO mol% for CaP and BaP glasses. The γ values were calculated to be $2.56 \leq \gamma \leq 3.07$ for CaP glasses and $1.96 \leq \gamma \leq 2.25$ for BaP glasses. In order to achieve negative dn/dT, the condition must be $\gamma < 3$. This corresponded to CaP glass compositions with CaO > 28 mol% and all BaP glass compositions.

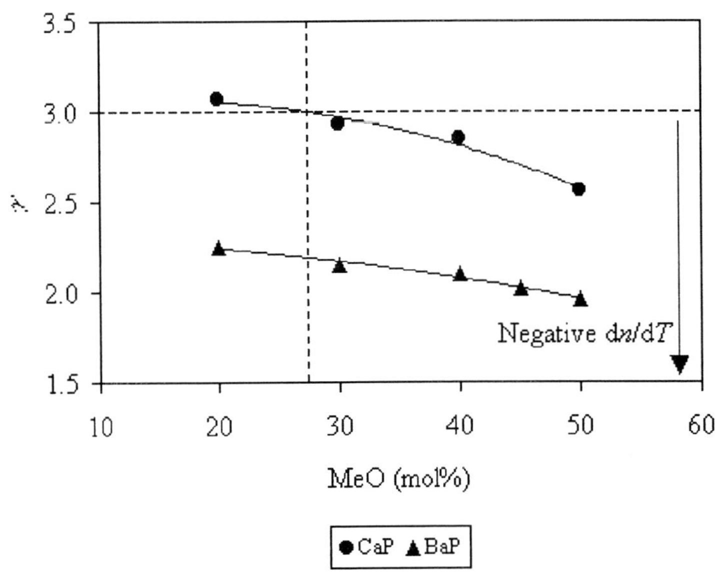

Figure 24. γ of CaP and BaP glasses as a function of MeO mol%

5.4. Summary

A range of binary calcium phosphate and barium phosphate at the ultraphosphate and metaphosphate compositions had been synthesised and characterised. The glass transition temperatures of the calcium phosphate and barium phosphate glasses exhibited a minimum point and such trend can be explained by the transition of a structural network containing modifying cations at isolated sites to a network with modifier sub-structures sharing terminal oxygens. The refractive indices of the glasses, however, displayed a different behaviour and this could be attributed to the presence of water in the phosphate network which depolymerised the network via the formation of (P-OH) bonds. The structural network transition, apparently did not affect the thermal expansion coefficients and thermo-optic coefficients of the glasses, where the systematic addition of modifiers led to the monotonic increase of the thermal expansion coefficient and the subsequent progress towards negative thermo-optic coefficient values.

dn/dT and α were inter-dependent and $\alpha > 9.0\times10^{-6}$/°C was required to obtain negative dn/dT for calcium phosphate glasses. Semi-empirical equations of dn/dT = -2.088α + 18.50 and dn/dT = -2.104α + 18.90 were defined for calcium phosphate and barium phosphate glasses, respectively, within specific compositional ranges. The similarity of the two semi-empirical equations suggested that the electronic polarisability was dominated by the phosphate matrix.

6. Borophosphate

This section presents the preparation and compositions of the ternary barium borophosphate and sodium borophosphate glasses examined in this study and the results and discussion of the optical and thermal properties of these glasses.

6.1. Introduction

Borophosphate glasses are interesting due to the fact that they incorporate a mixture of properties of borate, phosphate and borophosphate units, due to the presence of two glass formers, B_2O_3 and P_2O_5. Borophosphate glasses are suitable for athermalisation applications because of the inherent negative dn/dT displayed by pure phosphate and pure borate glasses at -92.2×10^{-6}/°C and -35.0×10^{-6}/°C [8], respectively. Different cations can be added to borophosphate glasses to modify their properties and barium and sodium were chosen for this study because barium metaphosphate and sodium metaphosphate glasses exhibit negative dn/dT of -10.6×10^{-6}/°C and -15.0×10^{-6}/°C, respectively (Table 2).

Studies so far have shown that boron atoms form either trigonal, B(3), or tetrahedral, B(4), sites within the borophosphate glass networks, with B(4) sites dominating at phosphate-rich compositions and B(3) sites dominating at borate-rich compositions [41-44]. The addition of B_2O_3 generally improves the chemical durability of the glasses, where the B(4) sites cross-link phosphate chains via the formation of (P-O-B) bonds [45]. However, further increases in B_2O_3 results in a structural change from a disordered borophosphate network to a borate network, as the boron coordination number changes from B(4) to B(3). This reduces the covalent cross-link density and weakens the glass network. The possible borophosphate [46] (analogous to $AlPO_4$ structure [30]) and borate units [47] present in the glass network are shown in Figure 25.

The objective for this section is to study the compositional effects of barium borophosphate glasses (designated as BaBP) and sodium borophosphate glasses (designated as NaBP) on their optical and thermal properties. This is carried out in two simultaneous investigations to study, firstly, the effect of incorporating B_2O_3 into phosphate glasses and secondly, the property trends as the glass composition makes the transition from phosphate-rich to borate-rich.

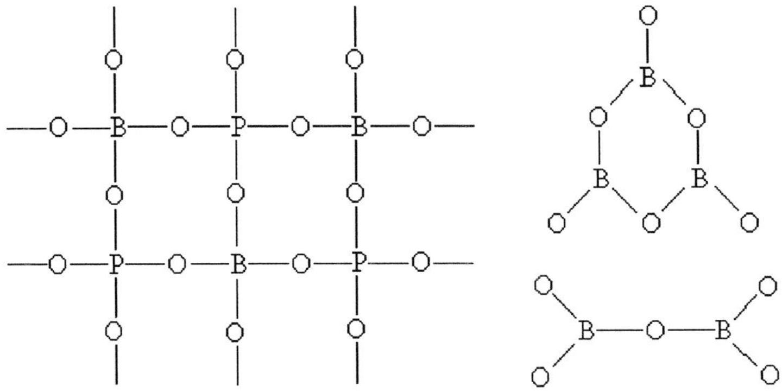

Figure 25. Borophosphate and borate groups in borophosphate glass network

6.2. Glass Preparation and Characterisation

Two series of BaBP glasses and a series of NaBP glasses were synthesised, corresponding to (mol%):

- BaBP Series 1 : $50BaO\text{-}xB_2O_3\text{-}(50\text{-}x)P_2O_5$ with $5 \leq x \leq 20$
- BaBP Series 2 : $(100\text{-}x)Ba(PO_3)_2\text{-}xB_2O_3$ with $10 \leq x \leq 60$
- NaBP : $(100\text{-}x)NaPO_3\text{-}xB_2O_3$ with $0 \leq x \leq 50$

The glasses were prepared from high purity powder of $BaCO_3$ (99.95%), Na_2CO_3 (99.5%), B_2O_3 (99.98%) and P_2O_5 (99.99%). The reaction mixtures were melted under an oxygen atmosphere for 60 minutes at temperatures between 1200°C and 1300°C and then annealed, using the synthesis route described in Section 3.1.

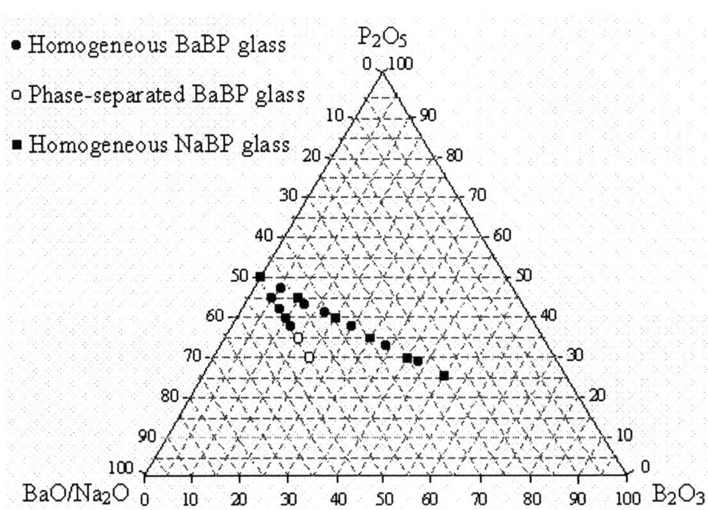

Figure 26. Barium borophosphate (BaBP) and sodium borophosphate (NaBP) glass melting

The melted glasses were then polished. Table 10 shows the nominal batch compositions for the BaBP and NaBP glasses that were prepared while Figure 26 shows the glass melting region for the borophosphate glasses. In the NaBP glass series, a sample of sodium metaphosphate (*NaBP-1a*) was melted for comparison purposes.

Table 10. Batch compositions of barium borophosphate (BaBP) and sodium borophosphate (NaBP) glasses

Sample	Batch composition (mol%)		
	BaO	B_2O_3	P_2O_5
BaBP-1a	50	5	45
BaBP-1b	50	8	42
BaBP-1c	50	10	40
BaBP-1d	50	12	38
BaBP-1e	50	15	35
BaBP-1f	50	20	30
Sample	$Ba(PO_3)_2$	B_2O_3	
BaBP-2a	90	10	
BaBP-2b	80	20	
BaBP-2c	70	30	
BaBP-2d	60	40	
BaBP-2e	50	50	
BaBP-2f	40	60	
Sample	$NaPO_3$	B_2O_3	
NaBP-1a	100	0	
NaBP-1b	90	10	
NaBP-1c	80	20	
NaBP-1d	70	30	
NaBP-1e	60	40	
NaBP-1f	50	50	

6.3. Results and Discussion

Tables 11-12 and Figures 27-32 (lines are drawn as guides to the eyes) show the optical and thermal properties of BaBP and NaBP glasses prepared. Most of the glasses prepared were transparent and homogeneous except for *BaBP-1e* ($50BaO-15B_2O_3-35P_2O_5$) and *BaBP-1f* ($50BaO-20B_2O_3-30P_2O_5$), which showed signs of phase-separation. The properties of barium metaphosphate (*BaP-1c* from Section 5) are also included in the results for comparison purposes. The refractive index and thermo-optic coefficient of *NaBP-1a* were obtained from References [9] and [8], respectively, as the glass prepared within this work was hygroscopic, where the coarse and nontransparent appearance of the glass surfaces prevented any measurements. The *NaBP-1f* d*n*/d*T* was also not measured due to the hygroscopic glass.

Table 11. Optical properties of BaBP and NaBP glasses

Sample	n (λ = 589.3 nm)	dn/dT ($\times 10^{-6}$/°C) (λ = 632.8 nm, 20 - 100°C)	r_{OH} (cm^{-1})
BaP-1c	1.585	-10.24	14.04
BaBP-1a	1.600	-9.27	9.88
BaBP-1b	1.604	-8.97	7.46
BaBP-1c	1.606	-7.82	6.93
BaBP-1d	1.610	-7.52	6.38
BaBP-1e	Phase-separated sample		
BaBP-1f	Phase-separated sample		
Sample	n (λ = 589.3 nm)	dn/dT ($\times 10^{-6}$/°C) (λ = 632.8 nm, 20 - 100°C)	r_{OH} (cm^{-1})
BaP-1c	1.585	-10.24	14.04
BaBP-2a	1.591	-8.45	17.88
BaBP-2b	1.593	-7.26	14.38
BaBP-2c	1.594	-5.01	13.79
BaBP-2d	1.596	-3.68	15.43
BaBP-2e	1.592	-0.69	21.96
BaBP-2f	1.589	+2.72	37.79
Sample	n(λ = 589.3 nm)	dn/dT ($\times 10^{-6}$/°C) (λ = 632.8 nm, 20 - 100°C)	r_{OH} (cm^{-1})
NaBP-1a	1.484 [9]	-15.00 [8]	nm
NaBP-1b	1.490	-12.75	18.05
NaBP-1c	1.501	-8.43	15.61
NaBP-1d	1.511	-3.61	14.28
NaBP-1e	1.512	-2.36	27.44
NaBP-1f	1.511	nm	38.14

nm = not measured

Table 12. Thermal properties of BaBP and NaBP glasses

Sample	T_g (°C)	T_x (°C)	T_p (°C)	α ($\times 10^{-6}$/°C) (50 - 120°C)
BaP-1c	478	651	699	13.67
BaBP-1a	507	706	740	13.23
BaBP-1b	545	729	762	12.52
BaBP-1c	565	748	787	12.23
BaBP-1d	586	749	809	11.58
BaBP-1e	Phase-separated sample			
BaBP-1f	Phase-separated sample			
Sample	T_g (°C)	T_x (°C)	T_p (°C)	α ($\times 10^{-6}$/°C) (50 - 120°C)
BaP-1c	478	651	699	13.67
BaBP-2a	497	705	778	12.63
BaBP-2b	533	783	825	11.63
BaBP-2c	571	802	848	10.48
BaBP-2d	596	848	871	9.42

Table 12. Thermal properties of BaBP and NaBP glasses (Continued)

BaBP-2e	589	810	864	7.53
BaBP-2f	564	806	844	5.68
Sample	T_g (°C)	T_x (°C)	T_p (°C)	α (×10^{-6}/°C) (50 - 120°C)
NaBP-1a	325	373	422	22.03
NaBP-1b	358	540	622	19.91
NaBP-1c	415	600	660	16.70
NaBP-1d	445	613	675	13.43
NaBP-1e	450	645	689	11.57
NaBP-1f	439	650	707	11.20

6.3.1. OH Content

The relative OH contents of BaBP and NaBP glasses are given in Table 11, which shows that generally, NaBP glasses had higher relative OH content compared to BaBP glasses with similar boron content, due to the lower field strength of Na^+ compared to Ba^{2+} [8]. The results for BaBP series 1 glasses show that the incorporation of B_2O_3 decreased the relative OH content. This was due to the formation of bridging bonds by the B^{3+} ions, which cross-linked the phosphate chains, thus improving the chemical durability of the glasses [45]. For BaBP series 2 and NaBP glasses, the relative OH contents of the glasses decreased on the initial addition of B_2O_3, reaching a minimum before increasing with further boron addition. As described in Section 6.1, B(4) units dominated at phosphate-rich compositions, forming (P-O-B) bridging bonds, while B(3) units dominated at borate-rich compositions. The formation of (P-O-B) bonds reduced the number of phosphate NBOs, thus minimising the formation of (P-OH) bonds.

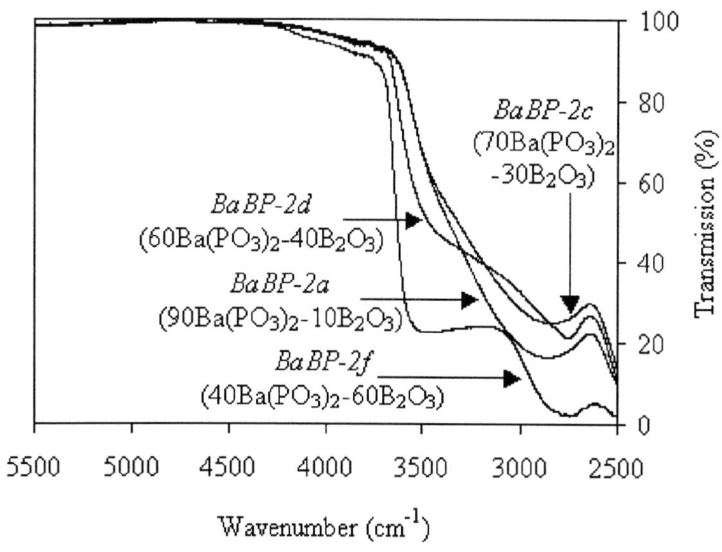

Figure 27. Infrared transmission spectra through (100-x)$Ba(PO_3)_2$-$x$$B_2O_3$ glasses. The spectra are normalised for a 1 mm thick sample and corrected for Fresnel loss

Hence the relative OH content of the glasses decreased. However, with increasing boron content, the glass network changed from a borophosphate network to a borate network. The number of bridging bonds was reduced and the network became less resistant to hydrolysis, which accounted for the increase in the relative OH content at high boron content. This is evident in Figure 27 where the transmission of *BaBP-2a* (with 10 mol% B_2O_3) is lower than *BaBP-2c* (with 30 mol% B_2O_3) but higher than *BaBP-2f* (with 60 mol% B_2O_3) at the OH absorption band around 3000 cm^{-1}. The step-like behaviour of *BaBP-2f* also suggested that it had a different structural network compared to glasses with phosphate-rich compositions, such as *BaBP-2a* and *BaBP-2c*.

6.3.2. Refractive Index and Glass Transition Temperature

Figures 28a and 28b show the refractive indices of BaBP and NaBP glasses, respectively, as a function of B/(B+P) ratio. For BaBP series 1 glasses, the addition of boron increased the glass *n* throughout the compositional range studied while for BaBP series 2 glasses, *n* increased upon the initial addition of boron, up to a maximum at around B/(B+P) = 0.4 and then decreased for higher boron concentrations. A maximum point was also observed for the NaBP glass results, occurring at around B/(B+P) = 0.6. Comparison of the results show that BaBP glasses had higher *n* than NaBP glasses, due to the more polarisable and larger cation size of Ba^{2+} ions. The glass transition temperature results for BaBP and NaBP glasses also exhibit similar behaviour with identical maximum points to that of their respective refractive index results, as shown in Figures 29a and 29b. The maximum points observed for both *n* and T_g also coincided with the positions of the minimum points observed for the relative OH contents of the glasses.

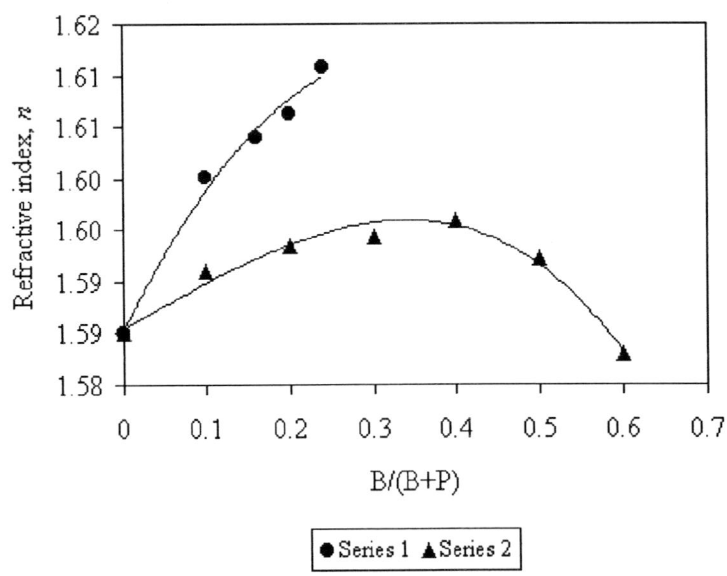

Figure 28a. Refractive index (n) of BaBP glasses, measured at λ = 589.3 nm

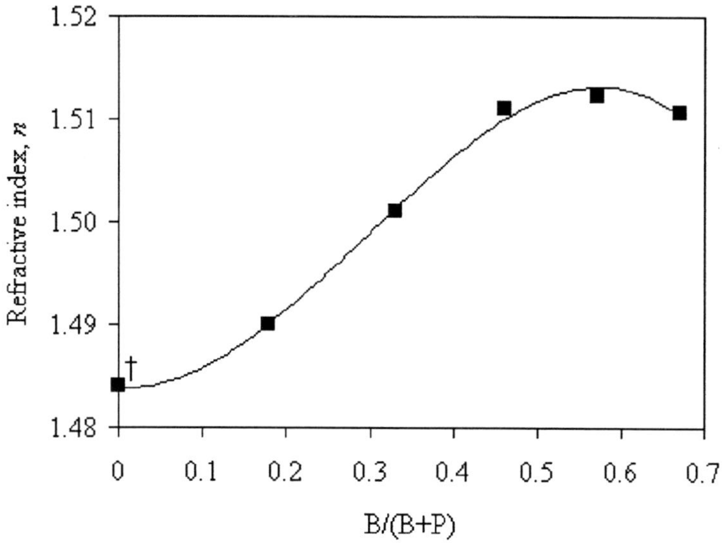

† n_e (measured at λ = 546.1 nm) [9]

Figure 28b. Refractive index (n) of NaBP glasses, measured at λ = 589.3 nm

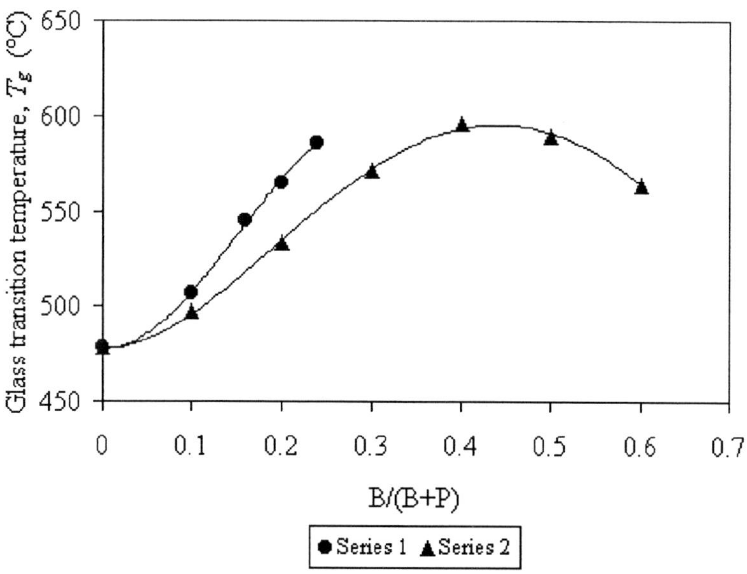

Figure 29a. Glass transition temperature (T_g) of BaBP glasses

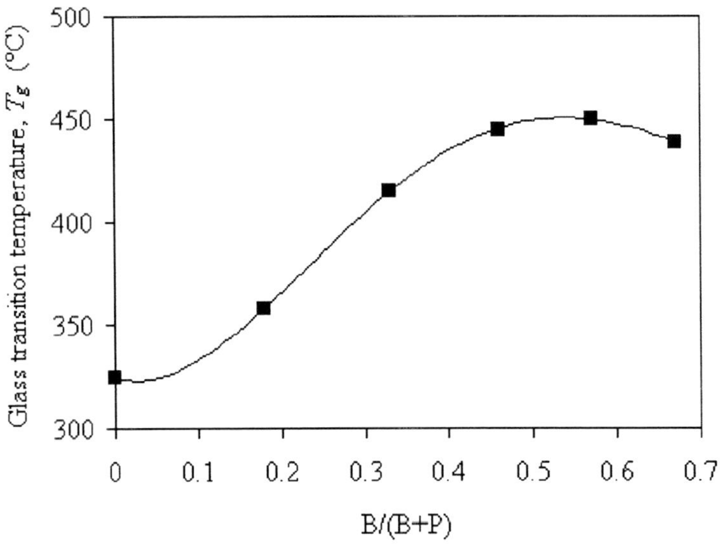

Figure 29b. Glass transition temperature (T_g) of NaBP glasses

At phosphate-rich compositions (low boron contents), B(4) units dominated within the structure and formed (P-O-B) bridging bonds which polymerised the glass network, creating a more compact borophosphate network and increasing the packing density and cross-linking density. This led to an increase in both n and T_g. As more boron was added, B(3) units became the preferred coordination in borate-rich compositions, resulting in a structural change from a disordered borophosphate network to a more open borate network and a reduction in the (P-O-B) bridging bonds. This led to a decrease in both n and T_g. The crossover points for this structural change are indicated by the maximum points observed for the BaBP series 2 and NaBP glasses.

6.3.3. Thermal Expansion Coefficient and Thermo-Optic Coefficient

Figures 30a and 30b show the compositional effects of boron on α and dn/dT of the borophosphate glasses. There seemed to be no evidence of any maximum or minimum point throughout the range of BaBP and NaBP glasses examined.

The addition of boron made the glass network highly cross-linked and hence led to a decrease in α of the glasses. Coupled with the increase in glass polarisability due to the coordination number change of boron from four to three, these combined effects skewed the corresponding dn/dT towards positive values, as expected from Equation (5). BaBP glasses had lower α values compared to NaBP due to the higher field strength of Ba^{2+}, which formed stronger bonds with the NBOs, compared to Na^+. In order to exhibit negative dn/dT, BaBP and NaBP glasses must exhibit a minimum α value of 7.5×10^{-6}/°C and 11.5×10^{-6}/°C, respectively.

The linear relationship between α and dn/dT for BaBP and NaBP glasses are depicted in Figures31a and 31b, respectively. A semi-empirical fit of dn/dT = $-1.599\alpha + 11.58$ can be defined to estimate dn/dT of BaBP glasses within the ranges of $1.585 \leq n \leq 1.610$ and $5.68\times10^{-6}/°C \leq \alpha \leq 13.67\times10^{-6}/°C$ while a fit of dn/dT = $-1.265\alpha + 12.74$ can be defined for NaBP glasses within $1.484 \leq n \leq 1.512$ and $11.20\times10^{-6}/°C \leq \alpha \leq 22.03\times10^{-6}/°C$.

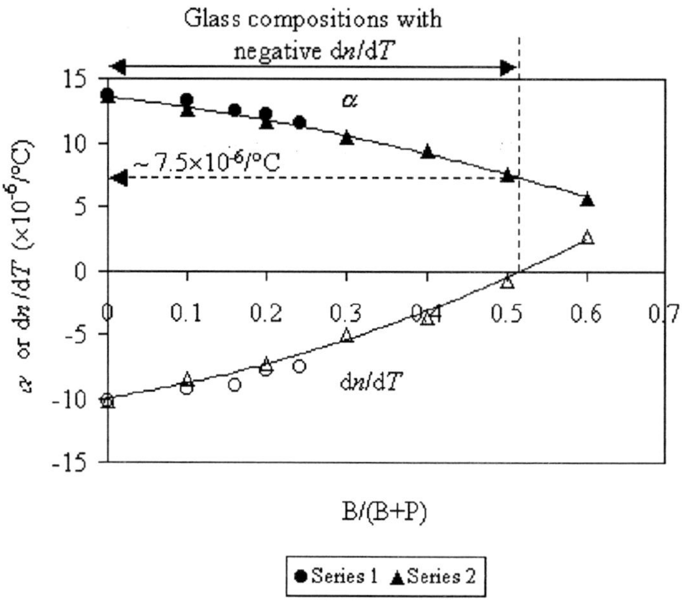

Figure 30a. α and dn/dT of BaBP glasses as a function of B/(B+P) ratio. α measured over 50 - 120°C and dn/dT measured at λ = 632.8 nm over 20 – 100 °C

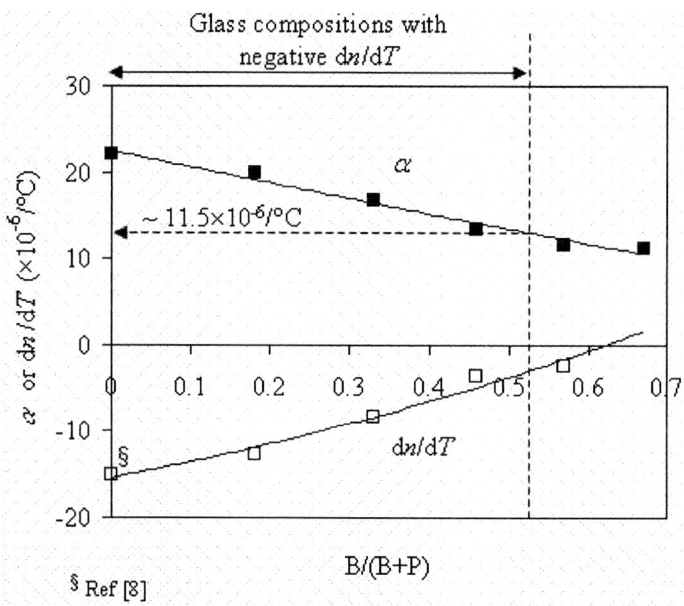

Figure 30b. α and dn/dT of NaBP glasses as a function of B/(B+P) ratio. α measured over 50 - 120°C and dn/dT measured at λ = 632.8 nm over 20 – 100 °C

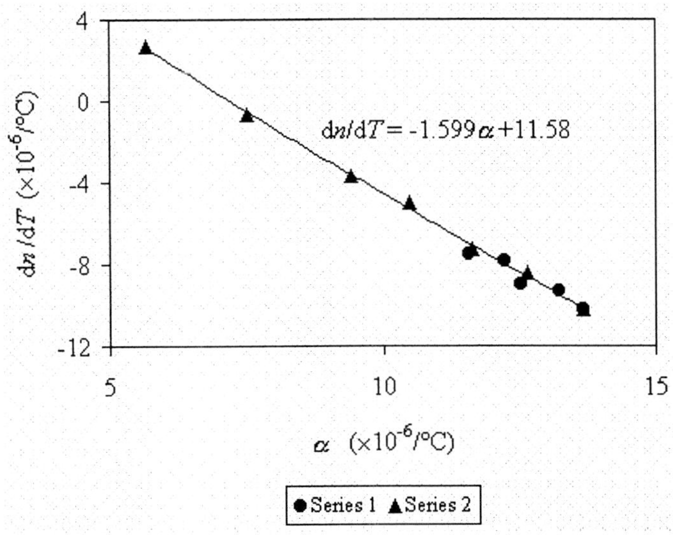

Figure 31a. dn/dT of BaBP glasses as a function of α

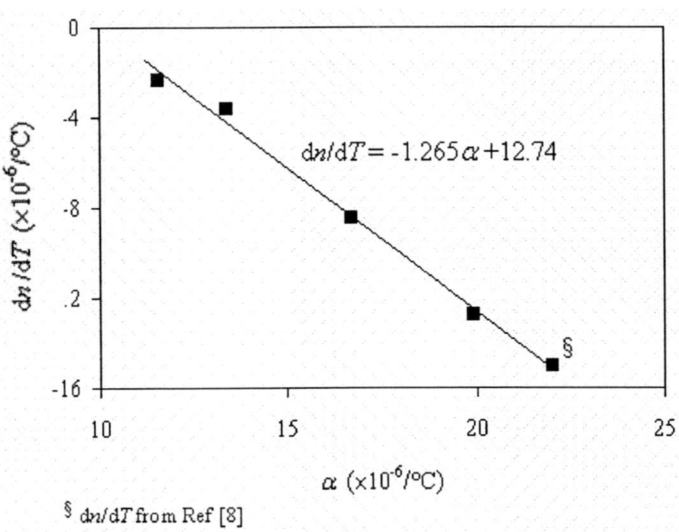

Figure 31b. dn/dT of NaBP glasses as a function of α.

Figure 32 shows a plot of γ against B/(B+P) ratio for BaBP and NaBP glasses. The γ values were calculated using Equation (6) and found to be $1.96 \leq \gamma \leq 3.67$ for BaBP and $1.80 \leq \gamma \leq 2.66$ for NaBP glasses. The condition to achieve negative dn/dT is $\gamma < 3$ and based on Figure 32, BaBP glass compositions must contain B/(B+P) < 0.52 (equivalent to < 52 mol% B_2O_3) and NaBP glass compositions must contain B/(B+P) < 0.66 (equivalent to < 49 mol% B_2O_3), after extrapolation of the NaBP′result.

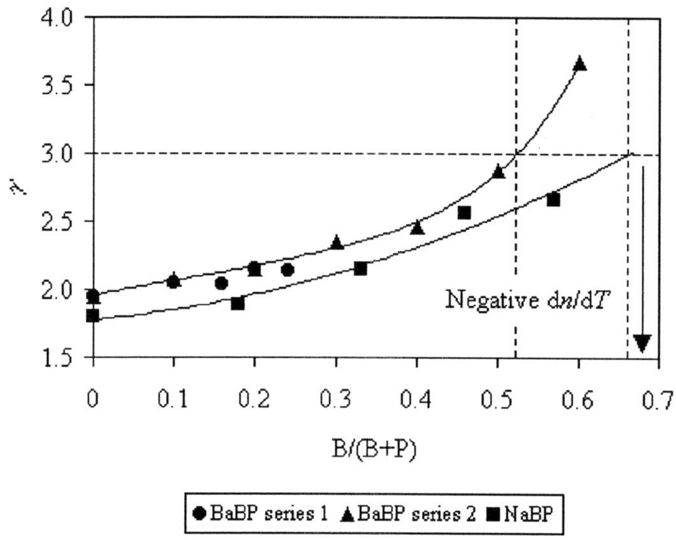

Figure 32. γ of BaBP and NaBP glasses as a function of B/(B+P) ratio

6.4. Summary

Different compositions of barium borophosphate and sodium borophosphate glasses had been studied and the changes in the properties were influenced by the effect and coordination number of B^{3+}. The presence of tetrahedral boron units at phosphate-rich compositions created a rigid borophosphate network due to the strong (P-O-B) bridging bonds between phosphate chains. Consequently, refractive index and glass transition temperature increased. However, the presence of trigonal boron units at borate-rich compositions replaced the borophosphate network with a more open borate network, leading to a decrease in refractive index and glass transition temperature. This structural change occurred at the maximum points observed in the results for the borophosphate glasses. However, the systematic addition of boron led to the monotonic decrease of thermal expansion and the subsequent change towards positive thermo-optic coefficient values.

Glasses with negative dn/dT were achieved, where the corresponding thermal expansion coefficient must exceed 7.5×10^{-6}/°C and 11.5×10^{-6}/°C for barium borophosphate and sodium borophosphate glasses, respectively. Semi-empirical equations were also obtained, enabling the estimation of dn/dT of borophosphate glasses within specific compositional ranges. For barium borophosphate glasses, a fit of $dn/dT = -1.599\alpha + 11.58$ was obtained within $1.585 \leq n \leq 1.610$ and $5.68\times10^{-6}/°C \leq \alpha \leq 13.67\times10^{-6}/°C$. For sodium borophosphate glasses, a fit of $dn/dT = -1.265\alpha + 12.74$ was obtained within $1.484 \leq n \leq 1.512$ and $11.20\times10^{-6}/°C \leq \alpha \leq 22.03\times10^{-6}/°C$.

7. Conclusion

Different families of phosphate glasses were synthesised and investigated within this work, namely the ternary potassium aluminophosphate glasses, the binary calcium phosphate and barium phosphate glasses and the ternary barium borophosphate and sodium borophosphate glasses. This study was carried out to ascertain the suitability of these glasses as the basis for which optical devices can be developed for athermalisation applications, especially in dense wavelength division multiplexing (DWDM) technology to minimise the sensitivity of lightwave devices such as gratings and filters to temperature.

Various compositions of ternary potassium aluminophosphate glasses encompassing ultraphosphate, metaphosphate and polyphosphate stoichiometries examined showed that their optical and thermal properties were influenced by the strengthening effect of Al^{3+} on the glass structural networks through the formation of (P-O-Al) bonds which cross-linked phosphate chains. These bridging bonds resulted in the increase in the structural polymerisation of the glass network and packing density of the atoms, resulting in a more compact network and the subsequent increase in refractive index and glass transition temperature, the decrease in thermal expansion coefficient and the change from negative thermo-optic coefficient to positive values. These behaviours were reversed when potassium was incorporated, as K^{+}, being a low field strength cation, depolymerised the network and formed weak ionic bonds with the phosphate non-bridging oxygens.

Binary calcium phosphate and barium phosphate glasses at ultraphosphate and metaphosphate compositions were also studied, where the glass structural network and hence properties depended on the availability of terminal oxygens to meet the coordination requirement of each modifying cation. When sufficient terminal oxygens were available, the modifying cations can exist at isolated sites. This resulted in the decrease in refractive index and glass transition temperature, due to the depolymerised network. When insufficient terminal oxygens were available, the cations formed clusters to share oxygens, creating a more compact network and increased the cross-linking density of the network, leading to an increase in refractive index and glass transition temperature. However, the systematic addition of modifying cations led to a monotonic increase of thermal expansion coefficient and the subsequent progress towards negative thermo-optic coefficient. In addition, the water content in the glasses must be minimised to reduce the formation of (P-OH) bonds which can depolymerise the glass network.

The properties of barium borophosphate and sodium borophosphate glasses ranging from phosphate-rich to boron-rich compositions were assessed and can be explained based on the effect and coordination number of B^{3+}. Tetrahedral boron units dominated at phosphate-rich compositions and formed (P-O-B) bridging bonds between phosphate chains. This resulted in the increase in the polymerisation of the glass structure and the covalent cross-link density, leading to an increase in refractive index and glass transition temperature. At borate-rich compositions, trigonal boron units dominated and a more open borate network replaced the rigid borophosphate network. This resulted in the decrease in the structural density and strength of the glass network, leading to a decrease in refractive index and glass transition temperature. However, the systematic addition of boron monotonically decreased the thermal expansion and changed the thermo-optic coefficient to positive values.

The results of the glasses suggested that dn/dT and α of the glasses were inter-dependent. Each glass family had a specific minimum α value beyond which α would overcome the effect of polarisability change with temperature so that the glasses would exhibit negative dn/dT. Semi-empirical equations relating dn/dT and α had been defined for the various phosphate glass systems, allowing the estimation of dn/dT of any glass within the same family within specific validity ranges, as summarised in Table 13.

Table 13. dn/dT semi-empirical equations

Glasses	α_{min} ($\times 10^{-6}$/°C)	dn/dT equation	Validity range : n, α ($\times 10^{-6}$/°C)
KAP	9.5	dn/dT = -1.514α + 14.82	$1.469 \le n \le 1.519$ $6.84 \le \alpha \le 21.25$
CaP	9.0	dn/dT = -2.088α + 18.50	$1.526 \le n \le 1.551$ $8.60 \le \alpha \le 10.22$
BaP	9.0 (hypothetical)	dn/dT = -2.104α + 18.90	$1.547 \le n \le 1.585$ $11.62 \le \alpha \le 13.67$
BaBP	7.5	dn/dT = -1.599α + 11.58	$1.585 \le n \le 1.610$ $5.68 \le \alpha \le 13.67$
NaBP	11.5	dn/dT = -1.265α + 12.74	$1.484 \le n \le 1.512$ $11.20 \le \alpha \le 22.03$

Overall, glasses with negative dn/dT had been achieved in all the various phosphate glass systems studied and this shows that phosphate glasses are promising candidates for the development of athermal devices.

ACKNOWLEDGEMENTS

The authors would like to acknowledge the financial support of the Engineering and Physical Sciences Research Council (EPSRC) UK, under the PHOTON project : 'Physical-Layer High-Speed Optoelectronics for Tomorrow's Optical Networks'.

REFERENCES

[1] Kokubun, Y.; Funato, N.; Takizawa, M. *IEEE Photonics Technol. Lett*. 1993, 5, 1297-1300.

[2] Kokubun, Y.; Takizawa, M.; Taga, S. *Electron. Lett.* 1994, 30, 1223-1224.

[3] Kokubun, Y.; Yoneda, S.; Tanaka, H. *European Conference on Optical Communications* 1996, Oslo.

[4] Kokubun, Y.; Yoneda, S.; Tanaka, H. *Electron. Lett.* 1996, 32, 1998-2000.

[5] Kokubun, Y.; Yoneda, S.; Matsuura, S. *Electron. Lett.* 1998, 34, 367-369.

[6] Weber, M. J. *J. Non Cryst. Solids* 1990, 123, 208-222.

[7] Campbell, J. H.; Suratwala, T. I. *J. Non Cryst. Solids* 2000, 263and264, 318-341.

[8] Vogel, W. In *Optical Properties of Glass*; Uhlmann, D. R.; Kreidl, N. J.; Eds.; The American Ceramic Society Inc.: OH, 1991; pp 18-19.
[9] Ehrt, D.; Ebeling, P.; Natura, U. *J. Non Cryst. Solids* 2000, 263and264, 240-250.
[10] Metwalli, E.; Brow, R. K. *J. Non Cryst. Solids* 2001, 289, 113-122.
[11] Park, H. D.; Kreidler, E. R. *J. Am. Ceram. Soc.* 1984, 67, 23-26.
[12] Sales, B. C.; Boatner, L. A. *Science* 1984, 226, 45-48.
[13] Peng, Y. B.; Day, D. E. *High Glass Technol.* 1991, 32, 166-173.
[14] Yu, X.; Day, D. E.; Long, G. J.; Brow, R. K. *J. Non Cryst. Solids* 1997, 215, 21-31.
[15] Lippma, E.; Magi, M.; Samoson, A.; Englehardt, G.; Grimmer, A. R. *J. Am. Ceram. Soc.* 1980, 102, 4889-4893.
[16] Uchino, T.; Ogata, Y. *J. Non Cryst. Solids* 1995, 181, 175-188.
[17] Sidek, H. A. A.; Collier, I. T.; Hampton, R. N.; Saunders, G. A.; Bridge, B. *Phil. Mag. B* 1989, 59, 221-232.
[18] Van Wazer, J. R. *The Phosphorus and its Compounds 1 and 2*; Publisher: Interscience, New York, NY, 1958.
[19] Bunker, B. C.; Arnold, G. W.; Wilder, J. A. *J. Non Cryst. Solids* 1984, 64, 291-316.
[20] Born, M.; Wolf, E. *Principles of Optics*; Pergamon Press: Oxford, UK, 1984; Ch. 2.
[21] Prod'homme, L. *Phys. Chem. Glasses* 1960, 1, 119-122.
[22] Izumitani, T.; Toratani, H. *J. Non Cryst. Solids* 1980, 40, 611-619.
[23] Adams, R. V. *Phys. Chem. Glasses* 1961, 2, 39-49.
[24] Hayden, J. S.; Marker III, A. J.; Suratwala, T I.; Campbell, J. H. *J. Non Cryst. Solids* 2000, 263and264, 228-239.
[25] Jewell, J. M.; Askins, C.; Aggarwal, I. D. *Appl. Opt.* 1991, 30, 3656-3660.
[26] Abbate, G.; Bernini, U.; Maddalena, P.; De Nicola, S.; Mormile, P.; Pierattini, G. *Opt. Commun.* 1989, 70, 502-508.
[27] Harris, R. J.; Johnston, G. T.; Kepple, G. A.; Krok, P. C.; Mukai, H. *Appl. Opt.* 1997, 16, 436-438.
[28] Brow, R. K. *J. Am. Ceram. Soc.* 1993, 76, 913-918.
[29] Lin, Y.; Zhang, Y.; Huang, W.; Lu, K.; Zhao, Y. *J. Non Cryst. Solids* 1993, 112, 136-141.
[30] Brow, R. K.; Kirkpatrick, R. J.; Turner, G. L. *J. Am. Ceram. Soc.* 1993, 76 (4), 919-928.
[31] Metwalli, E.; Brow, R K.; Stover, F. S. *J. Am. Ceram. Soc.* 2001, 84, 1025-1032.
[32] Dietzel, A. Z. *Elektrochem.* 1942, 48, 9-23.
[33] Klein, L. C.; Gray, P. E. *Glastech. Ber.* 1983, 56K, 868-873.
[34] Minami, T.; Mackenzie, J. D. *J. Am. Ceram. Soc.* 1977, 60, 232-235.
[35] Suzuya, K.; Price, D. L.; Loong, C.; Martin, S. W. *J. Non Cryst. Solids* 1998, 232-234, 650-657.
[36] Hoppe, U.; Walter, G.; Kranold, R.; Stachel, D.; Barz, A. *J. Non Cryst.* Solids 1995, 192and193, 28-31.
[37] Hoppe, U. *J. Non Cryst. Solids* 1996, 195, 138-147.
[38] Hudgens, J. *J. PhD Thesis*, Iowa State University, 1994.
[39] Brow, R. K.; Click, C. A.; Alam, T. M. *J. Non Cryst. Solids* 2000, 274, 9-16.
[40] Gray, P. E.; Klein, L. C. *Glass Technol.* 1983, 24, 202-206.
[41] Yun, H.; Bray, P. J. *J. Non Cryst. Solids* 1978, 30, 45-60.
[42] Ray, N. H. *Phys. Chem. Glasses* 1975, 16, 75-80.

[43] Scagliotti, M.; Villa, M.; Chiodelli, G. *J. Non Cryst. Solids* 1987, 93, 350- 360.
[44] Brow, R. K.; Tallant, D. R. *J. Non Cryst. Solids* 1997, 222, 396-406.
[45] Ray, N. H. *Inorganic polymers*; Academic Press Inc.: London, UK, 1978; pp 79.
[46] Kumar, S.; Vinatier, P.; Levasseur, A.; Rao, K. J. *J. Solid State Chem.* 2004, 177, 1723-1737.
[47] Kreidl, N. J. Glass Science and Technology 1; *Materials Science and Technology*; Academic Press Inc.: London, UK, 1983; pp 162.

In: Trends in Optical Materials Research
Editor: Owen G. Gardens, pp. 185-217

ISBN: 1-60021-290-5

Chapter 6

AZO POLYMER SYNTHESES, ELECTROSTATIC LAYER-BY-LAYER SELF-ASSEMBLY AND PHOTOINDUCED SURFACE MODULATION PROPERTIES

Yaning He and Xiaogong Wang*

Department of Chemical Engineering, Laboratory for Advanced Material
Tsinghua University, Beijing, 100084, P. R. China

ABSTRACT

This chapter reviews recent research progress in the synthesis, electrostatic layer-by-layer self-assembly and surface-relief-grating (SRG) formation behavior of azo polymers. The azo polymers were synthesized by some well-established methods such as post-polymerization azo-coupling reaction, functionalization of poly(acryloyl chloride), and polymerization through azo-coupling reactions. According to the polymer chain architecture, the azo polymers can be classified as epoxy-based, acrylate-based, urethane-based linear polymers and hyperbranched polymers. Some of the polymers were designed to contain ionizable groups. Multilayer films were prepared through the electrostatic layer-by-layer deposition of the azo polyelectrolytes and PDAC. SRGs were photofabricated by exposing both the spin-coated polymer films and multilayer films to an interference pattern of coherent Ar^+ laser beams. The SRG formation behavior was characterized by the inscription rates and the saturation levels of SRGs on the polymer films. The results show that the rates of SRG formation significantly depend on the types of the electron-withdrawing groups in the *para*-positions of the azobenzene units. The inscription rates decrease with the increase of the number of methyl substitutes on the *ortho*-positions of the azo groups. Meanwhile backbone structures also have some effect on the photoprocessibility through the ways such as altering the chromophore loading density and changing main-chain rigidity. The electrostatic multilayer films can form SRGs, but the inscription rates and the saturation levels are lower than those of the spin-coated films. Those results can be used to explain the SRG forming mechanism and to develop new SRG materials with better properties.

* E-mail Address: wxg-dce@mail.tsinghua.edu.cn

1. Introduction

Azobenzene and many of its derivatives have been well studied since those colorful compounds were first successfully prepared in the 19th century.[1-4] One of well-known application of those azo compounds is to be used as dyestuffs for a variety of substrates, which show the fastness properties and a wide spectrum of shades based on a proper molecular design.[1] The most interesting property explored for functional materials preparation is the photochemical trans-cis isomerization of the aromatic azo chromophores, which can be induced by UV or visible light irradiation (Figure 1).[2-4] The effect was tried to be prevented by colorists for harmful tint changes in various azo colorants in the history, but today it can offer polymer chemists a lot of chances to develop various photoresponsive polymers.

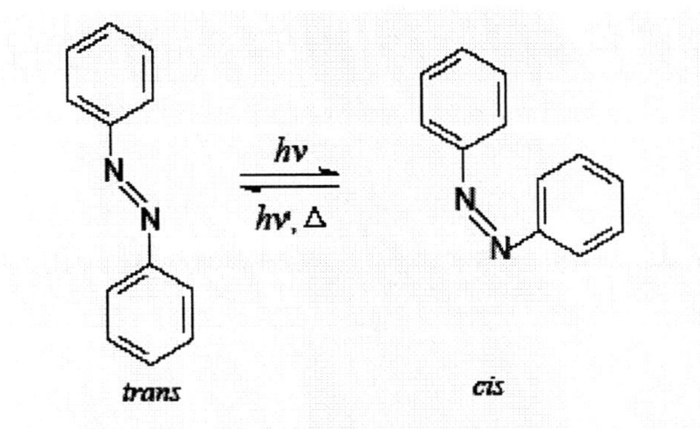

Figure 1. *Trans* and *Cis* isomers of azobenzene-based chromophores

Azo polymers (polymers containing aromatic azo chromophores) can show a variety of photoresponsive properties caused by photoisomerization of the azo chromophores.[3, 4] The photoisomerization can induce the structure and property variations such as phase transition, dichroism and birefringence, surface-relief-gratings (SRGs) among many others.[5, 6] Polymers having such properties are promising for applications in holographic gratings, reversible optical storage systems, optical switching modulators, sensors and other devices.[5-10]

Photoinduced surface-relief-gratings (SRGs) is a large reversible surface modulation occurring on the azo polymer films, which is caused by the photoisomerization of azo chromophores.[11-13] In a typical process, SRGs are formed by exposing the polymer films to an interference pattern of Ar^+ laser beams at modest intensities.[11-13] The surface relief structures formed are stable below the glass transition temperatures (*Tg*s) of the polymers and can be removed by heating samples to a temperature above their *Tg*s or erased optically even below the *Tg*s.[14] By selecting suitable photofabrication conditions, it is easy to produce complex surface patterns that could be very useful in many future applications. [15, 16]

The forming mechanism of the photofabricated SRGs has been actively investigated. It is generally accepted that the SRGs are formed due to large-scale migration of the polymer chains induced by the low power laser irradiation, in which the photoinduced trans-cis-trans isomerization cycles of the azo chromophores play a key role. [5, 6, 17-21] The large-scale mass transport of the polymer chains has been attributed to internal pressure gradients produced by an isomerization-driven free volume expansion,[18] anisotropic intermolecular interactions resulted from photoinduced ordering of dye molecules,[19] worm-like movement in a direction parallel to former orientation of the azo molecules,[20] dipole interaction of azo chromophores with the electric field gradient present in the polymer materials.[21] However, the understanding of the exact relationship between the cis-trans isomerization and the mass transport is still evolving.

For the sake of simplicity, almost all of the models proposed in the past years have neglected the effect of polymer structures on the SRGs information. However, some recent studies have shown that the polymer structures can have significant effect on the SRG formation.[15, 22-27] The photoinduced dynamic process is affected by structural factors of azo polymers such as chemical structure of backbones,[22, 25] types of azo chromophores, [15, 24-27] and degree of functionalization among others.[23, 25] In order to fully understand this photoinduced dynamic process of azo polymers, an intensive study on the correlation between the polymer structures and the SRG formation properties is required. For this purpose, a series of azo polymers with various structures have been prepared and studied to understand the structure-property relationship.[25-37] In the meantime, those different kinds of azo polymers can also supply a good stock of the potential optical materials for different future applications.

Several excellent reviews have been written by Tripathy, Barrett, Rochon and Natansohn et al. which cover the early literatures, molecular design, experimental details, forming mechanism, and potential applications of the SRGs.[5, 6, 17, 38, 39] This chapter will only concentrate on the recent progress in the azo polymer preparation and SRG formation property study in authors group, and focus on the synthetic strategy and the structure effect on the SRG formation properties.

2. Synthesis of Azo Polymers

To prepare azo polymers, the task is how to incorporate the azo chromophores into the polymer backbone. One common way is first to prepare monomers containing azo chromophores, and then to prepare azo polymers by step polymerization or free radical polymerization.[3-5] In recent years, several new synthetic strategies, such as postpolymerization azo-coupling reaction,[25-30, 40, 41] functionalization of poly(acryloyl chloride),[30-33] and polymerization through azo-coupling reactions,[26, 37] have been developed and used to prepare azo polymers. Those new methods show some advantages, such as mild reaction conditions, easiness in the performance, and elimination of the azo chromophore deterioration during polymerization.

2.1. Postpolymerization Azo-Coupling Strategy

The postpolymerization azo-coupling scheme was first used to prepare nonlinear optical azo polymers by introducing various azo chromophores through the azo-coupling reaction between an epoxy-based precursor polymer and diazonium salts at the final stage of the polymer synthesis.[40, 41] The precursor polymer was prepared by the reaction between diglycidyl ether of Bisphenol-A and aniline. This reaction route can provide a convenient and often the simplest way to synthesize side-chain azo polymers containing various types of azo chromophores, which generally has high reaction yields and can reach high degrees of functionalization (DFs).[40, 41]

The postpolymerization azo-coupling reaction strategy was used to prepare various azo polymers. The first step was to prepare different precursor polymers containing anilino moieties. Then the precursor polymers were reacted with various diazonium salts to introduce the azobenzene chromophores (Scheme 1). Through this way, a variety of epoxy-based (Scheme 2, 3), acrylate-based (Scheme 4), and urethane-based azo polymers (Scheme 5) were synthesized.[25-30]

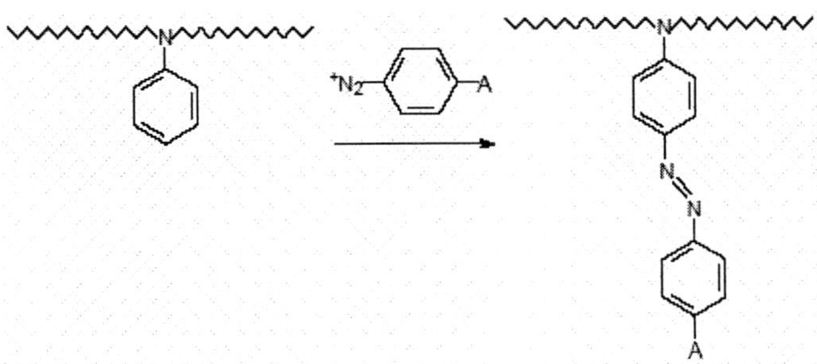

Scheme 1. Preparation of azo polymers through postpolymerization azo-coupling reaction

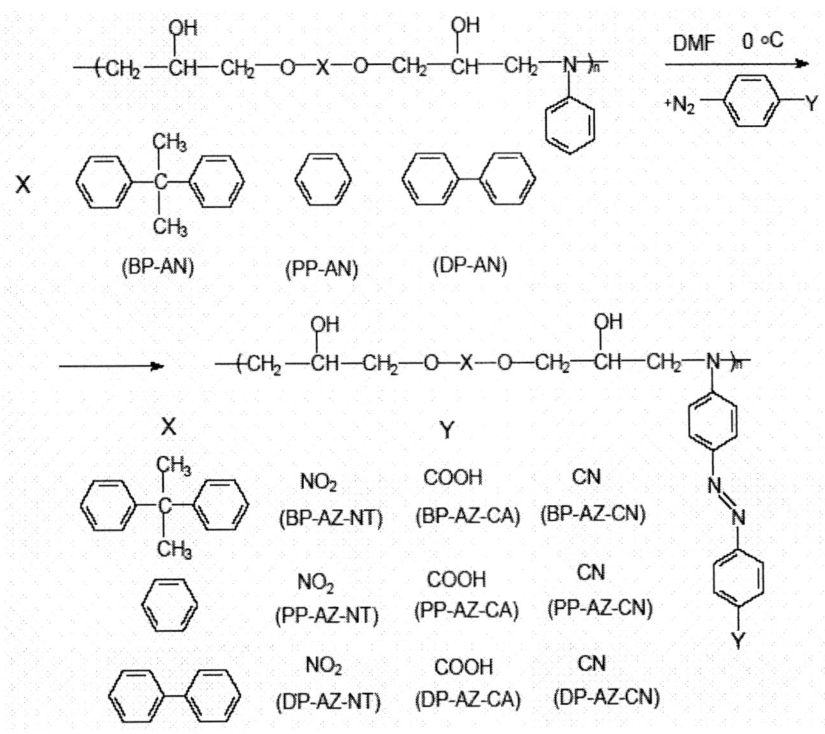

Scheme 2. Synthetic route and chemical structures of the epoxy based azo polymers prepared by postpolymerization azo-coupling reaction

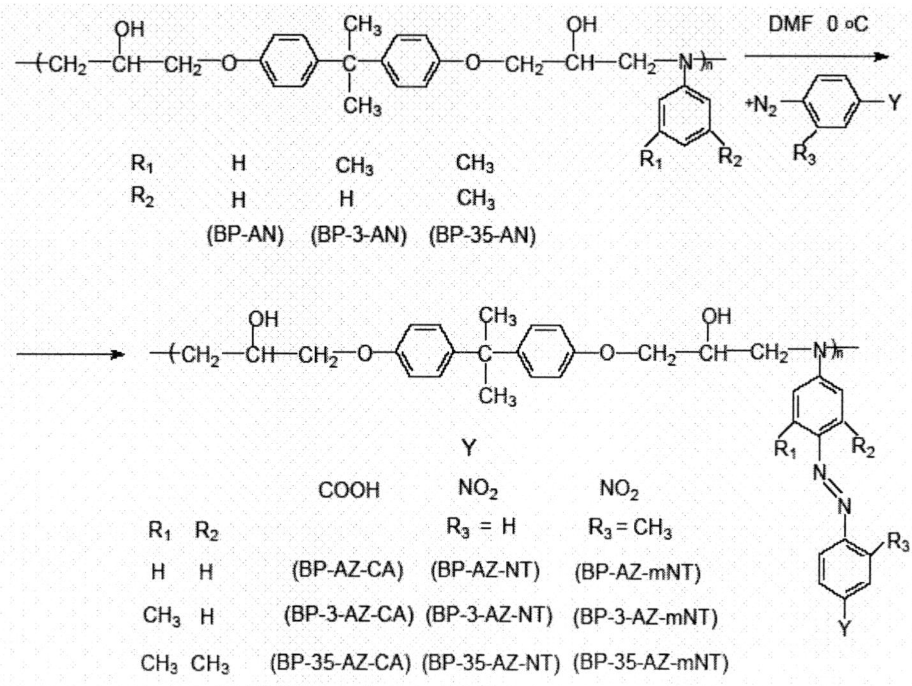

Scheme 3. Synthetic route and chemical structures of the epoxy-based polymers functionalized with methyl-substituted azobenzene chromophores

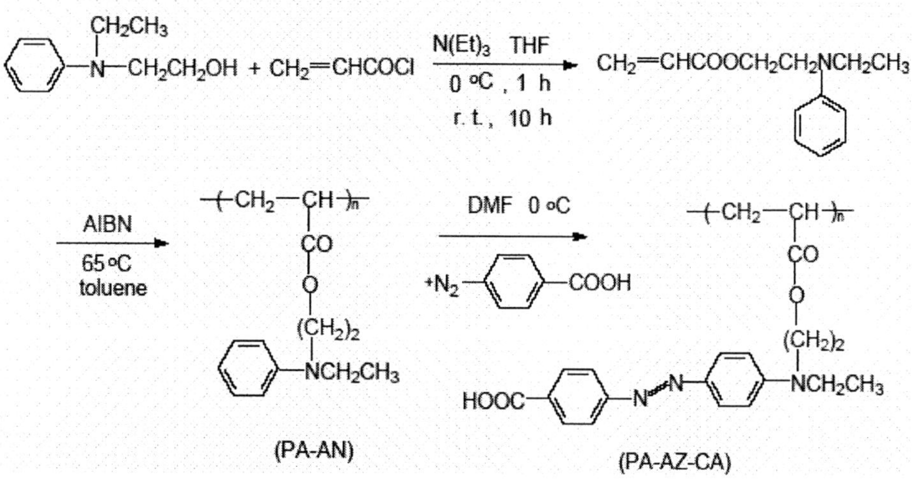

Scheme 4. Synthetic route and chemical structures of the acrylate-based azo polymers prepared by postpolymerization azo-coupling reaction

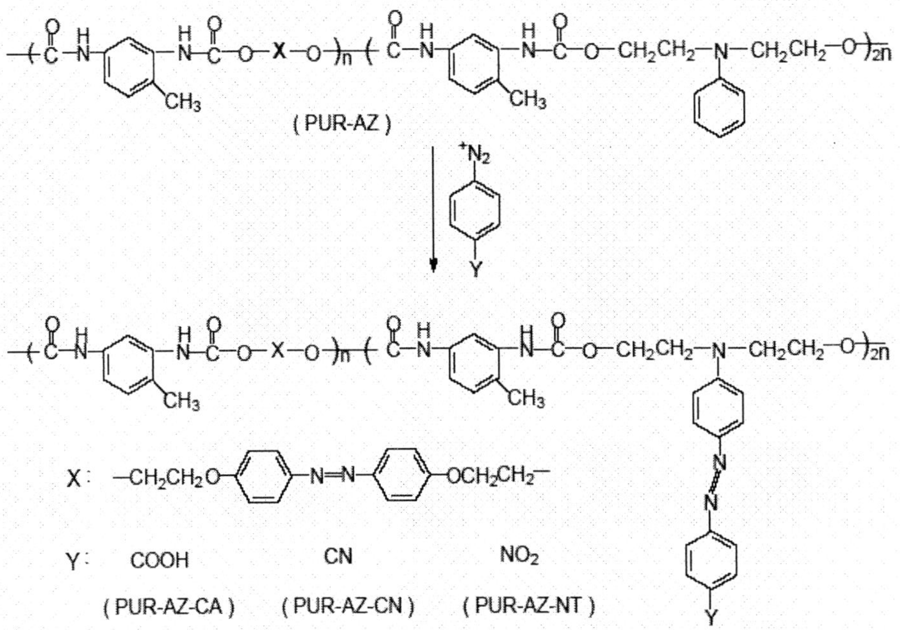

Scheme 5. Synthetic route and chemical structures of the urethane-based azo polymers prepared by postpolymerization azo-coupling reaction

The epoxy-based precursor polymers were synthesized from the different epoxide monomers and aniline or its derivatives.[25, 27] To avoid the reaction between the secondary OH groups formed during the polymerization and the unreacted epoxide groups, the polymerizations were carried out at a relatively low temperature (110 °C). BP-AN, BP-35-AN, BP-3-AN, and PP-AN are amorphous polymers with glass transition temperatures (*Tg*) at 90 °C, 72°C, 84°C, and 85 °C. DP-AN is a liquid crystal polymer with *Tg*, *Tm*, and *Ti* at 110 °C, 126 °C and 185 °C. The acrylate-based precursor polymer (PA-AN) was synthesized from 2-(phenylethylamino)ethyl acrylate by radical polymerization.[28] The monomer 2-(phenylethylamino)ethyl acrylate was prepared from the Schotten-Baumann reaction between acryloyl chloride and *N*-ethyl-*N*-hydroxyethylaniline. PA-AN is an amorphous polymer with *Tg* at 14 °C. The urethane-based precursor polymer PUR-AZ was synthesized by the polycondensation reaction of 4,4'-di(2-hydroxyethoxy) azobenzene, *N,N*-di(2-hydroxyethyl)aniline and 2,4-toluylene diisocyanate (TDI).[29] The *Tg* of the precursor polymer PUR-AZ is about 107 °C. All the precursor polymers showed good solubility in polar solvents like dimethylformamide (DMF).

Different type azo chromophores were introduced into the polymers by the post-polymerization azo coupling reactions.[25-30] The ^{1}H NMR analysis indicated that the azo-coupling reactions were taken place exclusively at the para-positions of anilino moieties with high yield for the precursor polymers (Figure 2). For example, for precursor polymer (BP-35-AN), the chemical shifts of unreacted aniline moieties appear at about 6.33 ppm and 6.18 ppm, attributed to the protons at ortho- and para-positions of the amino groups. When slight excesses of diazonium salts (molar ratios to aniline moieties of the precursor polymers) were used in all these reactions, the following changes in the ^{1}H NMR spectra can be observed for aniline moieties in the precursor polymers upon postpolymerization azo-coupling reaction.

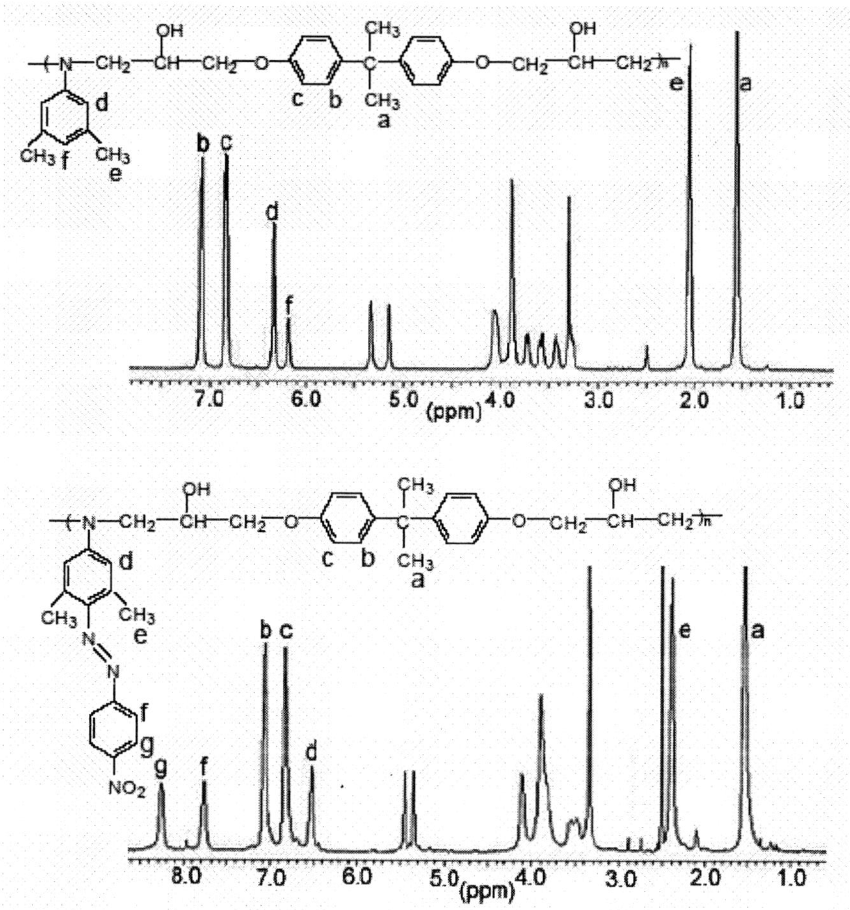

Figure 2. ^{1}H NMR spectra of BP-35-AN and BP-35-AZ-NT in DMSO-d_6

The resonance (6.18 ppm) corresponding to the protons at para-position of amino group disappears. The chemical shifts of protons ortho to the amino groups shift to lower magnetic field, due to the presence of electron withdrawing groups introduced by the azo-coupling reaction. Moreover, additional resonances corresponding to chemical shifts of the benzenoid ring protons introduced through the postpolymerization azo-coupling reaction appear at a lower magnetic field. It can be concluded that the azo-coupling reactions occur at the para positions of aniline moieties with high yields.

By the postpolymerization azo-coupling reaction, the degrees of functionalization (DFs) of azo polymers can be easily adjusted by selecting a suitable feed ratio. The feed ratio is defined as the molar ratio of diazonium salt to aniline moiety of the precursor polymer. DFs of the azo polymers were determined by ^{1}H NMR analysis. For instance of the BP-AN series of polymers, the DFs were estimated from ^{1}H NMR by comparing the peak areas for the protons meta to the amino groups in substituted aniline moieties with the peak areas at 1.55 ppm corresponding to the methyl protons of the Bisphenol-A moieties. The relationships between the molar feed ratios and the DFs for two series of the azo polymers are given in Figure 3. From the figure, it can be seen that the DFs increase almost linearly with the feed ratios. The DFs can be easily controlled by selecting a suitable feed ratio.

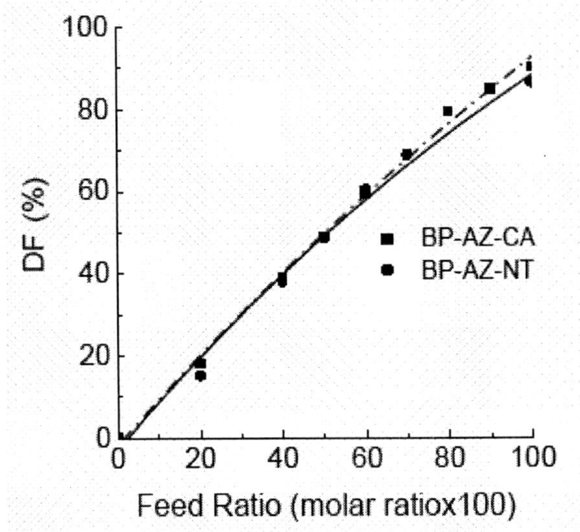

Figure 3. Relationship between the degrees of functionalization and the feed ratios for preparing BP-AZ-CA and BP-AZ-NT

2.2. Poly(Acryloyl Chloride) Functionalization Strategy

Polyelectrolytes (PEL) have aroused considerable research interest for their importance in academic exploration and practical applications.[42] The incorporation of ionizable segments in polymers is a very attractive molecular design option to prepare stimuli-responsive materials, which can exhibit drastic property changes in response to infinitesimal environmental variations such as temperature or pH.[43, 44] Polyelectrolytes are able to form PEL complex with oppositely charged macroions, surfactants or colloid particles.[45-47] The latest development by exploring this property is to produce self-assembled multilayer films through an electrostatic layer-by-layer assembly from two or more solutions containing oppositely charged polyions.[48, 49]

Azo polyelectrolytes, can combine most interesting properties of azo polymers and polyelectrolytes.[50-52] Polymers with such properties will be useful for preparing electrostatic multilayer films, photoresponsive aggregates among others.[30-34] The chemical modification of poly(acryloyl chloride) (PAC) offers a feasible way to prepare a variety of azo polyelectrolytes.[30-34] By this method, the precursor polymer PAC is prepared by the radical polymerization of acryloyl chloride. PAC is then functionalized by the Schotten-Baumann reactions with azo chromophores bearing hydroxyl groups (scheme 6) or with aniline moieties bearing hydroxyl groups (scheme 7). The latter can be further modified by azo coupling reactions. The degrees of functionalization of azo chromophores can be controlled by the feed ratios and then unreacted acyl chloride groups are hydrolyzed to obtain COOH groups. Through this way, a variety of acrylate-based azo polyelectrolytes have been synthesized.[30-34]

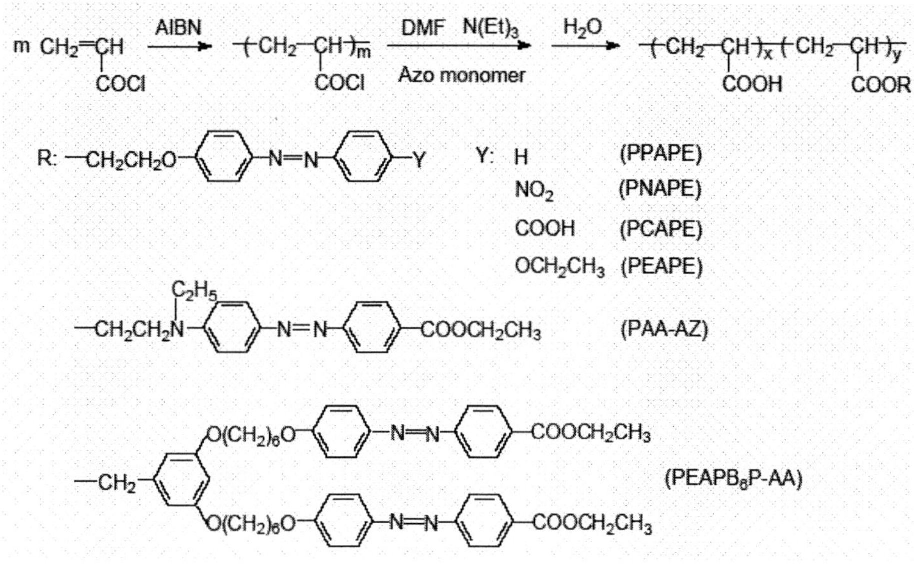

Scheme 6. Synthetic route and chemical structures of the azo polyelectrolytes prepared by poly(acryloyl chloride) functionalization

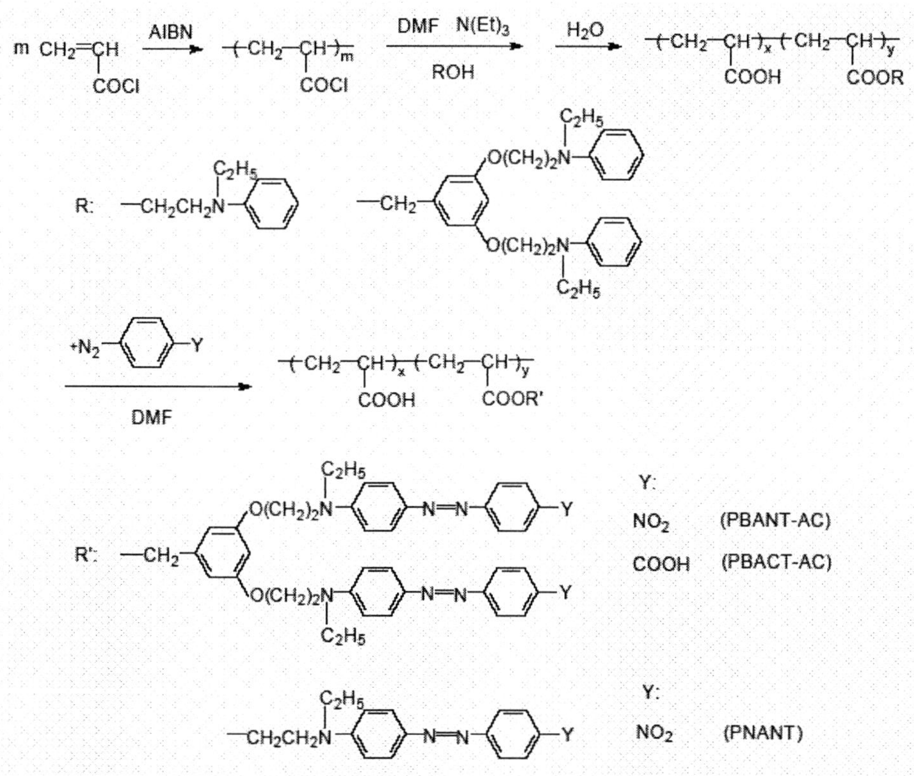

Scheme 7. Synthetic route and chemical structures of the azo polyelectrolytes prepared by poly(acryloyl chloride) functionalization and postpolymerization azo-coupling reaction

In the synthetic process, the radical polymerization of acryloyl chloride, precipitation, washing and drying were all carried out under strictly anhydrous condition to protect the acyl chloride groups from the hydrolysis. Compared to the IR spectrum of acryloyl chloride, the disappearance of 1610 cm^{-1} band (C=C stretching) in that of PAC confirmed the polymerization of acryloyl chloride and nonexistence of the monomer in the product. The precursor PAC was then post-functionalized with azo chromophores bearing hydroxyl groups. Unlike small molecular reactions, the conversion of reactive groups attached to polymeric chains is affected heavily by the conformation of polymers. Therefore, the solvent plays an important role in the reactions on the polymer. DMF was found to be a good solvent, which can result in a higher degree of functionalization due to its good solubility to PAC, the azo chromophores and the polyelectrolytes. The structure and purity of the final products were confirmed by ^{1}H-NMR spectra (see Figure 4). The sharp and clearly separated resonances, corresponding to low molecular weight azo reactants, disappear after the reaction, while those broad and overlapped resonances of azo chromophores in the macromolecules appear at nearly the same positions. The ^{1}H-NMR spectra of the azo polyelectrolytes and other spectroscopic analyses verify that the azo chromophores are incorporated in the polymeric chains. The broad and overlapped peaks (in δ 1-2.5 ppm range) corresponding to main-chain protons demonstrate a random copolymerization sequence and atactic configuration. The precursor PAC was functionalized with aniline moieties bearing hydroxyl groups by a similar way, which gives polymeric intermediates. By azo-coupling reaction of the polymeric intermediates, various azo polyelectrolytes were prepared. The relationship between the DF and the feed ratio is shown in Figure 5. The DF is linearly dependent on and slightly lower than the feed ratio.

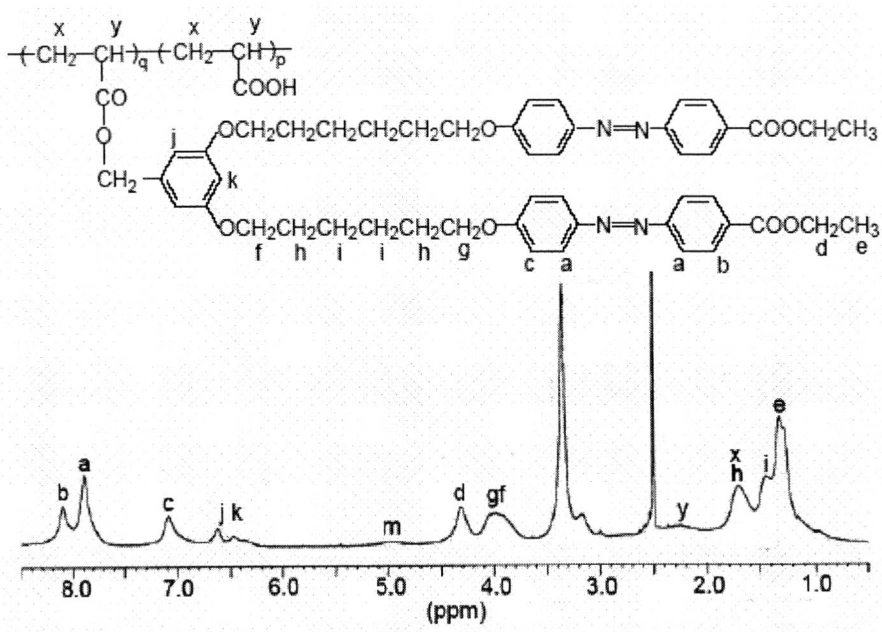

Figure 4. ^{1}H NMR spectrum of PEAPB6P-AA in DMSO-d_6

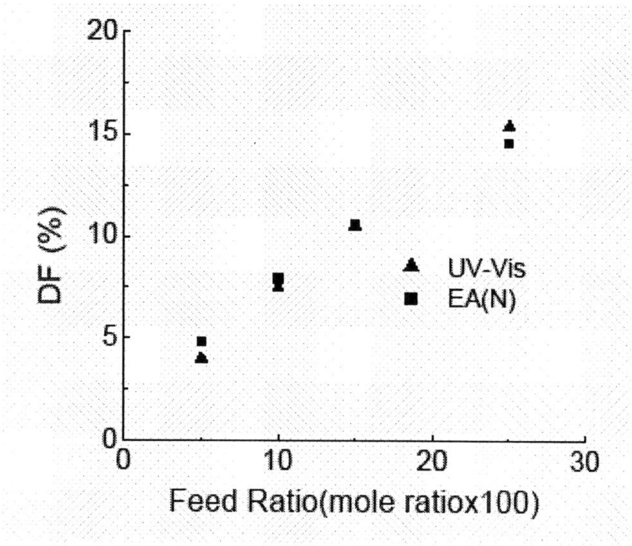

Figure 5. Relationship between the degree of functionalization of PEAPE and the feed ratio of EAPE (2-[4-(4'-ethoxyphenylazo)phenoxy]-ethanol) to PAC. The DFs were estimated by elemental analysis (EA) and UV-vis Spectroscopy (UV-Vis)

As a result, the DFs of azo chromophores can be easily adjusted by selecting suitable feed ratio. Although the acyl chloride groups of PAC are very reactive, the reactions with the azo reactants cannot undergo a complete conversion. One possible reason is due to the steric hindrance of the main chain to bulky azo reactants. Remaining certain percentage of unreacted acyl chloride groups on the main-chain, which can be hydrolyzed to produce carboxylic acid groups after the reaction, is necessary for the azo polymers to act as polyelectrolytes.

2.3. Polymerization Through Azo-Coupling Strategy

Hyperbranched polymers, as one type of dendritic polymers, have received great attention during the last decade.[53-56] Due to the structural characteristics, hyperbranched polymers show interesting properties of the dendritic polymers, such as good solubility, low melt and solution viscosity, the feasibility to be further modified through various chemical reactions of the peripheral groups.[57-59] In comparison with dendrimers, hyperbranched polymers have less regular structure and a broader molecular weight distribution. On the other hand, hyperbranched polymers can be prepared through some relatively simple methods, such as one-step polymerization of ABx type monomers (where x is generally 2 or 3). The preparation doesn't need the tedious isolation and purification procedures required for preparing dendrimers through the generation-by-generation scheme.[60, 61]

More recently, dendrimers containing azobenzene units have attracted considerable attention and have been synthesized to bear azo chromophores in the exterior, interior, or throughout the dendritic architecture.[62-64] Those azo dendrimers can exhibit some fascinating properties, such as harvesting low-energy photons, improving both the nonlinear optical properties and thermal stability of the azo polymers.[65-67] But until now, only few hyperbranched azo polymers have been synthesized and studied, which is partly due to the lack of a feasible way to prepare hyperbranched azo polymers under mild conditions.[35, 36]

Recently, a feasible way has been developed to prepare hyperbranched azo polymers through azo-coupling reaction of AB_2 monomer under an extremely mild condition.[37] For many years, azo-coupling reaction has been well developed to prepare various types of dyestuff and can be carried out in many different ways.[1-4] By exploiting the knowledge base, a variety of hyperbranched azo polymers can be designed and prepared. Through this scheme, the strict reaction conditions such as high temperature and vacuum can be avoided. In a typical process, the hyperbranched azo polymer was prepared through the route shown in scheme 8. The AB_2 monomer *N, N*-bis[2-(4-aminobenzoyloxy)ethyl]aniline was prepared by the nucleophilic substitution reaction between *N, N*-bis(2-chloroethyl)aniline and 4-aminobenzonic acid. The diazonium salt of *N, N*-bis[2-(4-aminobenzoyloxy)-ethyl]aniline was added into DMF to form hyperbranched polymer directly through azo-coupling reaction between the monomers. The hyperbranched polymer with a large number of peripheral diazonium salt groups can be further used as polyelectrolytes or reacted with some compounds such as *N, N*-bis(2-hydroxyethyl)aniline that will enable the remaining diazonium salt groups to be completely reacted through the azo coupling reaction.

This synthetic scheme has been further improved through two-stage azo coupling reactions of a new type AB_2 monomer.[26] A hyperbranched precursor azo polymer, which contained a large number of peripheral anilino groups, was prepared by step-growth polymerization of the monomer through azo-coupling reaction. The precursor polymer was further modified through post-polymerization azo-coupling reactions to introduce different peripheral azo chromophores.

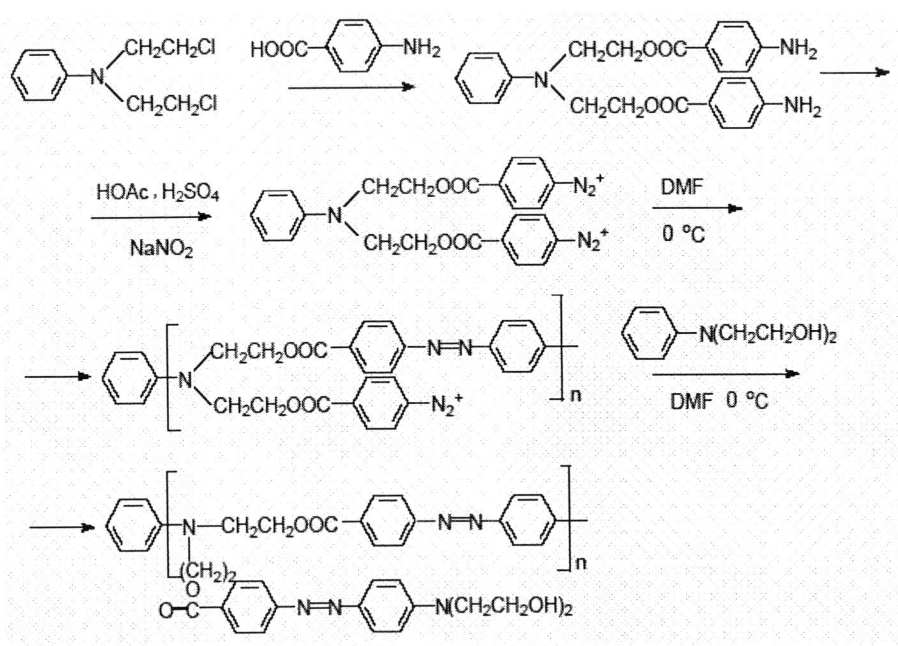

Scheme 8. Synthetic route and chemical structures of the hyperbranched azo polymers prepared by polymerization through azo-coupling reaction

The scheme can be used to prepare a variety of azo polymers that have a highly branched azo backbone and bear different types of azo chromophores at the peripheral positions.

The way to prepare the hyperbranched azo polymers is given in scheme 9. AB_2 monomer (DEEAP) was synthesized by the nucleophilic substitution reaction between *N*-ethyl-*N*-(2-chloroethyl)aniline and 3,5-biscarboxyl aniline in DMSO. The hyperbranched precursor azo polymer (HPAP) was prepared by step-growth polymerization of DEEAP through azo-coupling reaction. HPAP was further functionalized to incorporate three types of the peripheral azo chromophores by post-polymerization azo-coupling reactions. In these reactions, HPAP was reacted directly with the diazonium salts to introduce different azo chromophores at the peripheral groups.

The number average molecular weight (*Mn*) and polydispersity index of HPAP were obtained by gel permeation chromatography (GPC) to be 142000 and 1.44. The absorption band corresponding to the π-π* transition can be observed from the UV-Vis spectrum of HPAP. The results of ^{1}H NMR, ^{13}C NMR, UV-Vis and GPC analysis all suggested that the designed hyperbranched azo polymer with a reasonably high molecular weight was synthesized. The post-polymerization azo coupling reactions have been used to modify linear precursor polymers to incorporate azo chromophores. The current research shows that the same scheme can be successfully used to introduce various peripheral azo chromophores into a hyperbranched polymer.

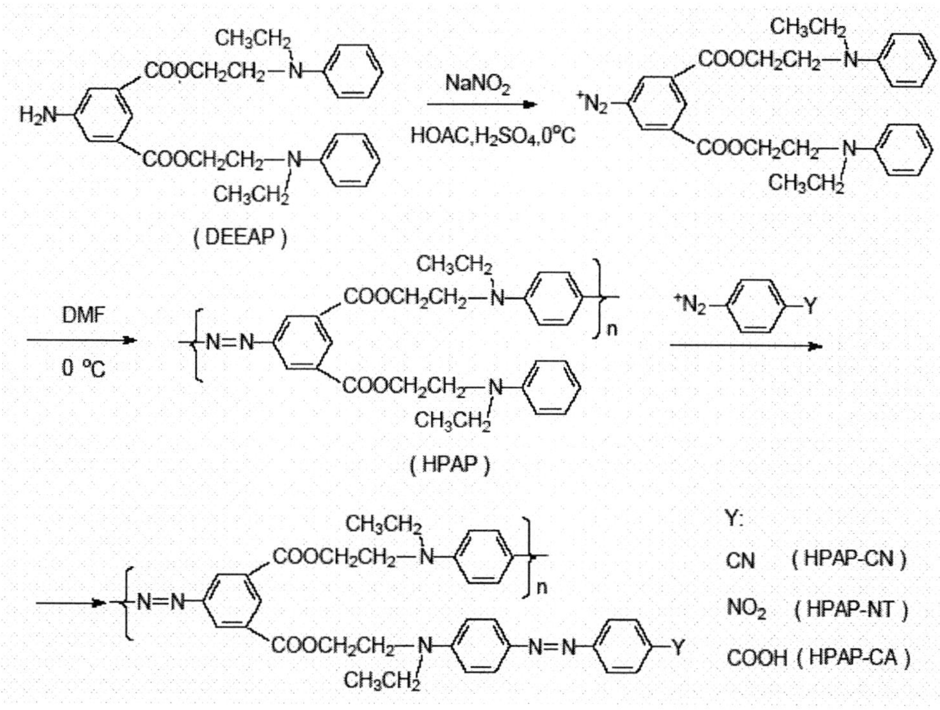

Scheme 9. Synthetic route and chemical structures of the hyperbranched azo polymers prepared by two-stage azo-coupling reactions

The polymerization through azo-coupling strategy can also be used to prepare linearly main-chain azo polymers. In a typical process, the linearly main-chain azo polymer was prepared through the route shown in scheme 10.

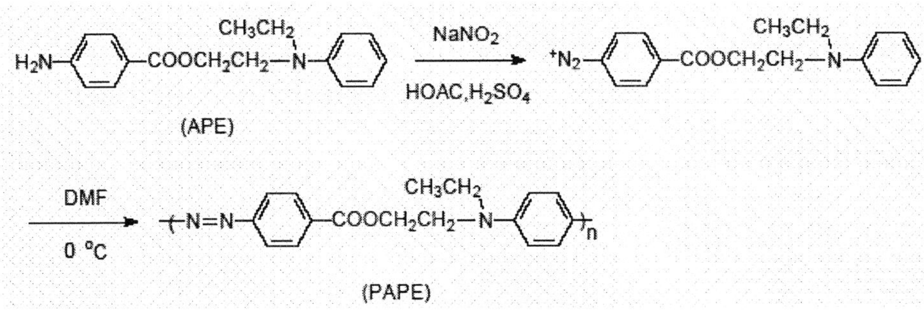

Scheme 10. Synthetic route and chemical structures of the linearly main chain azo polymer prepared by polymerization through azo-coupling reaction

The monomer (APE) was synthesized by the nucleophilic substitution reaction between *N*-ethyl-*N*-(2- chloroethyl)aniline and 4-aminobenzoic acid in DMSO. The azo polymer (PAPE) was prepared by step-growth polymerization of APE through azo-coupling reaction. The number average molecular weight determined by GPC is Mw=13,200 with a polydispersion index 2.2. The glass transition temperature (*T*g) of the polymer is 132 °C obtained from DSC thermogram. The results show that the azo polymer has been successfully synthesized through azo-coupling reaction. The synthetic scheme demonstrated here is also a feasible way to prepare main-chain azo polymers under an extremely mild condition.

2.4. Polystyrene Functionalization Strategy

Polystyrene (PS), as a common backbone for various functional polymers for its good mechanical properties and easy commercial availability, is also an appealing macromolecular carrier for azo chromophores because of its general transparency and photochemical stability. Polystyrene was functionalized through a nitration, reduction, diazotization, coupling four-step reactions (named as NRDC route) to prepare azo polymers.[68] Previous attempts to incorporate azo groups to a polystyrene chain suffered from a low degree of functionalization after several reaction steps or cross-linking side-reactions between PS chains, which resulted in a low chromophoric loading density or insolubility of the final polymers.[68, 69]

In our recent work, a PS precursor was functionalized through a modified NRDC (Nitration-Reduction-Diazotization-Coupling) scheme. Compared with original NRDC route, first three steps in this modified NRDC route gave very a high reaction yield. The ratios of nitration and reduction could reach nearly 100%. Water soluble poly(4-aminostyrene) was obtained and diazotization could be easily processed in aqueous solution to prepare almost 100% macromolecular diazonium salt (MDS). In this route, the degree of azo functioanlization was only controlled by the azo-coupling step. Soluble PS-based azo polymers with high degree of functionalization (DF) could be easily synthesized. Due to small amount of diazonium salt was unreacted in the polymer reaction, a thermolysis step after the azo coupling reaction was found to be essential to maintain the solubility of the azo polymers in the post reaction processing. Synthetic route and chemical structure of some PS-based azo polymers is given in scheme 11 and the details of the NRDC process are given in the following paragraphs.

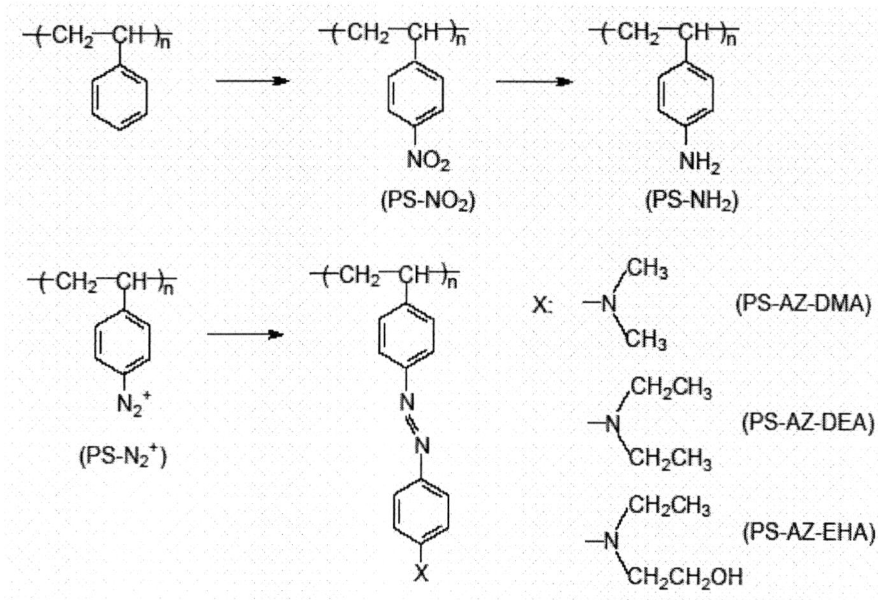

Scheme 11. Synthetic route and chemical structures of PS-based azo polymers prepared by NRDC route

In the nitration step, polystyrene (10 g, 96.2 mmol) was dissolved in 3-nitrotoluene (100 mL) and cooled to 15 °C. Nitric acid (80 mL) was added dropwise into the solution with vigorous stirring and followed by a slow addition of sulfuric acid (20 mL). The temperature was kept at 15 °C until the addition completed. The reaction mixture was stirred at 30 °C for 12 h. Then the mixture was separated and the upper oil phase was collected. The crude product, which included 3-nitrotoluene and nitrified PS (PS-NO_2), was dropped into 200 mL isopropyl alcohol under stirring. The white-yellow precipitate was filtered, washed with isopropyl alcohol and dried at 50 °C. The powder was then dissolved in 100 mL DMF and precipitated into 200 mL isopropyl alcohol. The final product was vacuum dried at 50 °C for 48 h.

For reduction of PS-NO_2, granulated Tin metal (20 g) was added into 100 mL hydrochloric acid (37%) and heated up to 70 °C rapidly. PS-NO_2 (5 g, 33.6 mmol) dissolved in DMF (50 mL) was added dropwise under stirring. The reaction was continued for 24 h at 100 °C until the solid in reaction mixture was completely dissolved. Then the solution was poured into 1000 mL ethanol and the precipitate was collected by filtration. The crude poly(4-amino styrene) (PS-NH_2) was dissolved in methanol and precipitated in ethanol twice. Then the white precipitate was collected and vacuum-dried at room temperature for 48 h. To prevent oxidation in air, PS-NH_2 was dissolved in water soon after purification and used as aqueous solution with determinate concentration in the following step.

Diazotization of PS-NH_2 was carried out in a hydrochloric acid solution. PS-NH_2 (20 mL) aqueous solution (containing 1 g, 8.4 mmol PS-NH_2) was mixed with 3 mL hydrochloric acid (37%) under stirring. Then the clear solution was cooled to 0 °C in an ice bath. $NaNO_2$ (1.2 g, 17.4 mmol) dissolved in 10 mL water was added dropwise into the above solution. The reaction was kept at 0 °C for 6 h. Then the diazonium salt (PS-N_2^+) solution was filtered rapidly and kept at 0 °C for further reaction.

The polymeric diazonium salt could react with different aromatic amino compounds. For a typical example, *N*, *N*-dimethylanline (DMA) (2g, 16.5 mmol) was dissolved in DMF (100 mL) and cooled to 0 °C in an ice bath. Polymeric diazonium salt of PS-NH_2 (1 g, 8.4 mmol) was added dropwise into above solution, which was stirred at 0 °C for 12 h and at 40 °C for 6 h. The last step was designed to eliminate the little amount of unreacted diazonium salts. The reaction mixture was precipitated into sodium carbonate water solution (5 g $NaCO_3$ in 300 ml H_2O). The product was purified by reprecipitation from formic acid into a sodium carbonate water solution, collected by filtration and washed with water repeatedly. The finally product (PS-AZ-DMA) was washed with acetone and vacuum-dried at room temperature for 48 h.

Via a procedure similar to that described above for the synthesis of PS-AZ-DMA, other PS-based azo polymers can be prepared by the reaction between PS-N_2^+ and different kinds of N-alkyl aromatic amines.

3. Surface-Relief-Grating Behavior

The SRG formation experimental setup is shown in Figure 6. In a typical process, a *p*-polarized Ar^+ laser beam (488 nm, 150 mW/cm^2) was used as the light source. SRGs were optically inscribed on the polymer films with interfering laser beams, where the Ar^+ laser beam was split by a mirror and the reflecting half beam coincided with the other half on the film surface. The surface profiles of the resulting gratings were recorded by using a Nanoscope atomic force microscope (AFM) in the tapping mode. The diffraction efficiency of the gratings was probed for the first-order diffracted beam of an unpolarized low power He-Ne laser beam (633 nm) in transmission mode. Although using two circularly polarized interfering beams can produce SRGs more efficiently, the *p*/*p*-polarization condition has also been frequently used to fabricate good quality SRGs and the result is relatively easier to be explained. [38]

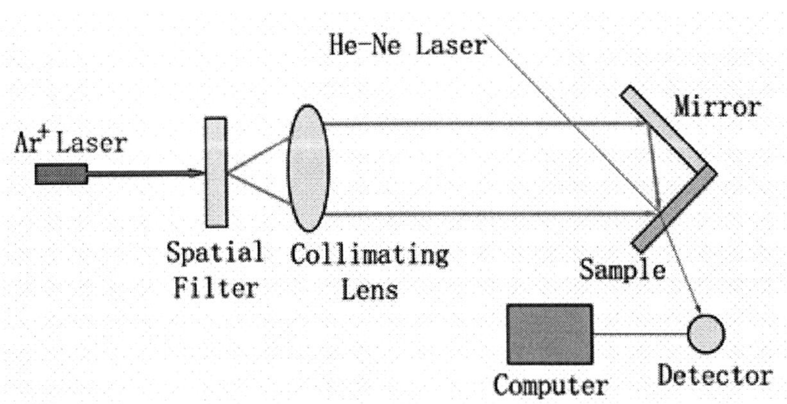

Figure 6. Experimental setup for SRG fabrication

3.1. SRG Behavior of Spin-Coated Films

The rates of grating formation can be probed by measuring the first order diffraction efficiency of the SRGs during the grating inscription process. The diffraction efficiency was recorded in-situ and used to characterize the rate of the surface modification. The rate of grating formation depends on both the structures of the polymers and the conditions such as film thickness, the light intensity, and the angle between the two interfering beams.[5, 6, 17, 38, 39] In order to attribute the difference of the SRG-forming rates to the polymer structures, the film thickness was controlled to be the same (about 1 μm). Relatively low light intensity (150 mW/cm_2) was used in the experiments to avoid the sample damage and possible side effect caused by high intensity laser irradiation. Figure 7(a) shows a typical AFM image of the sinusoidal surface relief structures with regular spaces formed on the spin-coated azo polymer film. The spatial period depends on the wavelength λ of the writing beam and the angle θ between the two interfering beams, which has a relationship as: $\Lambda = \lambda / 2 \sin (\theta / 2)$. Figure 7(b) shows the atomic force microscopy (AFM) images of the surface relief structures when two gratings are recorded orthogonally to each other at the same location.

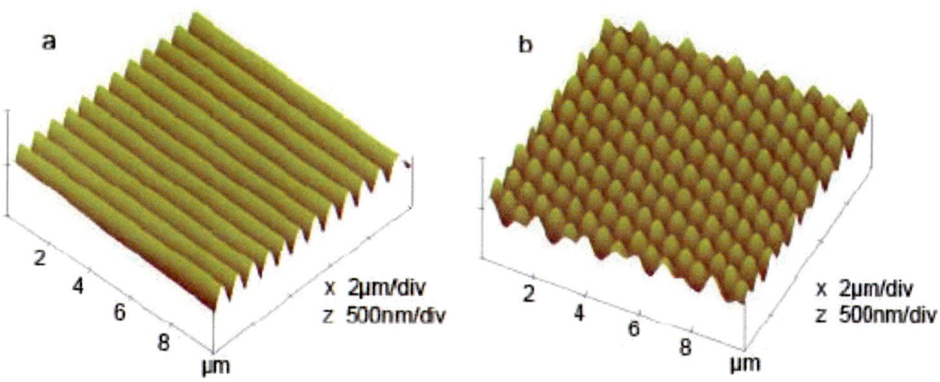

Figure 7. Atomic force microscopy (AFM) images of the surface relief gratings formed on spin-coated azo polymer films: (a) 3-D view of the SRG, (b) orthogonal gratings

3.1.1. Azobenzene Para-Substituent Effects

Figure 8 shows the diffraction efficiency as a function of irradiation time for three epoxy-based azo polymers containing different azo chromophores. The azo polymers BP-AZ-CA, BP-AZ-CN and BP-AZ-NT (see Scheme 2), which distinguishes with each other only by electron-withdrawing groups of the azo chromophores, show significantly different inscription rates of SRG formation.[25] With the same backbone, the polymer containing 4-nitroazobenzene chromophores and 4-carboxyazobenzene exhibit the slowest and fastest inscription rates for SRG formation among the three polymers. The increases of the diffraction efficiency of the SRGs with irradiation time are saturated for BP-AZ-CA, BP-AZ-CN and BP-AZ-NT at 3000s 3300 s and 5000 s respectively. The λ_{max} of BP-AZ-NT is in range 482-483 nm, which is nearest to the 488 nm irradiation of the Ar+ laser. However, the SRG forming rate is obviously slower than the rate of BP-AZ-CA or BP-AZ-CN bearing 4-cyanoazobenzene or 4-carboxyazobenzene chromophores (λ_{max} at 449-450 nm or 431-435 nm). For acrylate-based (see Scheme 4) and urethane-based azo polymers (see Scheme 5), the

SRG forming rates of polymer bearing 4-nitroazobenzene are also obviously slower than the rates of polymers bearing 4-cyanoazobenzene or 4-carboxyazobenzene chromophores. This observation contradicts the simple deduction that irradiation at λ_{max} could accelerate the trans-cis-trans circles and increase the inscription rates.

A plot of the diffraction efficiency as a function of irradiation time for the hyperbranched azo polymers is given in Figure 9.

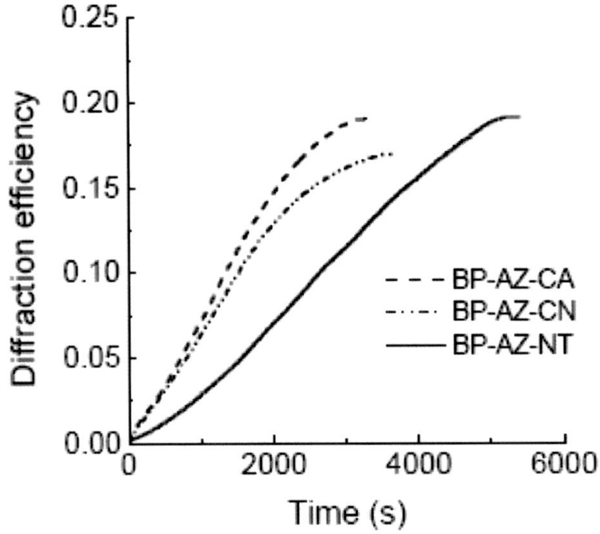

Figure 8. Diffraction efficiency as a function of irradiation time for three kinds of epoxy-based azo polymers BP-AZ-CA, BP-AZ-CN and BP-AZ-NT containing different electron withdrawing groups at *para*-positions of azobenzene units

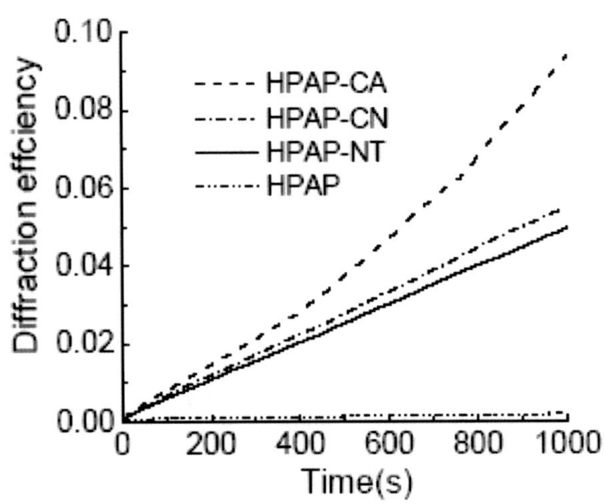

Figure 9. Diffraction efficiency of the surface gratings inscribed on the films of HPAP, HPAP-NT, HPAP-CN, and HPAP-CA as a function of irradiation time

The precursor polymer HPAP exhibits the slowest inscription rate for the SRG formation among the four polymers. The rates of SGR formation are much faster for HPAP-NT, HPAP-

CA, and HPAP-CN (see Scheme 9), which are functionalized by a large number of peripheral azo chromophores.[26] The amount of azo chromophores in the polymers increases as introducing terminal azobenzene chromophores. It is even more important that these peripheral azo chromophores have less steric hindrance and are easier to undergo the photo-induced trans-cis isomerization. HPAP-CA, HPAP-NT and HPAP-CN also show a considerable difference in the grating-forming rates, which distinguishes with each other only by electron-withdrawing groups of the terminal azo chromophores. With the same backbone structure, HPAP-CA shows the fastest inscription rate and HPAP-NT shows the slowest inscription rate among the series.

For both the linear and hyperbranched azo polymers, the rate of SRG formation is closely related with the electron-withdrawing groups in the azobenzene units. The linear polymer containing 4-carboxyazobenzene and 4-cyanoazobenzene groups shows an obviously faster SRG formation rate than that of polymer containing 4-nitroazobenzene groups. The similar tendency exists for the hyperbranched polymers with those types of azo chromophores in the terminal positions. Although hyperbranched azo polymers are obviously different with their linear analogs in backbone architecture and located positions of the azo chromophores, the influence of the electron withdrawing groups on the grating-forming rates appears to be the same. This rule can be confirmed for the amorphous molecular materials (the chemical structures are shown in Figure 10).[70] The molecule containing 4-cyanoazobenzene groups also shows an obviously faster SRG formation rate than that of molecule containing 4-nitroazobenzene groups. Those azo compounds only contain one azobenzene unit in each molecule and possess a much lower molecular weight than those of the azo polymers. From all above experimental results obtained from different systems, we believe that the effect of the electron-withdrawing groups on the SRG formation rate is a general rule. However, the explanation of this observation is still unclear to us at the present stage.

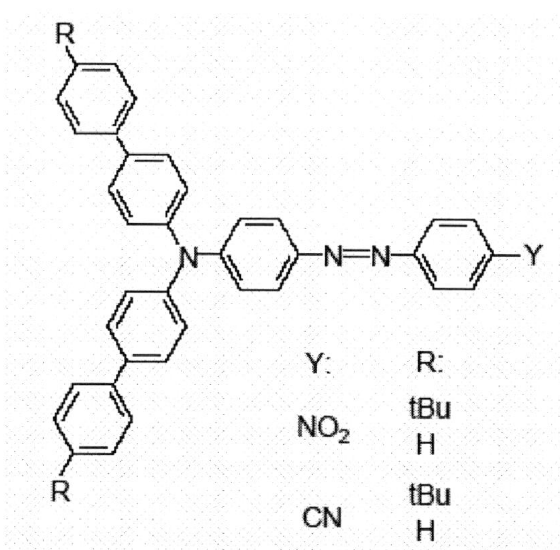

Figure 10. Chemical structures of azo amorphous molecular materials

3.1.2. Azobenzene Ortho-Substituent Effect

In addition to the effect of the *para*-substituents of the azobenzene units on the SRG formation behavior, the substitution in the *ortho*-positions of the azo groups also shows a significant influence on the SRG-forming properties. [27]

Figure 11 shows different inscription rates for BP-35-AZ-NT, BP-3-AZ-NT, and BP-AZ-NT (see Scheme 3), which have the same electron-withdrawing groups and are different only in the number of the methyl substituents in *ortho*-positions of the azo groups. BP-35-AZ-NT with two methyl substituents shows the slowest SRG-forming rate. BP-AZ-NT without the substitution shows the fastest formation rate. A similar result can be seen for BP-35-AZ-CA, BP-3-AZ-CA, and BP-AZ-CA. In this series, the *p*-position electron withdrawing groups are changed to the carboxyl groups, the methyl substitution shows the same effect to decrease the SRG-forming rate.

The relationship of the diffraction efficiency to the writing time for BP-35-AZ-NT and BP-3-AZ-mNT (see Scheme 3) is shown in Figure 12(a). BP-3-AZ-mNT shows an obviously faster inscription rate than that of BP-35-AZ-NT. Both polymers contain azo chromophores bearing the same type *p*-position electron-withdrawing groups and two methyl groups in the *ortho*-positions of the azo groups. In BP-35-AZ-NT, the two ortho methyl substituents are in the anilino moieties. On the other hand, the two ortho methyl substituents of BP-3-AZ-mNT are in the two benzene rings in the different sides of the azo groups. For the polymers containing azo chromophores with the same type electron-withdrawing group and one methyl group in the ortho-positions of azo groups, BP-AZ-mNT shows a faster SRG-forming rate than BP-3-AZ-NT (Figure 12(b)). The results indicate that the *o*-methyl substituents in the anilino moiety side show a more significant effect to decrease the SRG-forming rates. Compared to the strong electron donors and acceptors such as the amino and nitro groups, the electron donating effect of methyl groups is relatively weak. Therefore, it is believed that the methyl substitution effect on the SRG-forming rates is mainly resulted from the steric hindrance of the substituents to the trans-cis isomerization.

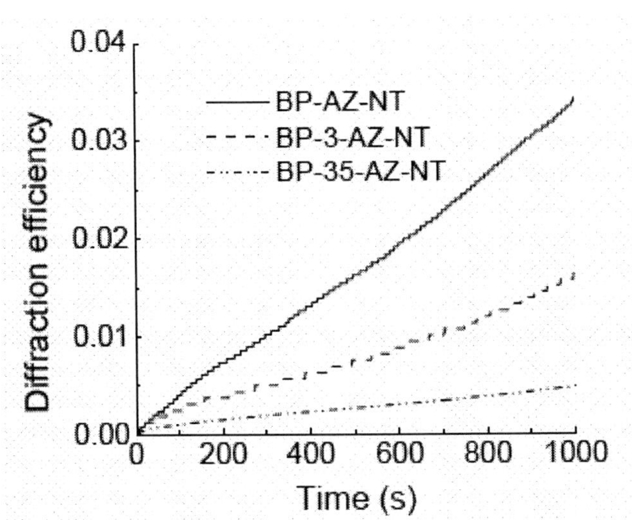

Figure 11. Diffraction efficiency as a function of the irradiation time for the azo polymers: BP-35-AZ-NT, BP-3-AZ-NT, and BP-AZ-NT

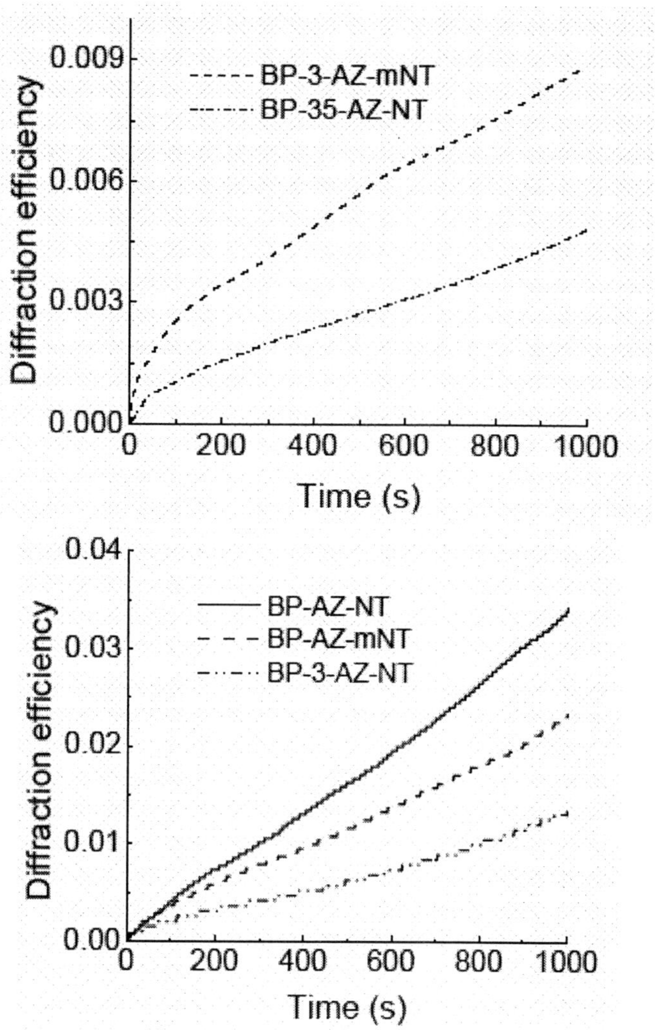

Figure 12. Diffraction efficiency as a function of the irradiation time for two series of azo polymers: (a) BP-35-AZ-NT and BP-3-AZ-mNT; (b) BP-AZ-NT, BP-AZ-mNT and BP-3-AZ-NT

3.1.3. Effect of Backbones and Degrees of Functionalization

Figure 13 shows the diffraction efficiency as a function of the irradiation time for BP-AZ-CA with different degrees of functionalization. For each different DF, the increase of the diffraction efficiency with irradiation time shows a different rate. Figure 14 gives the relationship between the surface modulation and the degree of functionalization for the same series of polymers characterized by atomic force microscopy (AFM). Both results show that the surface modulations at the same irradiation time increase with the increased amount of azobenzene chromophores attached to each polymeric chain.[25] Fukuda et al. has reported that the SRG formation rate exhibited an S-shaped dependence on the azo functionalization.[23] After azo chromophore density reaches 40-50 wt%, the inscription rate was almost independent of the degree of functionalization. Even for fully functionalized epoxy-based polymers, the azo chromophore density is lower than 40-50 wt%, therefore, the result can not be directly compared with those reported by Fukuda.

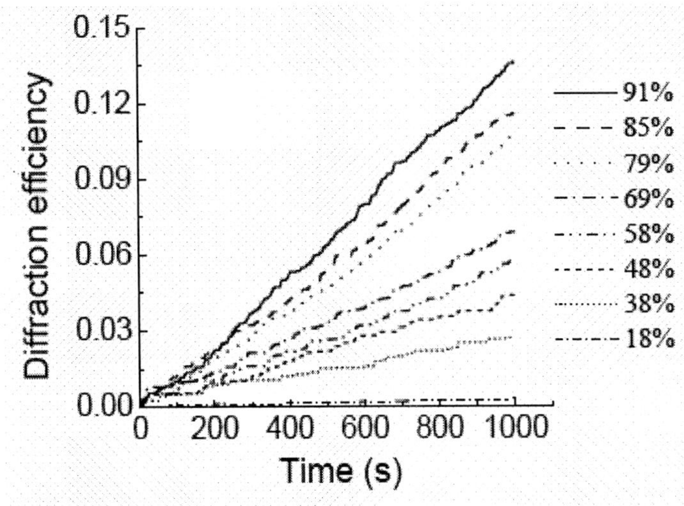

Figure 13. Diffraction efficiency as a function of the irradiation time for BP-AZ-CA with different degrees of functionalization

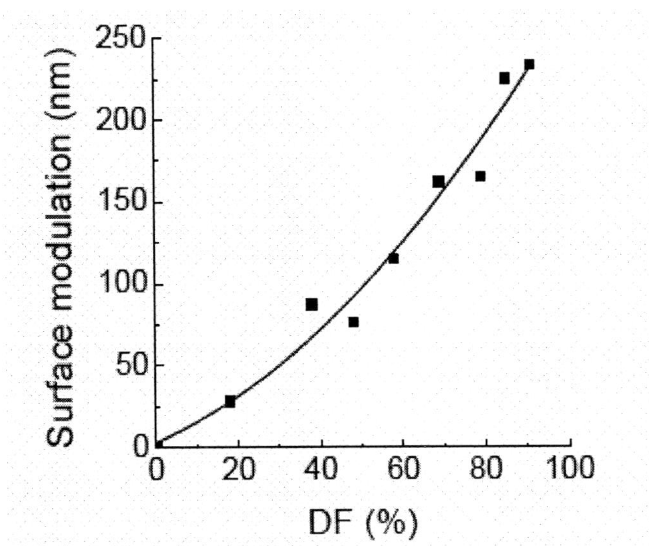

Figure 14. Relationship between surface modulation and degree of functionalization for BP-AZ-CA measured from AFM images after 1000 s irradiation

The effect of the backbones on the inscription rates of the azo polymers can be demonstrated by comparing the SFG forming rates of the azo polymers with the same type of chromophores and different type of backbones (Figure 15).[25] Although the type of azo chromophores and the way they attached to the backbones are the same, the inscription rates of PP-AZ-NT, DP-AZ-NT, and BP-AZ-NT (see Scheme 2) are obviously different. PP-AZ-NT exhibits the fastest inscription rate and BP-AZ-NT shows a slightly slower rate, while DP-AZ-NT possesses the slowest inscription rate. The fastest inscription rate of PP-AZ-NT can be attributed to the highest chromophore loading density among the three polymers. The slower rate of DP-AZ-NT compared with BP-AZ-NT can be attributed to the backbone rigidity of DP-AZ-NT that affects the effective viscosity of the system.

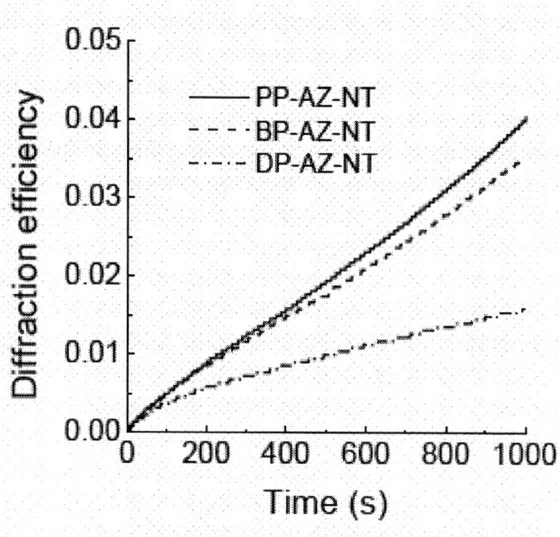

Figure 15. Diffraction efficiency as a function of the irradiation time for three kinds of epoxy-based azo polymers PP-AZ-NT, DP-AZ-NT, and BP-AZ-NT with the same type of azo chromophores

3.2. SRG Behavior of ELBL Self-Assembly Films

For SRG studies, the azo polymers are generally used in the form of thin films prepared by casting or spin-coating methods. The layer-by-layer growth methods, which allow structural control at the molecular level, can lead to thin films with controllable thickness and architecture. Electrostatic layer-by-layer (ELBL) self-assembly, which builds up thin films through alternate deposition of a substrate in dilute solutions of cationic and anionic polyelectrolytes, is a simple, versatile and effective technique to fabricate thin multilayer films.[48-50] The thickness and composition of the films can be easily controlled by this technique. Upon exposure to an interference pattern of Ar^+ laser beams at modest intensities, the significant surface modulation can be observed on the ELBL films.[39, 71] The ELBL self-assembly process can be readily extended to a wide variety of azo polyelectrolytes for optical applications.

In a typical process, the ELBL selfassembly is carried out in water solutions [48-50].However, the functional polyelectrolytes, such as azo polyelectrolytes, have a very poor solubility in water. Therefore, the ELBL selfassembly of azo polyelectrolytes is carried out in a polar organic solvent such as DMF or DMF-H_2O mixed solvents.[28, 30] This process can also be used for other azo polyeletrolytes or functional polyelectrolytes with poor solubility in water.

3.2.1. Multilayer Films Preparation

For multilayer preparation, the azo polyelectrolytes PA-AZ-CA, PBANT-AC, or PBACT-AC used as the polyanion (see Scheme 4 and Scheme 7) were dissolved in anhydrous DMF and a series of DMF-H_2O mixed solvents. Poly(diallyldimethyl- ammonium chloride) (PDAC, MW 20000~35000, 20% solution, Aldrich) was used as the polycation and diluted to a concentration of 0.1 mmol/L (repeated unit) with Milli-Q water (resistivity

>18MΩ). The Quartz slides and silicon wafers used as adsorption substrates and were treated by standard methods.[34-36] A freshly treated quartz slide was first dipped in the polycation solution for 10 min. The slide was rinsed with the Milli-Q water and blown dry with an air stream. And then the slide was dipped in the polyanion solution for 10 min and thoroughly rinsed with the solvent and blown dry. During the process, anhydrous DMF or DMF-H_2O mixed solvents with the same DMF to H_2O ratio were used to rinse the multilayers formed after the deposition of azo polyelectrolytes on the substrates (both quartz slides and silicon wafers). This process was repeated until the multilayers with the required number of layers were obtained. The UV-vis absorption spectra of the multilayers were measured after each monolayer (polycation or polyanion layer) was formed. In order to monitor the thickness of the multilayers formed, the silicon wafers were dipped and washed under the same condition. The thickness of the multilayer films formed on the silicon wafers was measured by ellipsometry.

3.2.2. Formation and Characterization of the Multilayer Films

All the azo polyelectrolytes, such as PBANT-AC, PBACT-AC, and PA-AZ-CA (see Scheme 4 and Scheme 7), can form uniform multilayer films through an electrostatic layer-by-layer adsorption process by using dipping solutions of the polyelectrolytes in anhydrous DMF and in DMF-H_2O mixed solvents.[28, 30] A linear increase of the maximum absorbance with the increase of the layer-pair number can be observed for these systems, which indicates that the multilayers on the substrates grew an equal amount after each dipping cycle. The azo polyelectrolytes can be self-assembled in a layer-by-layer manner by using a solution of anhydrous DMF and DMF-H_2O mixed solvents.

The solvents used for the polyanions have a significant influence on the aggregation states of the azo chromophores and film thickness of the multilayer films. Figure 16a and 16b show UV-vis spectra of PBANT-AC/PDAC multilayer films with the increasing number of deposition cycles, using anhydrous DMF and a DMF-H_2O mixed solvent (DMF:H_2O =2:1, v:v) as the solvents for the dipping solutions. A significant blue-shift of the maximum absorbance band can be seen for the latter, which indicates a high degree of H-aggregation existing in the multilayer films. Although a linear increase of the maximum absorbance can be observed as the number of layer-pairs increases no matter for the anhydrous DMF dipping solution and the DMF-H_2O dipping solutions, the incremental absorbance for each layer-pair is considerably dependent on the solvents (Figure 17). Thickness of the multilayers assembled on silicon wafers, measured by ellipsometry, also exhibits a linear increase with the number of adsorbed layer-pairs (Figure 18). The incremental rates are dependent on the solvents of dipping solutions, which suggests that the thickness of the layer-pairs containing the same azo polyelectrolyte can be significantly adjusted by the ratio of DMF to H_2O.

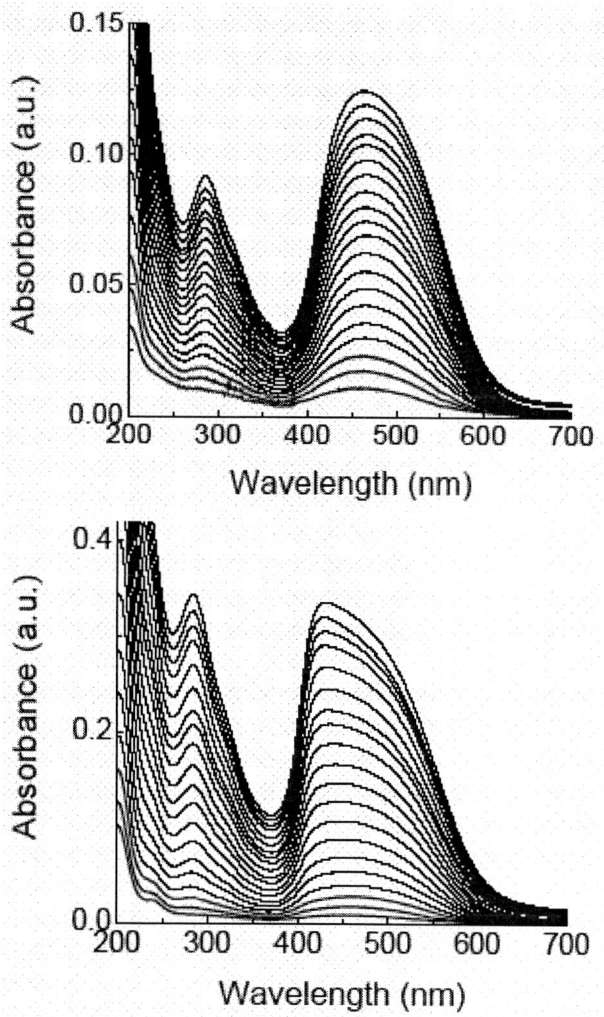

Figure 16. UV-vis spectra of PBANT-AC/PDAC multilayers changing with the deposition cycles: (a) PBANT-AC in anhydrous DMF; (b) PBANT-AC in a DMF-H_2O solution (DMF:H_2O = 2:1, v:v). The concentrations are 0.2 mg/mL for both. The chemical structures of PBANT-AC and PBANT-AC are shown in Scheme 7

The solvents for azo polyelectrolytes also influence the surface roughness and morphology of the multilayer films. The AFM images for the PBANT-AC/PDAC multilayers (20 layer-pairs) on quartz slides recorded in the tapping mode are given in a 3-D view (Figure 19). As it can be seen from the figure, the film surface of PBANT-AC/PDAC obtained by deposition in the DMF solution of PBANT-AC is relatively smooth with a small-scale fluctuation on the surface. On the contrary, for the multilayer obtained by deposition in the DMF-H_2O solution (DMF:H_2O=2:1, v:v) of PBANT-AC, the surface is rough. Multilayer films composed of PA-AZ-CA/PDAC show a similar result (Figure 20). The surface of the mutilayers assembled in DMF solution is much smoother than that of multilayers assembled in mixed solution.

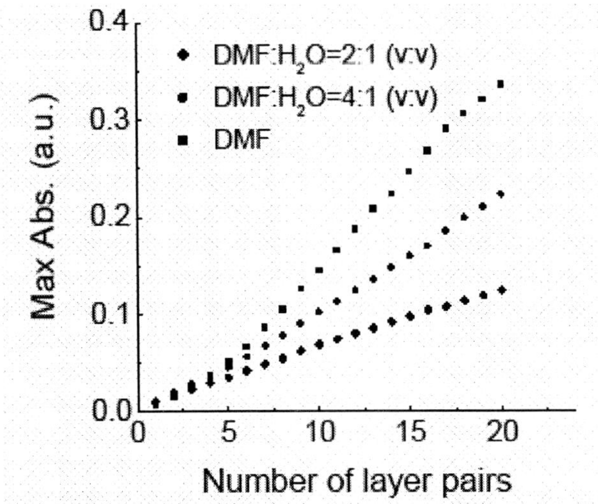

Figure 17. Relationship between the maximum absorbance of the PBANT-AC/PDAC multilayers and the number of layer-pairs. Concentrations of the polyelectrolytes in the dipping solutions are all 0.2 mg/mL. The solvents are anhydrous DMF (■), DMF:H_2O = 4:1 (v:v) (●), and DMF:H_2O = 2:1 (v:v) (♦), respectively

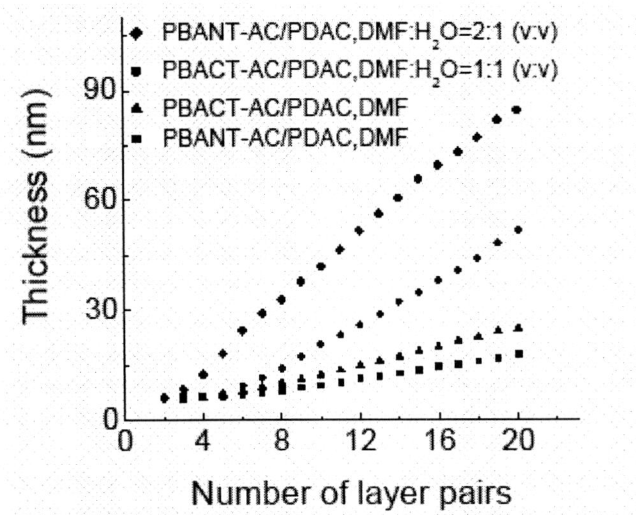

Figure 18. Ellipsometric thickness vs the number of layer-pairs on silicon wafers prepared from different dipping solutions: PBANT-AC in DMF-H_2O solution (DMF:H_2O = 2:1,v:v) (♦), PBANT-AC in anhydrous DMF solution (■), PBACT-AC in DMF-H_2O solution (DMF:H_2O = 1:1,v:v) (●), PBACT-AC in anhydrous DMF solution (▲). The concentrations of the polyelectrolytes in the solutions are all 0.2 mg/mL

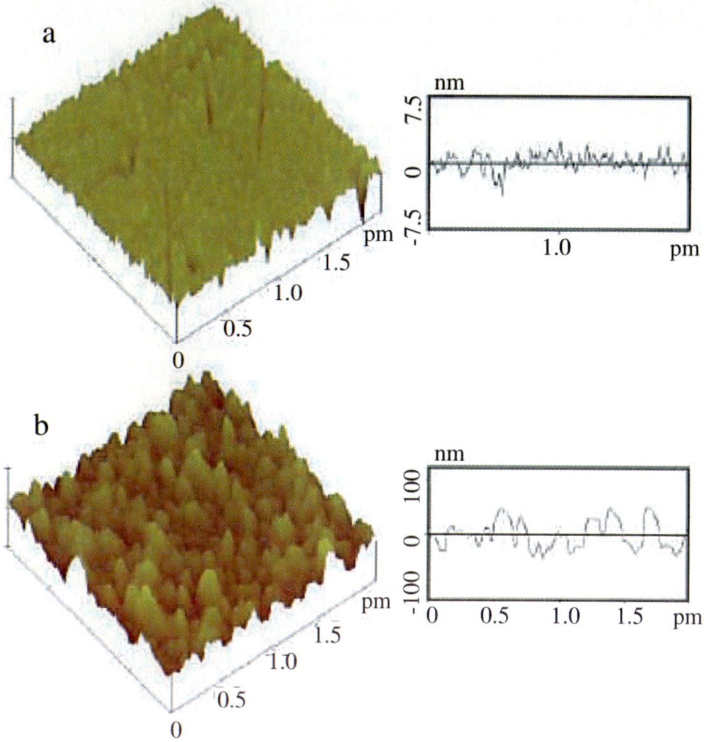

Figure 19. AFM images of the self-assembled films on quartz slides (20 layer pairs), obtained from (a) PBANT-AC in DMF; (b) PBANT-AC in DMF/H_2O (V_{DMF}/V_{water}= 2:1)

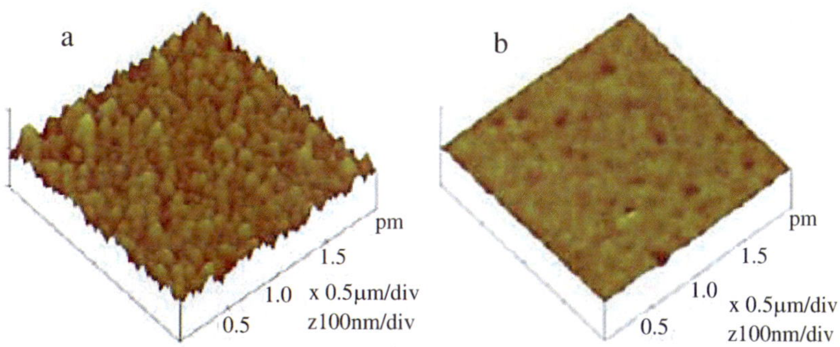

Figure 20. AFM images of the PDAC/PA-AZ-CA multilayers, obtained from (a) DMF/H_2O=1:10 mixed solution, (b) DMF solution

3.2.3. Photo-Induced Surface-Relief-Gratings

Photoinduced surface-relief-gratings can be observed for those self-assembly thin films under the same conditions as described in the above section. Figure 21 shows the AFM images of the 20-layer-pair PA-AZ-CA/PDAC multilayer films surface after exposing the sample to an interference pattern of Ar^+ laser beams at modest intensities (150 mW/cm^2). Figure 22 shows a typical surface-relief-grating (SRG) image for a spin-coated film of PA-AZ-CA. Comparing with the AFM image of SRG on the spin-coated film of PA-AZ-CA, the surface modulation on the multilayer surface is relatively small. The result is consistent with those reported before for fabricating SRGs on electrostatic layer-by-layer multilayer films of polyelectrolytes containing azo chromophores.[34] The result can be attributed to the strong electrostatic attraction between layers in the multilayer films, in which the light-driven mass transport process is more difficult to happen. Even the quality of SRGs is still not as good as spin-coated films, the electrostatic layer-by-layer self-assembly method can offer an easy way to control the thickness of thin film and a possible new route to explore the mechanism of SRG formation.

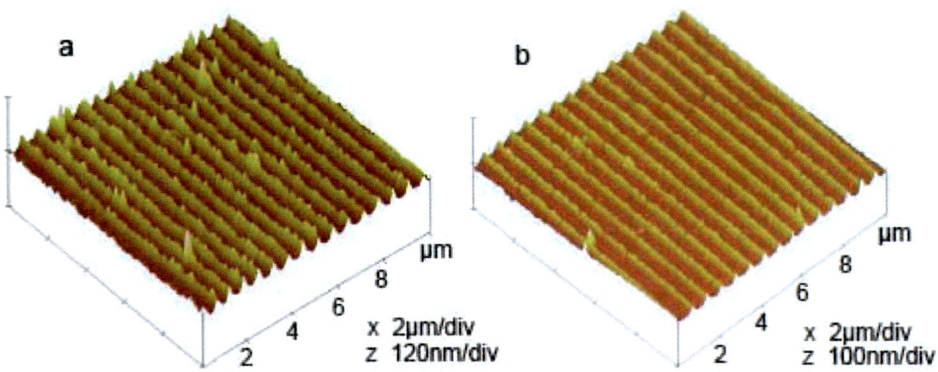

Figure 21. Atomic force microscopy (AFM) images of the surface-relief-gratings formed on the multilayers from (a) DMF/H_2O=1:10 mixed solution, (b) DMF solution

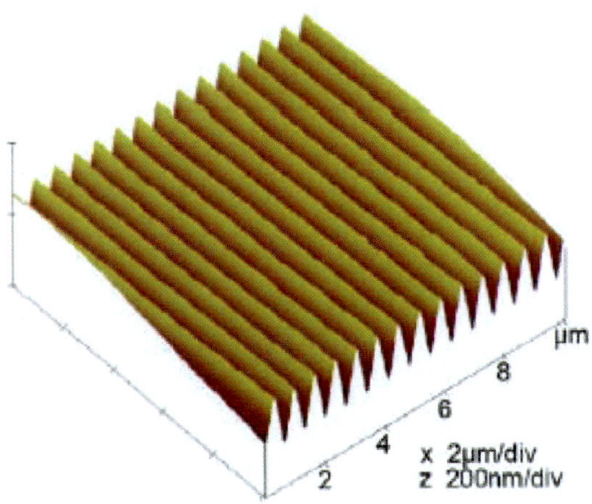

Figure 22. Surface-relief-grating AFM image for a spin-coated film of PA-AZ-CA

Under the similar condition, surface modifications can also be observed for the PBANT-AC/PDAC and PBACT-AC/PDAC multilayer films. For the spin-coated films, the rate of SRG formation is also closely related with the electron-withdrawing groups in the azobenzene units. The polymer containing 4-carboxyazobene groups shows an obviously faster SRG formation rate than that of polymer containing 4-nitroazobenzene groups. After Ar^+ laser beam irradiation with the same intensity and irradiation time, the SRGs depth of the PBACT-AC/PDAC multilayer films is obviously larger than that of PBANT-AC/PDAC multilayer films. This tendency is the same as those observed for the spin-coated thin films of azo polymers as described in the above section.

Conclusion

A variety of azo polymers were synthesized by methods such as postpolymerization azo-coupling reaction, functionalization of poly(acryloyl chloride), and polymerization through azo-coupling reactions. Surface-relief-gratings (SRGs) forming behavior was characterized by the SRG inscription rates and the saturation levels for both the spin-coated films and electrostatic selfassembled multilayer films. Results show that the rates and saturation levels of SRG formation can be significantly influenced by the polymer structures, which include types of the electron-withdrawing groups on the *pare*-positions of the azobenzene units, the number of methyl substitutes on the ortho-positions of the azo groups, the backbone structures of the polymers. The electrostatic layer-by-layer self-assembly films can form SRGs under a similar condition. The thickness and composition of ELBL films can be well controlled, but the surface modulation efficiency is still lower than the spin-coated films.

Acknowledgment

The financial support from NSFC under projects 50533040, 20504017, 20374033, 59925309, 50273018, 59873013 is gratefully acknowledged.

References

[1] Zollinger, H. *Colour Chemistry. Synthesis, Properties, and Applications of Organic Dyes*, VCH, Weinheim, 1987.

[2] Rau, H. *Photochemistry and Photophysics*; Rabek, J. F., Ed.; CRC Press: Boca Raton, FL, 1990; Vol. II. Chap. 4. p 119-141.

[3] Kumar, GS; Neckers, DC. Photochemistry of Azobenzene-Containing Polymers. *Chem. Rev.*, 1989, 89, 1915-1925.

[4] Xie, S; Natansohn A; Rochon P. Recent developments in aromatic azo polymers research. *Chem. Mater.*, 1993, 5, 403–411.

[5] Natansohn, A; Rochon, P. Photoinduced Motions in Azo-Containing Polymers, *Chem. Rev.,* 2002, 102, 4139-4175.

[6] Cojocariu, C; Rochon, P. Light-induced motions in azobenzenecontaining polymers, *Pure Appl. Chem*., 2004, 76, 1479–1497.

[7] Delaire, JA; Nakatani, K. Linear and Nonlinear Optical Properties of Photochromic Molecules and Materials. *Chem. Rev*., 2000, 100, 1817–1845.

[8] Ikeda, T; Tsutsumi, O. Optical Switching and Image Storage by Means of Azobenzene Liquid-Crystal Films. *Science*, 1995, 268, 1873–1875.

[9] Ichimura, K; Oh, SK; Nakagawa, M. Light-driven motion of liquids on a photoresponsive surface. *Science*, 2000, 288, 1624–1626.

[10] Hugel, T; Holland, NB; Cattani, A; Moroder, L; Seitz, M; Gaub, HE. Single-molecule optomechanical cycle. *Science*, 2002, 296, 1103–1106.

[11] Rochon, P; Batalla, E; Natansohn, A. Optically Induced Surface Gratings on Azoaromatic Polymer-Films. *Appl. Phys. Lett.*, 1995, 66, 136–138.

[12] Kim, DY; Tripathy, SK; Li, L; Kumar, J. Laser-Induced Holographic Surface-Relief Gratings on Nonlinear-Optical Polymer-films. *Appl. Phys. Lett.*, 1995, 66, 1166–1168.

[13] Ramanujam, PS; Holme, NCR; Hvilsted, S. Atomic force and optical near-field microscopic investigations of polarization holographic gratings in a liquid crystalline azobenzene side-chain polyester. *Appl. Phys. Lett.*, 1996, 68, 1329-1331.

[14] Jiang, XL; Li, L; Kumar J; Kim, DY; Tripathy SK. Unusual polarization dependent optical erasure of surface relief gratings on azobenzene polymer films. *Applied Physics Letters,* 1998, 72, 2502-2504.

[15] Barrett, C; Natansohn, A; Rochon, P. Mechanism of optically inscribed high-efficiency diffraction gratings in azo polymer films. *J. Phys. Chem.*, 1996,100, 8836-8842.

[16] Bian, S; Li, L; Kumar, J; Kim, DY; Williams, J. Tripathy SK. Single laser beam-induced surface deformation on azobenzene polymer films. *Applied Physics Letters*, 1998, 73, 1817-1819.

[17] Yager, KG; Barrett, CJ. All-optical patterning of azo polymer films, *Current Opinion in Solid State and Materials Science*, 2001, 5, 487–494.

[18] Barratt, CJ; Rochon, P; Natansohn, A. Model of laser-driven mass transport in thin films of dye-functionalized polymers. *J. Chem. Phys.*, 1998, 109, 1505-1516.

[19] Pedersen, TG; Johansen, PM; Holme, NCR; Ramanujam, PS; Hilvsted, S. Mean-field theory of photoinduced formation of surface reliefs in side-chain azobenzene polymers. *Phys. Rev. Lett.,* 1998, 80, 89-92.

[20] Lefin, P; Fiorini, C; Nunzi, JM. Anisotropy of the photo-induced translation diffusion of azobenzene dyes in polymer matrices. *Pure Appl. Opt.*, 1998, 7, 71-82.

[21] Kumar, J; Li, L; Jiang, XL; Kim, DY; Lee, TS; Tripathy, SK. Gradient force: The mechanism for surface relief grating formation in azobenzene functionalized polymers. *Appl. Phys. Lett.*, 1998, 72, 2096-2098.

[22] Lee, TS; Kim, DY; Jiang, XL; Li, L; Kumar, J; Tripathy, SK. Photoinduced surface relief gratings in high-Tg main-chain azoaromatic polymer films. *Journal of polymer science part A-polymer chemistry*, 1998,36,283-289.

[23] Fukuda, T; Matsuda, H; Shiraga, T; Kimura, T; Kato, M; Viswanathan, N; Kumar, J; Tripathy, SK. Photofabrication of surface relief grating on films of azobenzene polymer with different dye functionalization. *Macromolecules*, 2000, 33, 4220-4225.

[24] Helgert, M; Wenke, L; Hvilsted, S; Ramanujam PS. Surface relief measurements in side-chain azobenzene polyesters with different substituents. *Appl. Phys. B,* 2001, 72, 429-433.

[25] He, YN; Wang, XG; Zhou XQ. Epoxy-based azo polymers: synthesis, characterization and photoinduced surface-relief-gratings, *Polymer,* 2002, 43, 7325–7333.

[26] Che, PC; He, YN; Wang XG. Hyperbranched Azo-Polymers Synthesized by Azo-Coupling Reaction of an AB_2 Monomer and Postpolymerization Modification, *Macromolecules*, 2005, 38, 8657-8663.

[27] He, YN; Yin JJ; Che PC; Wang XG. Epoxy-based polymers containing methyl-substituted azobenzene chromophores and photoinduced surface relief gratings, *European Polymer Journal*, 2006, 42, 292–301.

[28] He, YN; Wang, HP; Tuo XL; Deng W; Wang XG. Synthesis, self-assembly and photoinduced surface-relief gratings of a polyacrylate-based Azo polyelectrolyte, *Optical Materials*, 2004, 26, 89-93.

[29] He, YN; Wang, XG; Zhou, QX, Photodynamic Polymers Containing Azo Chromophores Both in Main Chain and Side Chain, *Polymer Preprints* 2001, 42(2), 274-275.

[30] Wang, HP, He, YN; Tuo, XL; Wang, XG. Sequentially Adsorbed Electrostatic Multilayers of Branched Side-Chain Polyelectrolytes Bearing Donor-Acceptor Type Azo Chromophores, *Macromolecules*, 2004, 37, 135-146.

[31] Cheng, H; Tuo, XL; Wang, GJ; Wang, XG. A Novel Polyelectrolyte with Branched Azo Side Chains: Synthesis, Characterization and Self-Assembled Nanostructures, *Macromol. Chem. Phys*., 2001, 202, 3530-3535.

[32] Wang, GJ; He, YN; Wang, XG; Jiang, L. Self-assembly and optical properties of poly(acrylic acid)-based azo polyelectrolyte, *Thin Solid Films*, 2004, 458, 143-148.

[33] Wu, LF; Tuo, XL; Cheng, H; Chen, Z; Wang XG. Synthesis, Photoresponsive Behaviors and Self-assembly of Poly(acrylic acid) Based Azo Polyelectrolytes, *Macromolecules,* 2001, 34, 8005-8013.

[34] Li, YB; Deng, YH; He, YN; Tong, XL; Wang, XG. Amphiphilic Azo Polymer Spheres, Colloidal Monolayers, and Photoinduced Chromophore Orientation, *Langmuir*, 2005, 21, 6567-6571.

[35] Wang, GJ; Cheng, H; Wang, XG. Synthesis and properties of a photoresponsive azobenzene-containing hyperbranched polymer. *Chem. Lett*. 2002, 1, 78-79.

[36] He, YN; Wang, XG; Zhou, QX. Synthesis and Characterization of a Novel Photoprocessible Hyperbranched Azo Polymer. *Synthetic Metals* 2003, 132, 245-248.

[37] Che, PC; He, YN; Zhang, Y; Wang, XG. Synthesizing Hyperbranched Azo Polymer through Azo-coupling Reaction. *Chem. Lett*. 2004, 33(1), 22-23.

[38] Viswanathan, NK; Kim, DY; Bian, S; Williams, J; Liu, W; Li, L; Samuelson, L; Kumar, J; Tripathy, S. Surface relief structures on azo polymer films. *J. Mater. Chem*., 1999, 9, 1941-1955.

[39] Oliveira, JrON; Santos, JrDS; Balogh, DT; Zucolotto, V; Mendonca, CR. Optical storage and surface-relief gratings in azobenzene-containing nanostructured films, *Advances in Colloid and Interface Science*, 2005, 116, 179-192.

[40] Wang, XG; Chen, J; Marturunkakul, S; Li, L; Kumar, J; Tripathy, SK. Epoxy-based nonlinear optical polymers functionalized with tricyanovinyl chromophores. *Chem. Mater.,* 1997, 9, 45-50.

[41] Wang, XG; Kumar, J; Tripathy, SK; Li, L; Chen, J; Marturunkakul, S. Epoxy-based nonlinear optical polymers from post azo coupling reaction. *Macromolecules*, 1997, 30, 219-225.

[42] Mandel, M. Polyelectrolytes. In: Mark, FH; Bikales, NM; Overberger, CG; Menges, G(eds) *Encyclopedia of polymer science and engineering*, 2nd edn, vol 10. Wiley, NewYork, 1988, p 739.

[43] Matsuo, ES; Tanaka, T. Kinetics of discontinuous volume–phase transition of gels. *J. Chem. Phys*., 1988, 89(3), 1695-1703.

[44] Dusek, K. *Responsive Gels: Volume Transition* II, Springer: Berlin, 1993.

[45] Klitzing, RV; Tieke, B. Polyelectrolyte Membranes, *Adv. Polym. Sci*., 2004,165,177-210.

[46] Thunemann, AF; Muller, M; Dautzenberg, H; Joanny, JF; Lowen, H. Polyelectrolyte Complexes, *Adv. Polym. Sci.*, 2004,166,113-171.

[47] Forster, S; Abetz, V; Muller, AHE. Polyelectrolyte Block Copolymer Micelles, *Adv. Polym. Sci.*, 2004, 166, 173-210.

[48] Decher, G. Fuzzy nanoassemblies: toward layered polymeric multicomposites. *Science*, 1997, 277,1232-1237.

[49] Ferreira, M; Rubner, MF. Molecular-Level Processing of Conjugated Polymers. 1. Layer-by-Layer Manipulation of Conjugated Polyions. *Macromolecules*, 1995, 28, 7107-7114.

[50] Wang, XG; Balasubramanian, S; Li, L; Jiang, XL; Sandman, DJ; Rubner, MF; Kumar, J; Tripathy, SK. Self-assembled second order nonlinear optical multiplayer azo polymer. *Macromol. Rapid Commun.*, 1997, 18, 451-459.

[51] Wang, XG; Balasubramanian, S; Kumar, J; Tripathy, SK; Li, L.Azochromophore-functionalized polyelectrolytes. 1. Synthesis, characterization, and photoprocessing. *Chemistry of Materials*, 1998, 10, 1546-1553.

[52] Balasubramanian, S;Wang, XG; Wang, HC; Yang, K; Kumar, J; Tripathy, SK; Li, L.Azochromophore-functionalized polyelectrolytes. 2. Acentric self-assembly through a layer-by-layer deposition process. *Chemistry of Materials*,1998, 10, 1554-1560.

[53] Kim, Y; Webster, O. Hyperbranched polyphenylenes. *Macromolecules,* 1992, 25, 5561-5572.

[54] Inoue, K. Functional dendrimers, hyperbranched and star polymers. *Prog. Polym. Sci.,* 2000, 25, 453-571.

[55] Jikei, M; Kakimoto, M. Hyperbranched polymers: a promising new class of materials. *Prog. Polym. Sci.*, 2001, 26, 1233-1285.

[56] Gao, C; Yan, D. Hyperbranched polymers: from synthesis to applications. *Prog. Polym. Sci.*, 2004, 29, 183-275.

[57] Mori, H; Seng, D; Lechner, H; Zhang, M. Synthesis and characterization of branched polyelectrolytes. 1. Preparation of hyperbranched poly(acrylic acid) via self-condensing atom transfer radical copolymerization. *Macromolecules*, 2002, 35, 9270-9281.

[58] Kim, YJ. Hyperbranched polymers 10 years after. *Polym. Sci. A: Polym. Chem.*, 1998, 36, 1685-1698.

[59] Hong, CY; Pan, CY. Synthesis and characterization of hyperbranched polyacrylates in the presence of a tetrafunctional initiator with higher reactivity than monomer by self-condensing vinyl polymerization. *Polymer*. 2001, 42, 9385-9391.

[60] Miller, T; Neenan, T; Kwock, E; Stein, S. Dendritic Analogs of Engineering Plastics- A General One-Step Synthesis of dendritic Polyaryl Ethers. *J. Am. Chem. Soc.*, 1993, 115, 356-357.

[61] Turner, S; Voit, B; Mourey, T. All-aromatic hyperbranched polyesters with phenol and acetate end groups: synthesis and characterization. *Macromolecules*, 1993, 26, 4617-4623.

[62] Archut, A.; Azzellini, G.; Balzani, V.; Cola, L.; Vogtle, F. Toward Photoswitchable Dendritic Hosts. Interaction between Azobenzene-Functionalized Dendrimers and Eosin. *J. Am. Chem. Soc.,* 1998, 120(47), 12187-12191.

[63] Junge, D.; McGrath, D. Photoresponsive Azobenzene-Containing Dendrimers with Multiple Discrete State.s *J. Am. Chem. Soc.*, 1999, 121(20), 4912-4913.

[64] Yokoyama, S.; Nakahama, T.; Otomo, A.; Mashiko, S. Preparation and assembled structure of dipolar dendrons based electron donor/acceptor azobenzene branching. *Chem. Lett*., 1997, (11), 1137-1138.

[65] Yokoyama, S.; Nakahama, T.; Otomo, A.; Mashiko, S. Intermolecular Coupling Enhancement of the Molecular Hyperpolarizability in Multichromophoric Dipolar Dendrons. *J. Am. Chem. Soc*., 2000, 122(13), 3174-3181.

[66] Jiang, D; Aida, T. Photoisomerization in dendrimers by harvesting of low-energy photons. *Nature*, 1997, 388(6641), 454-456.

[67] Yokoyama, S.; Nakahama, T.; Otoma, A.; Mashiko, S. Second harmonic generation of dipolar dendrons in the assembled thin films. *Thin Solid Films*, 1998, 331(1-2), 248-253.

[68] Bachman GB; Hellman, H; Robinson, KR; Finholt, RW; Kahler, EJ; Filar, LJ; Heisey, LV; Lewis, LL; Micucci, DD. A new method of preparing substituted vinyl compounds. Depolymerization studies on vinyl polymers. *J. Org. Chem*. 1947, 12, 108-121.

[69] Chattopadhyay, P; Sinha, C; Pal, DK. Preparation and properties of a new chelating resin containing imidazolyl azo groups. *Fresenius Journal of Analytical Chemistry*, 1997, 357 (4), 368-372.

[70] Ishow, E.; Lebon,B.; He, Y.N.; Wang, X.G.; Bouteiller, L.; Galmiche, L.; Delaire, J.; Nakatani, K. Structural and photoisomerization cross studies of polar photochromic monomeric glasses forming surface relief gratings. *Chem. Mater*. 2006, 18,1261-1267.

[71] He, JA; Bian, S; Li, L; Kumar, J; Tripathy, SK; Samuelson, LA. Surface relief gratings from electrostatically layered azo dye films. *Appl. Phys. Lett*., 2000, 76, 3233-3235.

In: Trends in Optical Materials Research
Editor: Owen G. Gardens, pp. 219-255 ISBN: 1-60021-290-5

Chapter 7

STIMULATED EMISSION IN PHOTONIC CRYSTAL

Xu Xingsheng* and Zhang Daozhong

State Key Laboratory of Integrated Optoelectronics, Institute of Semiconductor, Chinese Academy of Sciences, Beijing 100083, China

Optical Physics Laboratory, Institute of Physics and Center for Condensed Matter Physics, Chinese Academy of Sciences, Beijing 100080, China

ABSTRACT

By using of Maxwell equations and the Bloch equations, we derive an analytical expression for stimulated emission in arbitrary 2-D photonic crystals. Using the analytical expression for stimulated emission we analyzed the transient emission, optical gain and gain threshold of two-level atom embedded in arbitrary structure photonic crystals. The transient emission can be expressed by the photon density of the states, the Rabi frequency and the relaxation coefficient. It is found that the sharp structures of the density of states near the band edge leads to a drastically enhancement of the emission and an abrupt change of the oscillation of the emission field when the ratio of Rabi frequency to decay rate is a proper value. Moreover, a transient gain is found to exist if the pumping pulse area is a certain value and the transition frequency is tuned to the band edge. It is also deduced that the optical gain in photonic crystal is proportional to the population inversion, and is inversely proportional to the group velocity and the relaxation coefficient of the phase, which indicates that the gain can be induced by the population inversion and be enhanced drastically due to the small group velocity or anomaly group velocity. Correspondingly, the lasing threshold expressed by the population inversion in arbitrary structure photonic crystals is obtained by an investigation of steady-state laser behavior. It shows that the lasing threshold expressed by population inversion is proportional to the second order of the group velocity, and proportional to the relaxation coefficient. The stimulated emission in the quasi-periodic photonic crystals is also investigated. It is found that there are some interesting phenomenons in the quasi-periodic crystals. The lasing peaks are more abundant and stronger than that in periodic crystals. More stimulated emission peaks appear when the size and the gain increase, and some lasing peaks are independent on incident direction.

*E-mail: xsxu@red.semi.ac.cn, Fax: 86-10-82305052

These lasing peaks maybe come from the localization of the disorder in the quasi-periodic photonic crystals.

1. INTRODUCTION

Spontaneous emission is the most visible manifestation of the dynamical interaction between light and matter. The fact that spontaneous emissions not an intrinsic atomic property, but rather depends on the electromagnetic vacuum environment that the atom is subjected to, is well established [1]. When the atom is in the photonic crystals, its environment is different from the free-space. The interactions of light with atoms embedded in photonic crystals have been extensively studied [2]. The understanding of quantum electrodynamics behaviors of atoms in photonic crystals was an important subject of photonic crystals [3], and some peculiar properties were studied, such as the anomalous lamb shift [4], oscillatory behavior of spontaneous emission, enhanced quantum interference effects [5], and coherent phenomena in photonic crystals [6]. In Ref. [6], the authors discussed the transient behavior of the absorption, and they found a transient gain without population inversion for a three level system by the quantum theory analysis. Much attention was attracted to the spontaneous emission near the band edge of photonic crystals [7]. Konôpka analyzed stimulated emission from the atom when its transition frequency is tuned near the band gap edge in a fully quantized field way [8]. He described a model that the initial multimode filed is a superposition of single photon states with the spectrum of energy localizing near the band edge, and he assumed that the atom was in a fixed position. His intent was to demonstrate the influence of the initial field on the dynamics of the atom-field system. He judged the STE near the band edge by whether the atom predominantly emits into the initially occupied modes, which needs very complicated numerical calculation.

It was demonstrated that the small group velocities of the eigen-mode would bring about an enhancement of the effective gain or low-threshold lasing [9, 10, 11, 12]. The lasing threshold in photonic crystal was also discussed by carrying out a multi-scale analysis of the appropriate Maxwell-Bloch equations [13]. In which the authors focused on band-edge lasers in one-dimensional and two-dimensional (2-D) photonic crystal (PhC) by numerical calculation. These methods relied heavily on numerical simulation or evaluation of the laser action by generalized parameter. K. Sakoda derived an analytical expression for the light amplification in arbitrary PhCs, and he obtained an expression of lasing threshold expressed by the imaginary part of the dielectric constant [14]. The lasing threshold based on the enhancement of electric fields in PhCs was also derived and supplemented by numerical studies [15].

Using the semi-classical theory, we analyzed the oscillation of the emission and the transient gain for a two-level atom embedded in 2-D PhCs when coherent transients dominated the interaction between the light and the atom. Moreover, the stimulated emission from atoms in PhCs was studied with considering background decay. The analytical expressions of the emission fields were deduced. We found that the emission fields were proportional to the density of states, oscillated with the Rabi frequency, and exponentially decreased with background decay. The transient gain was found to exist when the pump pulse area and the decay rate are proper value. When the transition frequency of the atom is resonant with the frequency near the band gap edge, the transient gain and the oscillation of the electric field can be largely enhanced. These effects are due to strongly localized states of light near the band edge and the coherent transient process. Based on the results obtained above, we studied the optical gain and the lasing threshold in 2-D PhC. The obtained gain was proportional to the population inversion, and was inversely proportional to the group velocity and the relaxation coefficient, while the population-inversion threshold is proportional to the relaxation coefficient of the phase and the second order of group velocity. Therefore, the small group velocity will lead to drastically decreasing of lasing threshold.

The quasi-periodic photonic crystal (QPC) possesses rotating period but loses the transferring period. The interaction of light with atom in such a structure with gain is a very interesting topic [16]. It can be found that the theory of stimulated emission obtained for periodic PhC was also applicable to quasi-periodic PhCs. The studies of localization of classical waves in random disordered media have been well understood [17,18], and the wave propagation in amplifying random media has been pursued intensively [19,20]. But the periodically correlated amplifying disordered systems that are generated by introducing the disorder and the gain medium into a periodic system [21]. Between the periodic structure and the disorder structure, the quasi-periodic PhC with gain medium is also an interesting topic. So the stimulated emission in the quasi-periodic PhC was also investigated. It was found that there were some interesting phenomenons in that structure. The lasing peaks were more abundant and stronger than that in periodic crystals. More stimulated emission peaks appeared when the size and the gain increased, and some lasing peaks were independent on incident direction. These lasing peaks maybe come from the localization of the quasi-periodic structure.

The contents in following are arranged as: Section 2, transient gain in the photonic crystals; section 3, transient emission of two-level atoms in photonic crystals; section 4, the enhancement of stimulated emission near a photonic band edge; section 5, lowered laser threshold in photonic crystals; section 6, stimulated emission in dodecagonal quasi-periodic photonic crystals; section 7, conclusion.

2. Transient Gain in the Photonic Crystals

Using the optical response formula of the PhCs and the Bloch equations we analyzed the coherent emission in arbitrary 2-D PhCs. A transient gain was found to exist if the pumping pulse area is a certain value and the transition frequency is tuned to the band edge. It was found that the transient emission not only depended on the photon density of the states, but also on the intensity and the duration of the pumping laser pulse. It showed that the sharp structures of the density of states near the band edge would lead to a drastically enhancement of the transient emission intensity and an abrupt changing of the oscillation of the emission field.

The interaction between the light and the atom can be described by Maxwell and Bloch equations. The Maxwell equations can be changed as [14]:

$$\begin{cases} -\left(\frac{1}{c^2}\frac{\partial^2}{\partial t^2}+\zeta\right)E(\mathbf{r},t)=\frac{1}{c^2\varepsilon_0\varepsilon(\mathbf{r})}\frac{\partial^2}{\partial t^2}P(\mathbf{r},t) \\ \zeta E(\mathbf{r},t)=\frac{1}{\varepsilon(\mathbf{r})}\nabla\times\{\nabla\times E(\mathbf{r},t)\} \end{cases} \tag{1}$$

These equations describe the propagation of electromagnetic radiation through the optical medium. Where $P(\mathbf{r},t)$ is an extrinsic polarization field of the impurity atoms induced by external field, $E(\mathbf{r},t)$ is the electric field in PhCs, and $\varepsilon(\mathbf{r})$, is a periodic dielectric function of the position $\mathbf{r}$. The structure of arbitrary 2-D PhCs is illustrated in Fig.1. We assume that the electromagnetic wave propagates in the plane of periodicity (*xy* plane) with the electric field polarized perpendicular to this plane. In this case, the scalar equations can be adopted to describe the radiation of impurity atom in PhCs [22].

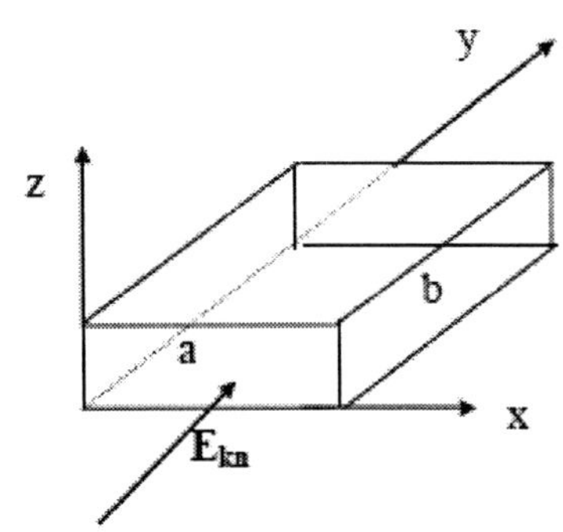

Fig. 1. Illustration of the photonic crystal structure. The pump pulse and the probe light input from facet *a* along direction *y*, propagating in the periodicity plane (*xy* plane), and the signal can be observed by a detector from facet *b*

For an arbitrary structure 2-D phtonic crystal, using equations (1), the emission field of the impurity atom induced by external field can be deduced as [14]:

$$E(\mathbf{r},t)+\frac{P(\mathbf{r},t)}{\varepsilon_0\varepsilon(\mathbf{r})}=\frac{1}{\varepsilon_0 V}\sum_{\mathbf{k}n}E_{\mathbf{k}n}(\mathbf{r})\int d\mathbf{r}'\int_{-\infty}dt' E^*_{\mathbf{k}n}(\mathbf{r}')P(\mathbf{r}',t')\omega_{\mathbf{k}n}\sin\left(\omega_{\mathbf{k}n}(t-t')\right) \tag{2}$$

Where $\omega_{\mathbf{k}n}$ is the eigen-angular frequency, $E_{\mathbf{k}n}(\mathbf{r})$ is the eigen-function of the wave equations, which is characterized by a wave vector $\mathbf{k}$ in the first Brillion zone and a band index *n*.

On the other hand, the Bloch equations that describe the atomic radiation kinetic process can be expressed as:

$$\begin{cases} \dfrac{dU}{dt} = -(\varpi-\omega)V - \dfrac{U}{T_2} \\ \dfrac{dV}{dt} = (\varpi-\omega)U + \dfrac{\mu E_{ex}(r,t)}{\hbar}W - \dfrac{V}{T_2} \\ \dfrac{dW}{dt} = -\dfrac{\mu E_{ex}(r,t)}{\hbar}V - \dfrac{(W-W(0))}{T_1} \end{cases} \tag{3}$$

Where ω, ϖ, μ, $\hbar$ are the frequency optical field, two-level atomic transition frequency, transition dipole moment and the Planck's constant, respectively. U is the real part of the slowly-varying amplitude of the atom polarization, representing the dispersion; V is the imagine part of the slowly-varying amplitude of the atom polarization, representing absorption or gain of the light; W is related to the population inversion with NW (N is the number of the atoms), $W(0)$ corresponds to the initial population inversion, and $1/T_1$ and $1/T_2$ are the longitudinal and the transverse relaxation time, representing the population and the phase relaxation processes, respectively.

When the light interacts with the atoms, various kinds processes may be generated, for example, Rabi oscillations, free induction decay, adiabatic population transfer, and transient phenomenon. When adiabatic criterion is fulfilled, a significant population transfer can be observed. If not, the final population depends crucially on the pulse integral and coherent transients dominates the interaction [23]. It is to say that when coherent light interacts with the atoms and the interaction time is shorter than the decay time of the electric field, the interaction can be regarded as coherent transient. For coherent transient, both the relaxation of the population (T_1) and that of the phase (T_2) can be neglected. The coherent transients result in oscillating population.

Optical Rabi oscillations are temporal oscillations of the population inversion in a two-level system driven by a strong resonant optical field on a time scale shorter than the relaxation time, with an oscillation frequency proportional to the transition dipole moment μ and the electric field amplitude $E_{ex}(\mathbf{r},t)$. Under pulsed excitation, Rabi oscillations manifest themselves as a sinusoidal dependence of the population inversion on the pulse area $\int_{-\infty}^{t} dt\, \mu E_{ex}(\mathbf{r},t)/\hbar$ (the time integrated Rabi frequency) [24].

We assume that the PhC can be regarded as thin sample, or the field is at a particular space point $\mathbf{r}$, then the external field $E_{ex}(\mathbf{r},t)$ can be regarded as constant electric field E_{ex}. We also assume that the laser pulse is a square pulse with pulse duration δt, where the variation of electric field with time and space can be neglected, and then the laser pulse area is

$$A = \frac{\mu E_{ex}}{\hbar}\delta t,\ 0 < t < \delta t. \tag{4}$$

Under above approximations, the optical Bloch equations can be written as:

$$\frac{dV}{dt} = (\varpi - \omega)U + \frac{\mu E_{ex}}{\hbar}W \tag{5}$$

$$\frac{dU}{dt} = -(\varpi - \omega)V \tag{6}$$

$$\frac{dW}{dt} = -\frac{\mu E_{ex}}{\hbar}V \tag{7}$$

Taking differential calculation by t to Eq.(5), substituting Eq.(6) and Eq.(7), then we obtain:

$$\frac{d^2V}{dt^2} + \Omega^2 V = 0 \tag{8}$$

where

$$\Omega = \sqrt{\left(\frac{\mu E_{ex}}{\hbar}\right)^2 + (\varpi - \omega)^2} \tag{9}$$

is the Rabi frequency.

Because $V(0) = 0$, then Eq.(8) can be solved as:

$$V = C\sin(\Omega t), \tag{10}$$

where C is a constant. Substituting Eq.(10) into Eq.(5), Eq.(6) and Eq.(7), taking integration, we obtain:

$$U = W(0)\frac{(\varpi - \omega)\Omega_r}{\Omega}[\cos(\Omega t) - 1], \tag{11}$$

$$V = W(0)\frac{\Omega_r^2}{\Omega}\sin(\Omega t), \tag{12}$$

$$W = W(0)\left\{1 + \left(1 + \frac{\Omega_r}{\Omega}\left[\cos(\Omega t - 1)\right]\right)\right\}, \tag{13}$$

where $W(0)$ is the amplitude of W, $\Omega_r = \left|\frac{\mu E_{ex}}{\hbar}\right|$ is the Rabi frequency as the light resonance with the two level atoms, *i.e.* $\varpi = \omega$. In this case, we have:

$$\begin{cases} U = 0 \\ V = W(0)\sin(\Omega_r t) \\ W = W(0)\cos(\Omega_r t) \end{cases} \tag{14}$$

The polarization $P(\mathbf{r},t)$ of the impurity atom can be expressed as:

$$P(\mathbf{r},t) = N\mu(U - iV). \tag{15}$$

For considering the emission of atom, the polarization can be abbreviated as:

$$P(\mathbf{r},t) = -iN\mu V = -iN\mu W(0)\sin(\Omega_r t). \tag{16}$$

If $V < 0$, *i.e.*, $\sin(\Omega_r t) < 0$, there is a gain.

Substituting $P(\mathbf{r},t)$ into Eq.(2), the emission field in the 2-D PhC can be written as:

$$E(\mathbf{r},t) + \frac{P(\mathbf{r},t)}{\varepsilon_0 \varepsilon(\mathbf{r})} = \frac{1}{i\varepsilon_0 V}\sum_{\mathbf{k}n} E_{\mathbf{k}n}(\mathbf{r})\int_V d\mathbf{r}'\int_{-\infty} dt' E^*_{\mathbf{k}n}(\mathbf{r}')N\mu W(0)\sin(\Omega_r t)\omega_{\mathbf{k}n}e^{-i\omega t'}\sin(\omega_{\mathbf{k}n}(t-t')) \tag{17}$$

Pay attention to that Eq.(16) only represents the slowly-varying amplitude of the atom polarization. The polarization in Eq.(2) and Eq.(17) that varys with time t' should be fast-variation part, and the slowly-varying amplitude does not vary with time t'. When we adopt strict resonant approximation, slowly-varying amplitude approximation, and uniform-material approximation [25], the right hand of Eq.(17), representing the propagation field, can be written as:

$$
\begin{aligned}
E(\mathbf{r},t) &\simeq \frac{NW(0)\mu E_{\mathbf{k}n}(\mathbf{r})}{i\varepsilon_0 V}\sum_{\mathbf{k}n}\omega_{\mathbf{k}n}\sin(\Omega_r t)\int E^*_{\mathbf{k}n}(\mathbf{r}')d\mathbf{r}' \\
&= \frac{NW(0)\overline{E}\mu E_{\mathbf{k}n}(\mathbf{r})}{i\varepsilon_0}\sin(\Omega_r t)\int_0^\infty \sum_{\mathbf{k}n}\omega_{\mathbf{k}n}\delta(\omega-\omega_{\mathbf{k}n})d\omega \\
&\simeq \frac{NW(0)\overline{E}\mu E_{\mathbf{k}n}(\mathbf{r})\omega_{\mathbf{k}n}}{i\varepsilon_0}\sin(\Omega_r t)\rho(\omega)
\end{aligned}
\tag{18}
$$

Where $\rho(\omega)=\sum_{\mathbf{k}n}\delta(\omega-\omega_{\mathbf{k}n})$ is the density of state, and $\overline{E}=\frac{1}{V}\int E^*_{\mathbf{k}n}(\mathbf{r})d\mathbf{r}$. Eq.(18) can be changed as:

$$
E(\mathbf{r},t)=\frac{\beta E_{\mathbf{k}n}(\mathbf{r})}{i}\rho(\omega)\sin(\Omega_r t) \tag{19}
$$

where

$$
\beta=\frac{NW(0)\mu\omega_{\mathbf{k}n}\overline{E}}{\varepsilon_0} \tag{20}
$$

is the amplitude amplification factor for electric field. It is found that the emission is proportional to the density of states and oscillates with the Rabi frequency. There is a phase difference π between the population W and emission field $E(\mathbf{r},t)$, but the transient behavior of V and $E(\mathbf{r},t)$ is qualitatively the same, and they all oscillate with the same frequency Ω_r. The amplification of the light occurs in the time interval in which $\frac{dV}{dt}<0$ and the peak amplification coincides with the time at which V reaches the maximum values.

The optical power is

$$
I=\left|E(\mathbf{r},t)*E(\mathbf{r},t)^*\right|=\beta^2\rho^2(\omega)E^2_{\mathbf{k}n}(\mathbf{r})\left|\sin(\Omega_r t)\right|^2 \tag{21}
$$

It is can be found from Eq.(19) and Eq.(21) that, when the density of states is large, there is an enhancement effect to the emission field, as well as the oscillations.

We consider the density of state near photonic band edge [26]

$$
\rho(\omega)=\frac{1}{2\pi}\frac{1}{\sqrt{\omega-\omega_c}}\theta(\omega-\omega_c) \tag{22}
$$

$\theta(\omega)$ is the step function, ω_c is the upper band gap edge frequency. Substituting Eq.(22) into Eq.(20), we obtain

$$E(\mathbf{r},t)=\frac{\beta}{i}\frac{1}{2\pi\sqrt{\omega-\omega_c}}E_{\mathbf{k}n}(\mathbf{r})\sin(\Omega_r t). \tag{23}$$

Then the optical power of the emission field is

$$I=\frac{\beta^2}{4\pi^2}\frac{1}{\omega-\omega_c}\left|E_{\mathbf{k}n}(\mathbf{r})\right|^2\left|\sin\Omega_r t\right|^2. \tag{24}$$

From above equation, the emission field $E(\mathbf{r},t)$ not only depends on the sharp structure of the band gap edge, but also on the initial state of the atom, the intensity and duration of the pumping laser pulse. Supposing an atom initially in its excited state starts to emit a wave packet, which propagates away from the atom. When this light encounters the rods or the holes of the PhCs, it is reflected back toward or scattered away the atom and adds either constructively or destructively to the light instantly radiated by the atom, as the case may be, to either enhance or inhibit the further atomic emission, similar to the case the atom embed in the cavity [27]. Physically, the atom exchanges energy back and forth with its own radiation with the Rabi frequency Ω_r. It indicates that in Fig.2, if $\sin(\Omega_r t)<0$, the light will be amplified, and there is a transient gain. The light will be amplified at a scale time $t\sim\Omega_r^{-1}$.

It is well known that the emission will be enhanced drastically as the emission frequency resonance with the band edge frequency [28]. As we consider the transient process of the atom emission near the photonic band gap edge, because of quantum interference, as well as the coherent photon localization [29], the transient gain of the emission can be strongly enhanced. The oscillatory behavior is direct consequences of interference effects that result either in an enhancement the radiation rate or in a periodic exchange of excitation between the atom and the field, and it will be strongly enhanced when the atomic resonant frequency lies near the edge of a perfect photonic band gap. The emission intensity as function of $\Omega_r t$ and the detuning between the frequency of eigenmode and that of the band gap edge is illustrated in Fig.3 and Fig.4, respectively. It can be found from those two figures that, as the pump pulse area being a certain value, such as $\Omega_r t=\frac{\pi}{2}$, or $\frac{\pi}{4}$, when the transition frequency of the atom is tuned to the band gap edge frequency, transient emission as well as the oscillation can be drastically enhanced.

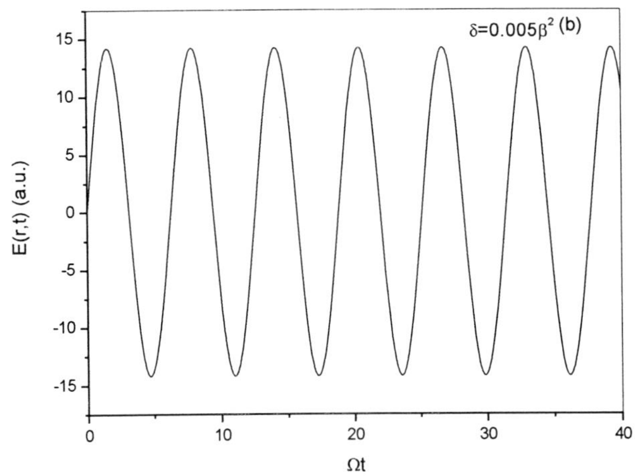

Fig. 2. Electric field of stimulated emission as a function of $\Omega_r t$. Where $\delta = \omega - \omega_c$ is the detuning of the emission frequency from the band gap edge frequency, and $\delta = 0.005\beta^2$. (After [30])

It should be stated that we describe the ideal case here that the relaxation process of the electromagnetic eigenmode are not considered and that the size of PhC is infinite. In fact, for a finite structure, near the band gap edge, the density of states as well as the emission field $E(\mathbf{r},t)$ is not a singular infinite, but a very large value.

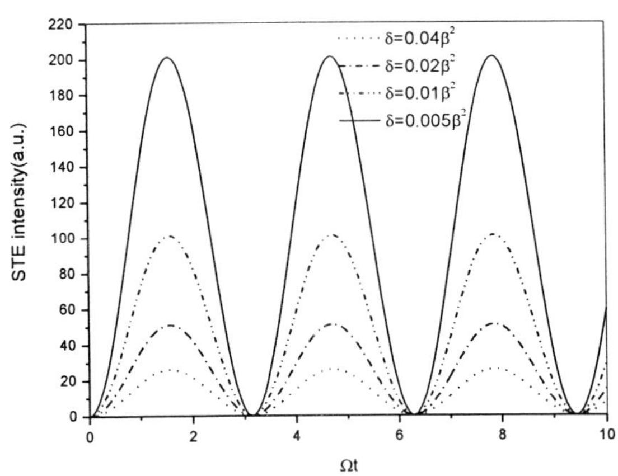

Fig. 3. Intensity of stimulated emission as a function of $\Omega_r t$. $\delta = \omega - \omega_c$ is the detuning of the emission frequency from the band edge frequency. The stimulated emission varies periodically with Rabi frequency with period π. (After [30])

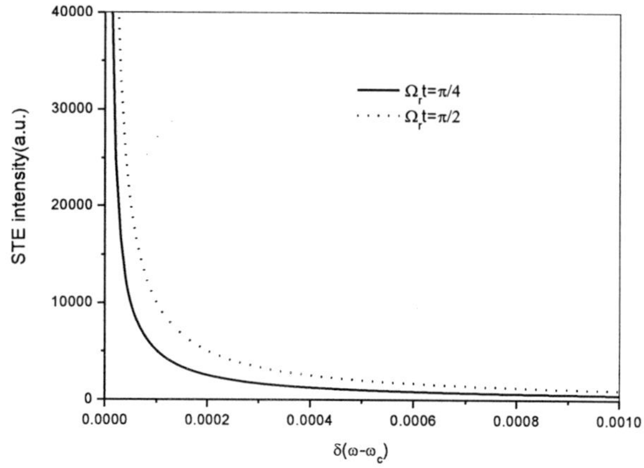

Fig. 4. Intensity of stimulated emission as a function of the detuning between the frequency of emission and that of the band gap edge. The solid represents the case that $\Omega_r t = \frac{\pi}{2}$, while the dotted represents the case $\Omega_r t = \frac{\pi}{4}$. (After [30])

We can explain the above phenomenon by virtue of a possible experiment. We can produce a slab-like active material by MOCVD as a compressively strained quantum wells (QW), such as an InGaAsP on Indium Phosphide (InP) substrate, and the QW emission wavelength is designed as $1.55\mu m$. Using FDTD or multiple-scattering program, the 2-D PhCs with square lattice, for example, can be designed, and the band edge of the 2-D PhCs is designed as $1.55\mu m$. By electron-beam lithography technology and reactive ion etching, we can fabricate a 2-D PhC. The pump light can choose a Ti-sapphire femotoseconds (fs) pulse laser with wavelength about 800nm. The pump laser is focused onto the 2-D PhC along *y* direction (shown in Fig.1), propagating in *xy* plane. Scanning the electric field of the pump laser through a programmable power supply allows the measurement of the gain or the intensity of the fluorescent radiation as a function of the electric field strength. Using the fast photo-electronic diode detector to observe the temporal behavior of the fluorescent or the stimulated emission. The signals can be averaged over the duration of the pump pulse with a Boxcar integrator. The electric field of the pump laser can deduce from $\Omega_r = \left|\frac{\mu E_{ex}}{\hbar}\right|$, and it is $E_{ex} = \frac{\hbar\Omega_r}{\mu}$. Assuming $W(0) = 1$, it means that the atom is initially on the upper level. It can be seen that from Fig.3, when the input pulse area $\Omega_r t = \pi/4$, *i.e.*, as the scanning electric field $E_{ex} = \frac{\pi}{4}\frac{\hbar}{\mu}\frac{1}{t}$, where t is the pulse duration, *W=0.707*, and

$E(\mathbf{r},t)=\dfrac{0.707\beta_0}{\sqrt{\omega-\omega_0}}$. When $\Omega_r t = n\pi$, n is integer, *i.e.*, $E_{ex}=n\pi\dfrac{\hbar}{\mu}\dfrac{1}{t}$ in the experiment, *W=1*, *V=0*, and $E(\mathbf{r},t)=0$, there is no emission. On the other hand, when $\theta = {n\pi}/{2}$, n is odd, *i.e.*, $E_{ex}=\dfrac{n\pi}{2}\dfrac{\hbar}{\mu}\dfrac{1}{t}$, *W=0*, *V=1*, and $E(\mathbf{r},t)=\dfrac{\beta_0}{\sqrt{\omega-\omega_0}}$, emission is strongly enhanced, and the transient gain reach maximum. It is to say that in the experiment, scanning the electric field from $E_{ex}=n\pi\dfrac{\hbar}{\mu}\dfrac{1}{t}$ to $E_{ex}=\dfrac{n\pi}{2}\dfrac{\hbar}{\mu}\dfrac{1}{t}$, we will observe the transient gain oscillating with Rabi frequency from 0 to maximum. In addition, we can introduce a tunable probe light range from 1545nm to 1555nm to propagate in the plane of periodicity (*xy* plane), along *y* direction from *a* facet, and then probe it from *b* facet (see Fig.1), we will find that the signal is amplified, and the detuning from 1550nm (band gap edge frequency) is smaller, the probed signal is larger, which is due to the enhanced transient gain when the emission frequency is near the photonic band gap edge. The lifetime of the spontaneous emission from some semiconductor QW material are about nanoseconds, when the pump laser pulse is femotoseconds, which is satisfied for the coherent interference condition to observe the above transient emission.

3. Transient Emission of Atom with Relaxation in Photonic Crystals

In the second section, we considered the transient process of the emission from atom in the ideal case that there is no background decay to the atom. In this section, we investigate the transient behavior of an external field induced transient gain in two-level atomic systems embedded in a PhC with considering relaxation process. The depending of the transient process on the band gap edge, the background decay rate and the Rabi frequency is discussed. It shows that the transient emission of two-level atom embedded in PhC can be effectively controlled.

Fig.5 displays a two-level atomic system. The population is assumed to be initially in state $|1>$. $\Delta=\omega_{10}-\omega_P$ is the detuning of the pump laser from the resonant transition $|0\rangle\leftrightarrow|1\rangle$, where ω_{10} is the resonant frequency of the transition $|0\rangle\leftrightarrow|1\rangle$ (corresponding to ϖ in Eq.(3)) and ω_P is the frequency of the pump light. The transition $|1\rangle\leftrightarrow|0\rangle$ is taken to be near resonant with a photonic band edge, and the spontaneous emission comes from the transition $|1\rangle\rightarrow|0\rangle$. Our interest is in coherent phenomena that occur in the PhCs reservoirs having relatively weak modal densities where the Born and Markovian approximations can be applied [31]. γ denotes the decay of state $|1>$ to basic state and the backgroud. As the laser field is sufficiently weak, γ can also account for the radiative decay of state $|1>$ to state $|0>$.

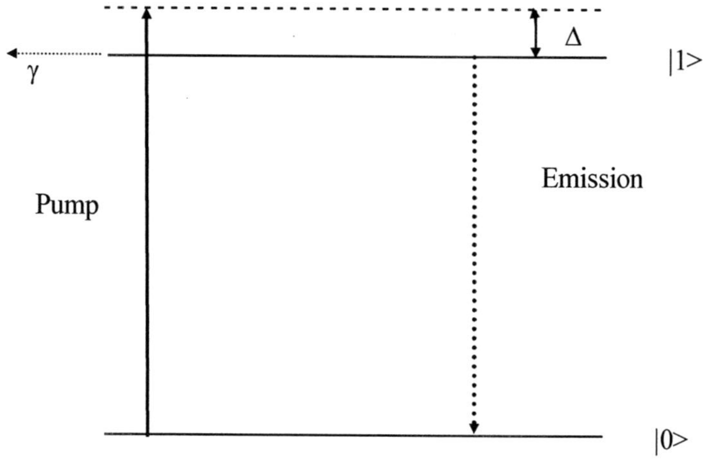

Fig. 5. A two-level atomic system. The solid line denotes the pump laser, and the dotted line denotes the emission light coupled to the PhC and the thin dotted line denotes the background decay

We assume that the emission light from transition $|0\rangle \leftrightarrow |1\rangle$ will all be coupled to the PhCs at the band-edge frequency. Bloch equations shown in Eq.(3) can describe the emission kinetic process.

Assuming $1/T_1 = 1/T_2 = \gamma$, then Eq.(3) can be solved:

$$U = U_0 + e^{-\gamma t}\left\{U(0) - \Delta\left(V(0) - V_0\right)\frac{\sin(\Omega t)}{\Omega} + \Delta\left[U(0)\Delta + \frac{\mu E}{\hbar}W(0) - V_0\gamma\right]\frac{\cos(\Omega t) - 1}{\Omega^2} - U_0\right\}, \tag{25}$$

$$V = V_0 + e^{-\gamma t}\left\{\left[\left(V(0) - V_0\right)\cos(\Omega t)\right] + \left[U(0)\Delta + \frac{\mu E}{\hbar}W(0) - V_0\gamma\right]\frac{\sin(\Omega t)}{\Omega}\right\}, \tag{26}$$

$$W = W_0 + e^{-\gamma t}\left\{\left[W(0) - W_0 - \frac{\mu E}{\hbar}\left(V(0) - V_0\right)\right]\frac{\sin(\Omega t)}{\Omega} + \frac{\mu E}{\hbar}\left[U(0)\Delta + \frac{\mu E}{\hbar}W(0) - V_0\gamma\right]\frac{\cos(\Omega t) - 1}{\Omega^2} + \frac{1}{\gamma}\frac{\mu E}{\hbar}V_0\right\} \tag{27}$$

Where the steady-state solutions to Eq. (3) read

$$U_0 = \frac{\mu E}{\hbar}W(0)\frac{-\Delta}{\gamma^2 + \Delta^2 + \left(\frac{\mu E}{\hbar}\right)^2}, \tag{28}$$

$$V_0 = \frac{\mu E}{\hbar}W(0)\frac{\gamma}{\left(\gamma^2 + \Delta^2 + \left(\frac{\mu E}{\hbar}\right)^2\right)}, \tag{29}$$

$$W_0 = W(0)\frac{\gamma^2+\Delta^2}{\gamma^2+\Delta^2+\left(\frac{\mu E}{\hbar}\right)^2}. \tag{30}$$

$\Omega_r = \left|\frac{\mu E}{\hbar}\right|$ is the Rabi frequency when the pump light is resonant with the transition $|0\rangle \to |1\rangle$, that is, $\Delta = \omega_{10} - \omega_P = 0$, while $\Omega = \sqrt{\left(\frac{\mu E}{\hbar}\right)^2 + \left(\omega_{10}-\omega_P\right)^2}$ is the nonresonance Rabi frequency.

Assuming that the population is initially in the upper level, the initial conditions

$$U(0)=0,\ V(0)=0,\ W(0)=1. \tag{31}$$

From above equations, we can obtain

$$\begin{aligned} V &= \frac{\mu E}{\hbar}W(0)\frac{1}{\gamma^2+\Delta^2+\left(\frac{\mu E}{\hbar}\right)^2} + \frac{\mu E}{\hbar}W(0)e^{-\gamma t}\left[\frac{-\gamma}{\gamma^2+\Delta^2+\left(\frac{\mu E}{\hbar}\right)^2}\cos(\Omega t) + \left(1-\frac{\gamma^2}{\gamma^2+\Delta^2+\left(\frac{\mu E}{\hbar}\right)^2}\right)\frac{\sin(\Omega t)}{\Omega}\right] \\ &= \left\{\frac{\gamma}{\gamma^2+\Delta^2+\Omega_r^2} + \left[\left(1-\frac{\gamma^2}{\gamma^2+\Delta^2+\Omega_r^2}\right)\frac{\sin(\Omega t)}{\Omega} - \frac{\gamma}{\gamma^2+\Delta^2+\Omega_r^2}\cos(\Omega t)\right]e^{-\gamma t}\right\}\Omega_r W(0) \end{aligned} \tag{32}$$

where the relation $\Omega_r = \frac{\mu E}{\hbar}$ with $E > 0$ is used.

The polarization $P(\mathbf{r},t)$ of the impurity atom can be expressed as:

$$P(\mathbf{r},t) = N\mu(U - iV). \tag{33}$$

The polarization related to optical gain can be written as:

$$P(\mathbf{r},t) = -iN\mu V = -iN\mu W(0)\Omega_r\left\{\frac{\gamma}{\gamma^2+\Delta^2+\Omega_r^2} + \left[\left(1-\frac{\gamma^2}{\gamma^2+\Delta^2+\Omega_r^2}\right)\frac{\sin(\Omega t)}{\Omega} - \frac{\gamma}{\gamma^2+\Delta^2+\Omega_r^2}\cos(\Omega t)\right]e^{-\gamma t}\right\} \tag{34}$$

Assuming

$$\chi(t) = \frac{\Omega_r\gamma}{\gamma^2+\Delta^2+\Omega_r^2} + \left[\left(1-\frac{\gamma^2}{\gamma^2+\Delta^2+\Omega_r^2}\right)\frac{\sin(\Omega t)\Omega_r}{\Omega} - \frac{\Omega_r\gamma}{\gamma^2+\Delta^2+\Omega_r^2}\cos(\Omega t)\right]e^{-\gamma t} \tag{35}$$

If $V < 0$ or $\chi(t) < 0$, there is a gain.

Substituting Eq.(34) into Eq.(2), the emission field in the PhC can be written as:

$$E(\mathbf{r}',t)+\frac{P(\mathbf{r}',t)}{\varepsilon_0\varepsilon(\mathbf{r}')} \tag{36}$$
$$=\frac{1}{\varepsilon_0 V}\sum_{\mathbf{k}n}E_{\mathbf{k}n}(\mathbf{r})\int_V d\mathbf{r}'\int_{-\infty}^{t} dt' E_{\mathbf{k}n}^{*}(\mathbf{r}')N\mu W(0)\chi(t)e^{-i\omega t'}\omega_{\mathbf{k}n}\sin\left(\omega_{\mathbf{k}n}(t-t')\right)$$

The part of the polarization that varies with time t' in Eq.(2) and Eq.(36) should be fast-varying amplitude, and the slowly-varying amplitude does not vary with time t'. When we neglect the non-resonant term, adopt slowly-varying amplitude approximation and uniform material approximation [25], the right hand of Eq.(36), representing the propagation field, can be written as:

$$\begin{aligned} E(\mathbf{r},t) &\simeq \frac{N\mu W(0)E_{\mathbf{k}n}(\mathbf{r})}{i\varepsilon_0 V}\sum_{\mathbf{k}n}\omega_{\mathbf{k}n}\chi(t)\int E_{\mathbf{k}n}(\mathbf{r}')d\mathbf{r}' \\ &= \frac{N\mu W(0)\overline{E}E_{\mathbf{k}n}(\mathbf{r})}{i\varepsilon_0}\chi(t)\int_0^{\infty}\sum_{\mathbf{k}n}\omega_{\mathbf{k}n}\delta(\omega-\omega_{\mathbf{k}n})d\omega \\ &\simeq \frac{N\mu W(0)\overline{E}E_{\mathbf{k}n}(\mathbf{r})\omega_{\mathbf{k}n}}{i\varepsilon_0}\chi(t)\rho(\omega) \end{aligned} \tag{37}$$

Where $\rho(\omega)=\sum_{\mathbf{k}n}\delta(\omega-\omega_{\mathbf{k}n})$ is the density of states, and $\overline{E}=\frac{1}{V}\int E_{\mathbf{k}n}(\mathbf{r})d\mathbf{r}$.

Eq.(37) can be abbreviated as:

$$E(\mathbf{r},t)=\frac{\beta E_{\mathbf{k}n}(\mathbf{r})}{i}\rho(\omega)\chi(t) \tag{38}$$

Where

$$\beta=\frac{N\mu W(0)\omega_{\mathbf{k}n}\overline{E}}{\varepsilon_0} \tag{39}$$

is the amplitude amplification factor for the electric field. The transient behavior of $\chi(t)$ and $E(\mathbf{r},t)$ is qualitatively the same, and they all oscillate with the same frequency Ω_r. The amplification of the light occurs in the time interval in which $V<0$ or $\chi(t)<0$ and the peak amplification coincides with the time at which V or $\chi(t)$ reaches the maximum values.

The optical power is

$$I = \left| E(\mathbf{r},t) * E^*(\mathbf{r},t) \right| = \beta^2 \rho^2(\omega) E_{\mathbf{k}n}^2(\mathbf{r}) |\chi(t)|^2 \tag{40}$$

It can be found from Eq.(38) and Eq.(40) that, the transient emission is affected strongly by the density of states, the pump light characters (determined by Rabi frequency) and the background decay. When the density of states is large, there is an enhancement effect to the emission, as well as the oscillations of the emission.

We consider the density of states near the band edge [26]

$$\rho(\omega) = \frac{1}{2\pi} \frac{1}{\sqrt{\omega - \omega_c}} \theta(\omega - \omega_c) \tag{41}$$

$\theta(\omega)$ is a step function, ω_c is the upper band gap edge frequency. Substituting Eq.(41) into Eq.(38), we obtain

$$E(\mathbf{r},t) = \frac{\beta}{i} \frac{1}{2\pi\sqrt{\omega - \omega_c}} E_{\mathbf{k}n}(\mathbf{r}) \left\{ \frac{\Omega_r \gamma}{\gamma^2 + \Delta^2 + \Omega_r^2} + \left[\left(1 - \frac{\gamma^2}{\gamma^2 + \Delta^2 + \Omega_r^2} \right) \frac{\sin(\Omega t)\Omega_r}{\Omega} - \frac{\Omega_r \gamma}{\gamma^2 + \Delta^2 + \Omega_r^2} \cos(\Omega t) \right] e^{-\gamma t} \right\} \tag{42}$$

Then the optical power of the emission field is

$$I = \frac{\beta^2}{4\pi^2} \frac{1}{\omega - \omega_c} \left| E_{\mathbf{k}n}(\mathbf{r}) \right|^2 \left| \frac{\Omega_r \gamma}{\gamma^2 + \Delta^2 + \Omega_r^2} + \left[\left(1 - \frac{\gamma^2}{\gamma^2 + \Delta^2 + \Omega_r^2} \right) \frac{\sin(\Omega t)\Omega_r}{\Omega} - \frac{\Omega_r \gamma}{\gamma^2 + \Delta^2 + \Omega_r^2} \cos(\Omega t) \right] e^{-\gamma t} \right|^2 \tag{43}$$

From above equations, the emission field not only depends on the density of states near the band gap edge, but also on the initial state of the atom, the intensity and duration of the pumping laser pulse (determined by Rabi frequency Ω_r), and the background decay rate. A time-dependent transient coherent process between states |0> and |1> is shown in Fig.6. If $\gamma \geq \Omega_r$, the system exhibits only absorption, however, small oscillations are visible at the beginning. As γ decreases, then the oscillations become more pronounced. The oscillatory behavior is direct consequences of interference effects that result either in an enhancement the radiation rate or in a periodic exchange of excitation between the atom and the field. If the background decay decreases further and reaches the regime that $\gamma << \Omega_r$ the oscillations increase further, the transient gain occurs and increases, the interaction becomes more nonadiabatic and the steady state value is reached for very large time. It can also be found that the decay rate γ is smaller, and then the needed time for the emission becoming steady state is longer. It indicates that in Fig.6, if $E(\mathbf{r},t) < 0$, that is, $\chi < 0$ (V<0), the stimulated emission occurs, and the transient gain occurs. The light will be amplified at a scale time $t \sim \Omega_r^{-1}$. The atom exchanges energy back and forth with its own radiation with the Rabi frequency Ω_r.

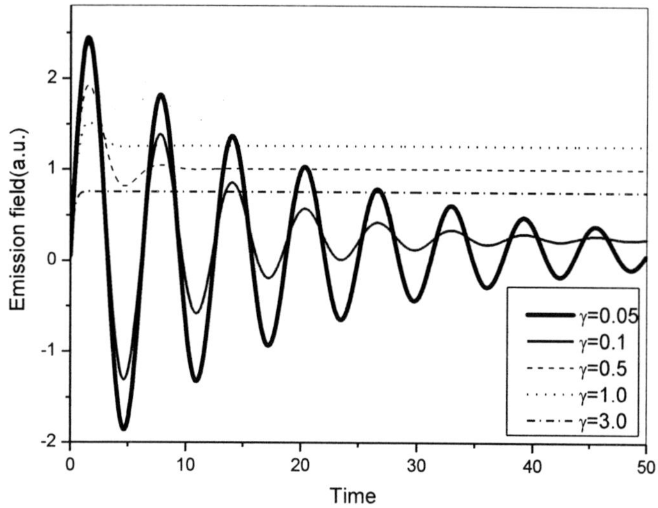

Fig. 6. The time evolution of electric field of stimulated emission. Where $\delta_e = \omega - \omega_c$ is the detuning of the emission frequency from the band gap edge frequency. $\delta_e = 0.001$, and $\Delta = \omega_{10} - \omega_P = \delta_e$. γ, Δ, and δ_e are in unit of Ω_r.

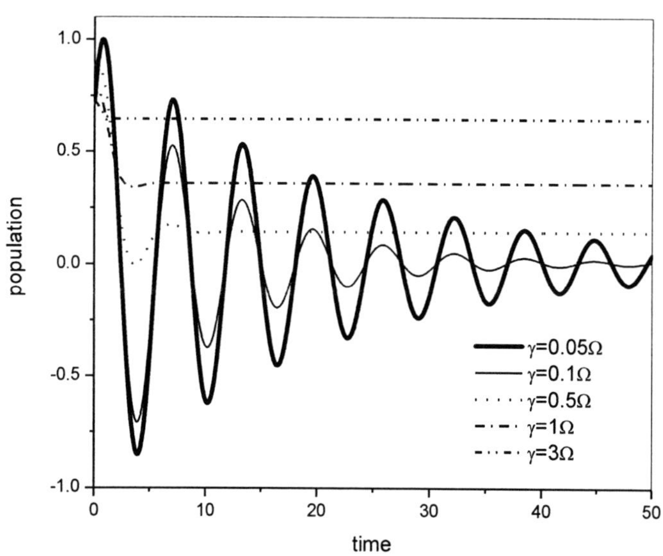

Fig. 7. The time evolution of the population in level |1>. $\delta_e = \omega - \omega_c$ is the detuning of the emission frequency from the band edge frequency, and $\Delta = \delta_e$. γ, Δ, and δ_e are in unit of Ω_r.

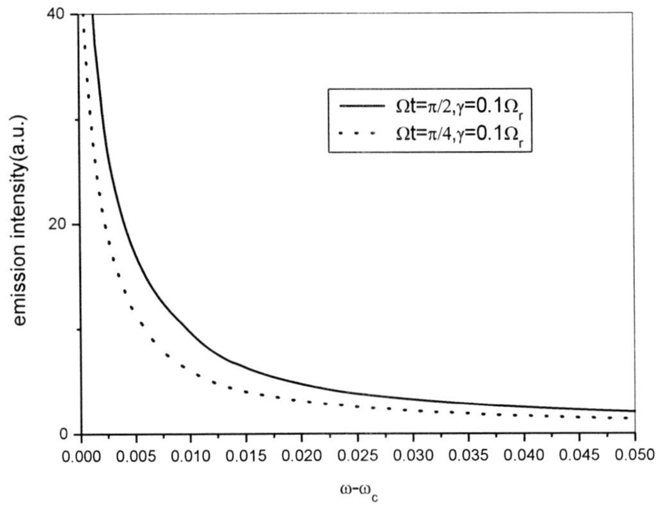

Fig. 8. Intensity of stimulated emission as a function of the detuning between the frequency of the emission and that of the band gap edge. The solid represents the case that $\Omega t = \frac{\pi}{2}$, while the dotted represents the case $\Omega t = \frac{\pi}{4}$.

The time evolution of population is displayed in Fig.6. It shows that, the population from the state |1> can either decay smoothly to the steady state or evolve by undergoing damped Rabi oscillations due to reversible decay which arises via the interferer interaction with the modified reservoir of PhCs [6]. These oscillations increase as the background decay decreases compared to the pump strength that transferred to the frequency-dependent radiation reservoir. The steady state of W is determined by Eq.(30), from which it can be explained that the steady states shown in Fig.7 are different with different γ / Ω_r. The transient emission and its oscillation can be strongly enhanced when the atomic resonant frequency is tuned to the band edge of a perfect photonic band gap [12]. The emission intensity as a function of the detuning between the frequency of emission field and that of the band edge is illustrated in Fig.8. It shows that, with the pump pulse area being a certain value, such as $\Omega t = \frac{\pi}{2}$, or $\frac{\pi}{4}$, when the transition frequency of the atom is tuned to the band gap edge frequency, transient emission as well as the oscillation can be drastically enhanced. It should be stated that we describe the ideal case here that the size of PhC is infinite. In fact, for a finite structure, at the band gap edge, the density of states as well as the emission field $E(\mathbf{r},t)$ is not a singular infinite value, but a large value.

4. THE ENHANCEMENT OF STIMULATED EMISSION NEAR A PHOTONIC BAND EDGE

We derive an analytical expression of light amplification for stimulated emission in arbitrary 2-D PhCs using the response formula of the PhCs and the Bloch equations. It shows that the gain is proportional to the population inversion, and is inversely proportional to the group velocity and the relaxation coefficient of the phase, which indicates that the gain can be induced by the population inversion and enhanced drastically due to the small group velocity or anomaly group velocity.

A periodic dielectric medium doped with two-level atoms is studied here. We consider the light propagates in one-dimensional PhC or 2-D PhC. For intense optical pulses containing many photons a semi-classical treatment of the radiation field is adequate. We assume that the polarized direction is parallel to periodic plane. The propagation of electromagnetic radiation from an impurity atom embedded in PhC can be described by Eq.(1), and the induced electric field in PhC can be expressed as Eq.(2).

On the other hand, the dynamics of the atoms in PhC can be described by Bloch equations:

$$\frac{dP(\mathbf{r},t)}{dt} = \left(-i\varpi - \gamma_p\right)P(\mathbf{r},t) + \frac{1}{i\hbar}\left(E(\mathbf{r},t)d^2\right)D(\mathbf{r},t) \tag{43}$$

$$\frac{dD(\mathbf{r},t)}{dt} = -\gamma_D\left(D(\mathbf{r},t) - D_0(\mathbf{r},t)\right) - \frac{4}{i\hbar}E(\mathbf{r},t)P^*(\mathbf{r},t) \tag{44}$$

$$D(\mathbf{r},t) = \rho(\mathbf{r})D(t) \tag{45}$$

where $D(t)$ represents the population inversion, it is a function of time t. $D(\mathbf{r},t)$ represents the population inversion at position $\mathbf{r}$, $D_0(\mathbf{r},t)$ is the steady-state equilibrium inversion at position $\mathbf{r}$, $\rho(\mathbf{r})$ characteristics the atoms distribution within the PhC, *i.e.*, the atoms number density, d is the dipole matrix element of the atomic transition, ϖ is the atomic resonance frequency, and γ_D, γ_p is the relaxation coefficient of the population and that of the phase, respectively.

Letting $E = \widetilde{E}(\mathbf{r},t)e^{-i\omega t}$, $P = \widetilde{P}(\mathbf{r},t)e^{-i\omega t}$, substituting them into Eq.(43) and Eq.(44), we can obtain

$$\frac{d\widetilde{P}(\mathbf{r},t)}{dt} = \left(-i(\varpi - \omega) - \gamma_p\right)\widetilde{P}(\mathbf{r},t) + \frac{1}{i\hbar}\left(\widetilde{E}(\mathbf{r},t)d^2\right)D(\mathbf{r},t) \tag{46}$$

$$\frac{dD(\mathbf{r},t)}{dt} = -\gamma_D\left(D(\mathbf{r},t) - D_0(\mathbf{r},t)\right) - \frac{4}{i\hbar}\widetilde{E}(\mathbf{r},t)\widetilde{P}^*(\mathbf{r},t) \tag{47}$$

Where $\widetilde{P}(\mathbf{r},t)$ is the slowly-varying amplitude of the polarization, $\widetilde{E}(\mathbf{r},t)$ is the slowly-varying amplitude of the electronic field, and ω is the frequency of the electric field. In the following, the wave signal is omitted.

Assuming $\gamma_P >> \gamma_D$, the variation of the polarization can follow the variation of the electric field in the period of $1/\gamma_P$, so we can let $\frac{d\widetilde{P}(\mathbf{r},t)}{dt} = 0$ in Eq.(46), and then we can obtain:

$$P(\mathbf{r},t) = \frac{d^2}{(\omega - \varpi + i\gamma_p)\hbar} E_{kn}(\mathbf{r},t)\rho(\mathbf{r})D(t) \tag{48}$$

When we consider the case that the frequency of optical field is resonant with the two-level atom, Eq.(48) can be expressed as:

$$P(\mathbf{r},t) = \frac{d^2}{i\gamma_p\hbar}\rho(\mathbf{r})D(t)E_{\mathbf{k}n}(\mathbf{r})\exp\{(-i\omega+\delta)t\} \tag{49}$$

To assure adiabatic switching of the polarization, a positive infinitesimal δ is introduced, and γ_P is assumed as a very small finite positive quantity. Substituting Eq.(49) into Eq.(2), the propagating part of the induced electric field is:

$$E(\mathbf{r},t) = \frac{d^2 D(t)}{2\varepsilon_0 V}\frac{1}{i\hbar\gamma_p}\sum_{\mathbf{k}'n'}\omega_{\mathbf{k}'n'}E_{\mathbf{k}'n'}(\mathbf{r})\int_V d\mathbf{r}'\rho(\mathbf{r}')E^*_{\mathbf{k}'n'}(\mathbf{r}')\int_{-\infty}^{t} dt' E_{\mathbf{k}n}(\mathbf{r}',t')\left(\frac{1}{\omega_{\mathbf{k}'n'}+\omega+i\delta} - \frac{1}{\omega-\omega_{\mathbf{k}'n'}+i\delta}\right) \tag{50}$$

In the deduction, the structure of PhC is assumed as a cubic lattice, but the following result is applicable for arbitrary structure PhC. Similar to the analysis made by K. Sakoda [14], the total induced electric field of the impurity atoms can be approximated to:

$$E(\mathbf{r},t) \approx E_{\mathbf{k}n}(\mathbf{r})\exp(\beta l)\exp(-i\omega t) \tag{51}$$

where

$$\beta = \frac{f\omega_{\mathbf{k}n}D(t)d^2}{2\varepsilon_0 v_g \hbar\gamma_p} \tag{52}$$

is the amplitude amplification factor of the induced electric field per unit length, l is PhC length along the light propagating direction, and $f = \frac{1}{V_0}\int_{V_0} d\mathbf{r}\rho(\mathbf{r})|\mathbf{E}_{\mathbf{k}n}(\mathbf{r})|^2$ is the effective number density of the impurity atoms. It should be mentioned that the electric fields in Eq.(19) and (38) are the first order of induced fields, not total induced field, while that in Eq.(51) is the total induced electric field.

From Eq.(51), the gain of the intensity of stimulated emission can be obtained (the gain of the optical intensity is 2 folds of that of the electrical field):

$$G = 2\beta = \frac{f\omega_{kn} D(t) d^2}{\varepsilon_0 v_g \hbar \gamma_p} \tag{53}$$

On the other hand, we can deduce the population inversion $D(t)$ from Eq.(47). The steady-state process can determine many properties such as the laser threshold, the gain and the laser line width. When we consider the steady-state case, that is, $\frac{dD(\mathbf{r},t)}{dt} = 0$, and $\frac{dP(\mathbf{r},t)}{dt} = 0$, using Eq.(46), Eq.(47) and Eq.(45), we can get:

$$D(\mathbf{r},t) = \frac{D_0(\mathbf{r},t)}{1 + \frac{|E(\mathbf{r},t)|^2}{I_{sat}}} \tag{54}$$

$$D(t) = \frac{D_0(t)}{1 + \frac{|E(\mathbf{r},t)|^2}{I_{sat}}} \tag{55}$$

where $I_{sat} = \frac{\hbar^2 \gamma_p \gamma_D}{4d^2}$ is the line-center saturation field intensity, it represents the field intensity at which the nonlinear response of the atoms becomes important, and $D_0(t)$ is the steady-state equilibrium inversion. For the field intensity $I >> I_{sat}$, the atomic response is saturated. In PhC, I_{sat}, which may be modified by the local density of state [32], is different from its value in free space.

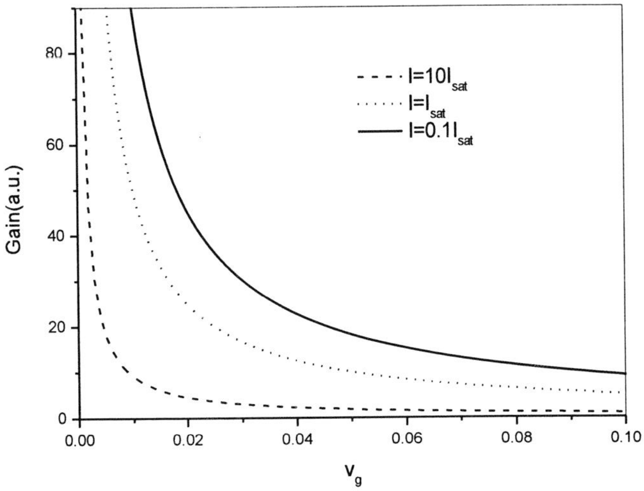

Fig. 9. Normalized gain as a function of the group velocity v_g when the eigenmode intensity $I=\left|E\left(\mathbf{r},t\right)\right|^2=0.1I_{sat}$ (solid), $I=I_{sat}$ (dotted) and $I=10I_{sat}$ (dashed). v_g is measured in units of α.. $\alpha=\dfrac{f\omega_{kn}d^2}{\varepsilon_0\hbar\gamma_p}\dfrac{D_0\left(t\right)}{1+\left|E\left(\mathbf{r},t\right)\right|^2/I_{sat}}$. (After [31])

Substituting Eq.(55) into Eq.(53), we thus obtain

$$G=\frac{f\omega_{kn}d^2}{\varepsilon_0\hbar\gamma_p v_g}\frac{D_0(t)}{1+\dfrac{\left|E(\mathbf{r},t)\right|^2}{I_{sat}}} \tag{56}$$

The gain is proportional to the steady-state equilibrium inversion, but it is inversely proportional to the group velocity of the eigenmode and the relaxation coefficient of the phase. Given the group velocity, the steady-state equilibrium inversion and the relaxation coefficient, the optical gain of stimulated emission in arbitrary PhC can be calculated by Eq.(56). It indicates that the gain in PhC could be induced by the external field (reflected by population inversion), and then can be enhanced by the small group velocity in PhC. Near the band gap edge of a PhC with infinite size, the group velocity approaches to zero [28], and then the gain will reach to infinite, which is described in Fig.9. Furthermore, some scholars have demonstrated that the group velocity anomaly can occur in the photonic pass band [33], which will cause a fairly large enhancement of stimulated emission. Reference [28] showed that the enhancement comes from the long interaction time between the radiation field and the matter system due to the small group velocity. Near the band-gap edge or in the pass band where the group velocity is anomalistic, the small group velocity dominantly contributes the gain, no matter how small population inversion $D_0(t)$ is. Especially, the gain even can occur when there is no population inversion [34]. On the other hand, if the group velocity is not small, such as in some pass band, but the population inversion is larger enough, the gain is mainly induced by the population inversion. The above results are different from that in uniform material where the optical gain will only be induced by population inversion in general case. Moreover, in uniform material, when the electric field intensity $|E(\mathbf{r},t)|^2 > I_{sat}$, the optical gain becomes to saturate. But in PhC, even $|E(\mathbf{r},t)|^2 >> I_{sat}$, the gain still may be enhanced largely due to the small group velocity. It can be found from Fig.9 that, for a constant group velocity, the optical gain becomes to saturate under the condition that $|E(\mathbf{r},t)|^2 > I_{sat}$, therefore, the gain still increases with the group velocity decreasing.

From Eq.(51) and Eq.(55), the optical intensity of stimulated emission can be obtained:

$$I = I_0 \exp\left(\frac{f\omega_{kn} d^2}{\varepsilon_0 \hbar \gamma_p v_g} \frac{D_0(t)}{1 + \frac{|E(\mathbf{r},t)|^2}{I_{sat}}} l \right) \tag{57}$$

where I_0 represents the initial field intensity. When the population inversion and the group velocity in the PhC are appropriate values, the stimulated emission will increase exponentially with the PhC length increasing (See Fig.10). The intensity of stimulated emission is extremely sensitive to v_g, $D_0(t)$ and l. For example, assuming $\alpha = \frac{f\omega_{kn} d^2}{\varepsilon_0 \hbar \gamma_p} \frac{D_0(t)}{1 + |E(\mathbf{r},t)|^2 / I_{sat}}$, with $l = 50\mu m$, $v_g = 5.0 \times 10^{-6}\alpha$, $I = 53.5 I_0$ and for $v_g = 2.5 \times 10^{-6}\alpha$, $I = 2864 I_0$. This phenomenon also demonstrates that the stimulated emission can be drastically enhanced when the group velocity is very small. The intensity of stimulated emission can be practically evaluated by using Eq.(57). Because the other parameters in Eq.(57), such as I_0, $D_0(t)$, γ_p, I_{sat}, f, and d etc., can be determined by the condition of an experiment. This equation will benefit to design a photonic band gap edge laser.

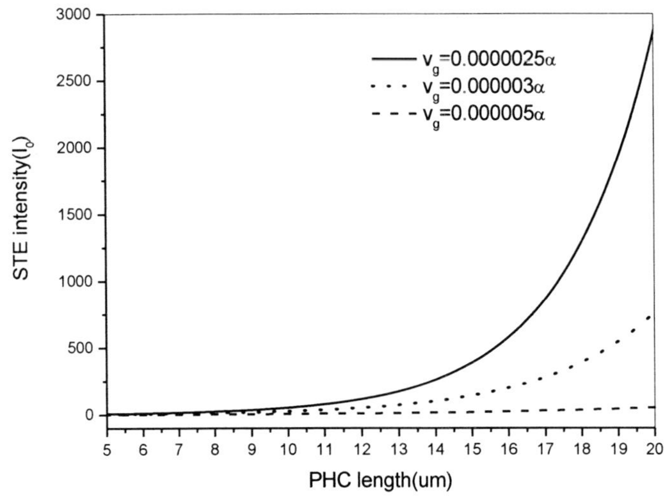

Fig. 10. The intensity of stimulated emission as a function of the photonic crystal length with group velocity $v_g = 2.5\times10^{-6}$ (solid line), $v_g = 3.0\times10^{-6}$ (dotted) and $v_g = 5.0\times10^{-6}$ (dashed). v_g is measured in units of $\alpha = \dfrac{f\omega_{kn}d^2}{\varepsilon_0\hbar\gamma_p}\dfrac{D_0(t)}{1+|E(\mathbf{r},t)|^2/I_{sat}}$. (After [36])

As an example, we proceed to the quantitative evaluation of the enhancement of stimulated emission for a 2-D PhC with square lattice. Fig.11(a) shows the transmittance as a function of the frequency for an incident wave with S polarization propagated perpendicular to the 2-D PhC with dielectric constant of 11.4. Fig.11(b) shows the transmittance for S polarization calculated for the same crystal of Fig.11(a) with a dielectric constant of 11.4-0.02i. The crystal consists of 153 (9×17) dielectric rods arranged as square lattice. The negative imaginary part of the dielectric constant stands for the inverted population of the impurity atoms [15], corresponding with $D_0(t)$ in Eq.(57). Note that the transmittance can be greater than unity because of the stimulated emission occurred in the crystals. Here, a strongest peak shows a large enhancement of the stimulated emission at 765THz., which exactly coincides with the third upper band-edge frequency (see Fig.11(a)) where $v_g \to 0$. This phenomenon just demonstrates the result of Eq.(57). The enhancement factor, which was estimated as the ratio of the transmittance when there is impurity atoms (see Fig.11(a)) to that there is no impurity atoms (see Fig.11(b)) in the PhCs as large as 141. So the emission near the band gap edge was greatly enhanced due to small group velocity.

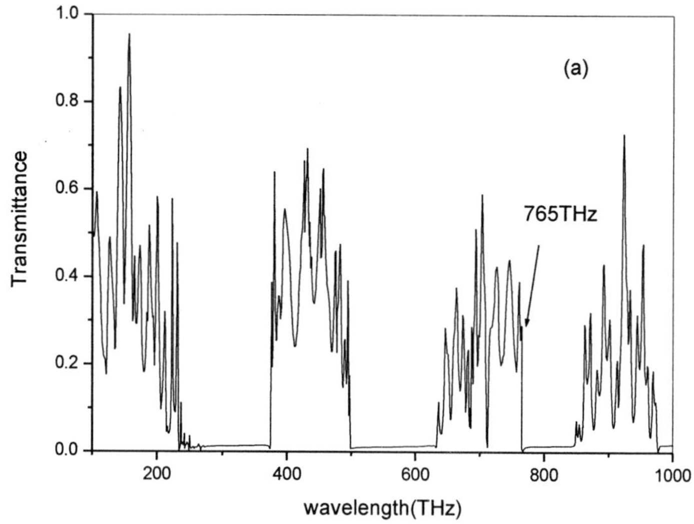

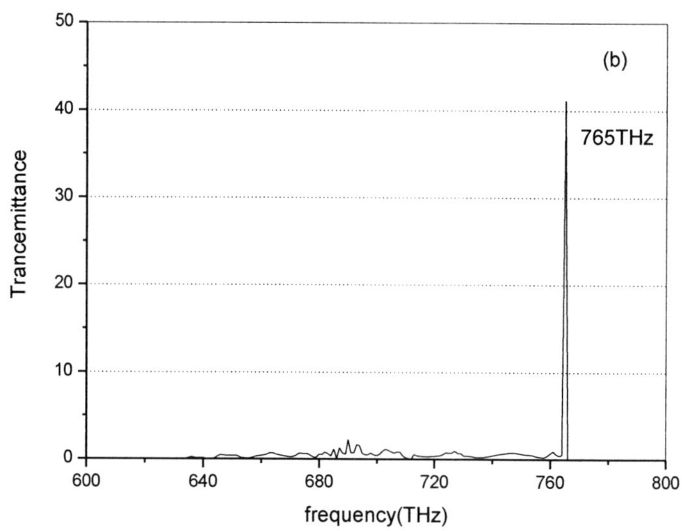

Fig. 11 (a). The transmittance for S polarization as a function of the frequency calculated for a square lattice crystal with 153 dielectric rods with dielectric constant of 11.4; (b) The transmittance for S polarization as a function of the frequency calculated for the same crystal of (a) with dielectric constant of 11.4-0.02i. The negative imaginary part of the dielectric constant stands for the population inversion of the impurity atoms. Note that the transmittance can be greater than unity because of the stimulated emission. (After [36])

5. Lowered Laser Threshold in Photonic Crystals

The lasing threshold in arbitrary 2-D PhCs is obtained by an investigation of steady-state laser behavior. It shows that the population inversion threshold is proportional to the second order of the group velocity, and proportional to the relaxation coefficient.

Eq.(56) shows that the gain is proportional to the steady-state equilibrium inversion and inverse proportional to the relaxation coefficient and the group velocity. When the group velocity of the eigenmode is small, there is an enhancement effect. Reference [28] pointed out that the enhancement came from the long interaction time between the radiation field and the matter system due to the small group velocity.

It can be demonstrated that the enhanced stimulated emission can reduce the lasing threshold. We designed a PhC slab with length L orienting along the pumping direction. The reflective amplitude coefficient of the eigenmode at the PhCs surface is set as R, and the lasing threshold is given by the balance between the loss at both surfaces and the optical gain in the path of 2L as:

$$R^2 \exp[2LG] = 1, \tag{58}$$

$$G = \frac{1}{2L}\ln\left(\frac{1}{R^2}\right), \tag{59}$$

where the cavity dissipation is neglected. When $R \approx 1$, $\ln R \approx 1 - R$, Eq.(59) can be changed as:

$$G \approx (1-R)/L. \tag{60}$$

In the PhC, the effective refractive index can be defined by $n_{eff} = c/v_g$, so the reflection coefficient can also be evaluated as: $R \approx \dfrac{n_{eff}-1}{n_{eff}+1}$ [35]. Substituting these relations into Eq.60, we can get threshold gain

$$G_{th} = \left(1 - \frac{c/v_g - 1}{c/v_g + 1}\right)\frac{1}{L} \approx \frac{2v_g}{cL} \tag{61}$$

This is the lasing threshold expressed by gain, where the case $v_g << c$ is taken. It shows that the gain threshold is proportional to the eigenmode group velocity. Using Eq.(53) and Eq.(61), we can obtain the population-inversion threshold:

$$D_{th} = \frac{2v_g^2 \varepsilon_0 \hbar \gamma_p}{cLf\omega_{kn} d^2}\left(1 + \frac{|E(\mathbf{r},t)|^2}{I_{sat}}\right) \tag{62}$$

If $|E(\mathbf{r},t)|^2 << I_{sat}$, then

$$D_{th} = \frac{2v_g^2 \varepsilon_0 \hbar \gamma_p}{cLf\omega_{kn}d^2} \tag{63}$$

If $|E(\mathbf{r},t)|^2 \doteq I_{sat}$, then

$$D_{th} = \frac{4v_g^2 \varepsilon_0 \hbar \gamma_p}{cLf\omega_{\mathbf{kn}}d^2} \tag{64}$$

It shows that the threshold population inversion is proportional to the second order of group velocity. Its large reduction is expected when group velocity is small. As Ref.[35] referred, the reduction of the lasing threshold was brought about by both the increase of stimulated emission and that of the amplitude of reflection coefficient.

It can be found from Eq.(62) that the population-inversion threshold not only determined by the group velocity, but also by the reflectivity of the PhC and the length of the PhC. A small v_g is beneficial to laser action, therefore, one can attain laser action in a very small cavity with very low pumping power. It is found that there is a very small group velocity v_g near the photonic band edge. For a 26 periods one-dimensional PhCs of GaAs/AlAs multi-layers plate, the group velocity near the band edge was observed as 0.07c [37], and the state density of photon modes was very large there [38]. The laser threshold is low near the photonic band edge, and the laser oscillation occurs firstly near the band edge. For example, a 2-D PhC surface-emitting laser with 2-D triangular lattice was realized near either of the photonic band edges [39]. In addition, lasing oscillation was observed at the four lowest band gap frequencies in a 2-D hexagonal organic photonic-crystal with lattice constants ranging from 0.18 to 0.44um [40]. Therefore, lasing threshold can be reduced when we take advantage of the small group velocity in some photonic band or band edge. It indicated that a PhC could be treated as an effective medium at frequencies not too close to a band gap. At the frequencies near the band edges, the effective index n_{eff} is much larger, which is caused by the multiple reflections at the internal interfaces in the PhCs. For one-dimensional PhC of GaAs/AlAs with 30 periods, the group index $n_{eff} = c/v_g$ reach to 17 at the frequency near the band edge [41]. Then the lasing threshold is low there.

The above phenomenon can also be demonstrated by the following calculations. We proceed to the quantitative evaluation of the enhancement of stimulated emission for a 2-D PhC with square lattice. Fig.12 shows the transmittance as a function of the frequency for an incident wave with S polarization (TM wave) propagated perpendicular to the 2-D PhC with dielectric constant of 11.4. Fig.13 shows the transmittance for S polarization calculated for the same crystal of Fig.12 with a dielectric constant of 11.4-0.02i. The crystal consists of 117 (9×13) dielectric rods arranged as square lattice with the lattice constant of 250nm and the rods radius of 74nm. The negative imaginary part of the dielectric constant stands for the inverted population of the impurity atoms [9], corresponding with $D_0(t)$ in Eq.(3) or D_{th} in

Eq.(62). Here, a strong peak shows a large enhancement of the stimulated emission at 497THz, which exactly coincides with the third upper band-edge frequency (see Fig.12) where $v_g \to 0$ and the lasing threshold is low. The enhancement factor, which was estimated as the ratio of the transmittance when there is impurity atoms (shown in Fig.12) to that there is no impurity atoms (shown in Fig.13) in the PhCs as large as 126. The other peaks, such as 685THz, 687THz, occur in the photonic band due to the small group velocity there. These strong stimulated emission peaks occur just show that there is a lower laser threshold at corresponding frequency than that at other frequency. These phenomena just demonstrate the result of Eq.(62), that is, a small group velocity will lead to a lower lasing threshold. The lasing threshold described by the population inversion and the group velocity can open out the microcosmic physical mechanism more manifest, which reflects directly the characteristics of the interaction between the light and the atom embedded in the PhC.

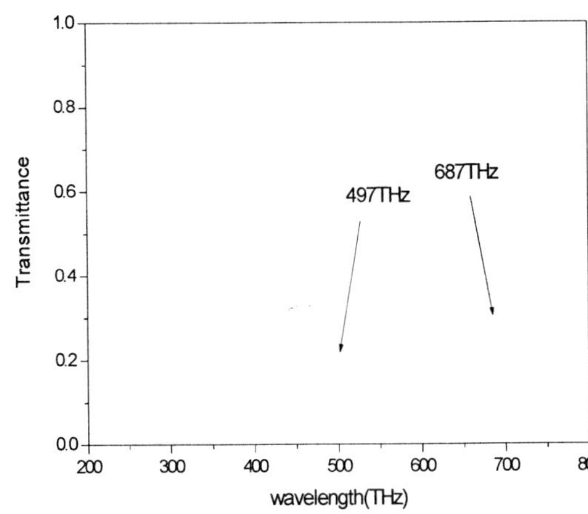

Fig. 12. The transmittance for S polarization as a function of the frequency calculated for a square lattice crystal with 117 dielectric rods with dielectric constant of 11.4. (After [42])

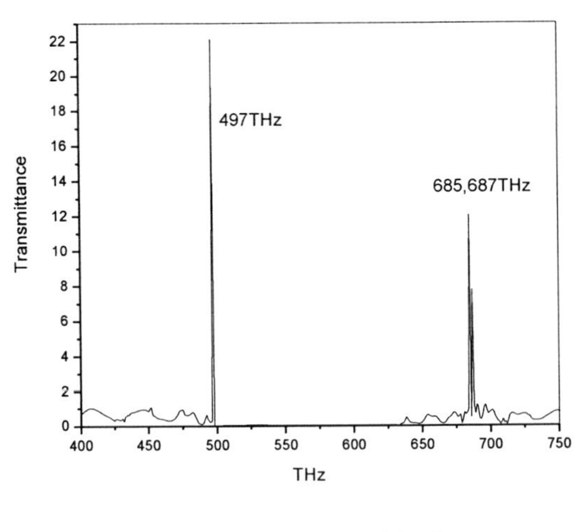

Fig. 13. The transmittance for S polarization as a function of the frequency calculated for the same crystal of Fig.12 with dielectric constant of 11.4-0.02i. The negative imaginary part of the dielectric constant stands for the population inversion of the impurity atoms. (After [42])

6. Stimulated Emission in Dodecagonal Quasi-Periodic Photonic Crystals

In this section, we mainly study the stimulated emission in quasi-periodic PhC with gain medium by numerical calculation, which is also a demonstration in quasi-periodic crystal for the theory discussed in previous sections. It is found that there is some interesting phenomenon in the quasi-periodic structure. The lasing peaks are more abundant and stronger compared with that in periodic crystals, and more stimulated emission peaks appear when the crystal size and the gain become larger. Sharp lasing peaks appear when the gain or length of the system was over a well-defined threshold value [43]. Some lasing peaks are independent on incident direction, which may be induced by localization due to the quasi-periodic structure. As K. Sakoda referred [14], in the deduction process of the electric filed of stimulated emission, for example, from Eq.(17) to Eq.(18) and from Eq.(50) to Eq.(51), an approximation for uniform materials is made. In this case, we believe that the theories in the above sections are still applicable for quasi-periodic PhC. Especially, Eq.(56) is still applicable for the quasi-periodic crystal. It means that the small group velocity will enhance the stimulated emission. Correspondingly, the large local density of states will contribute to the stimulated emission.

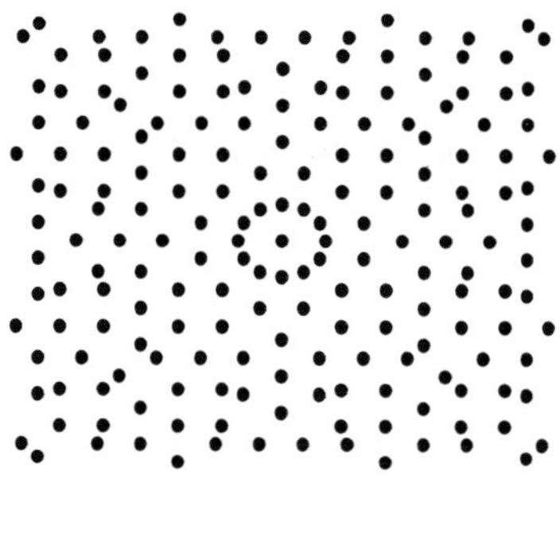

Fig. 14. Schematic of a two-dimensional dodecagonal quasi-periodic photonic crystal

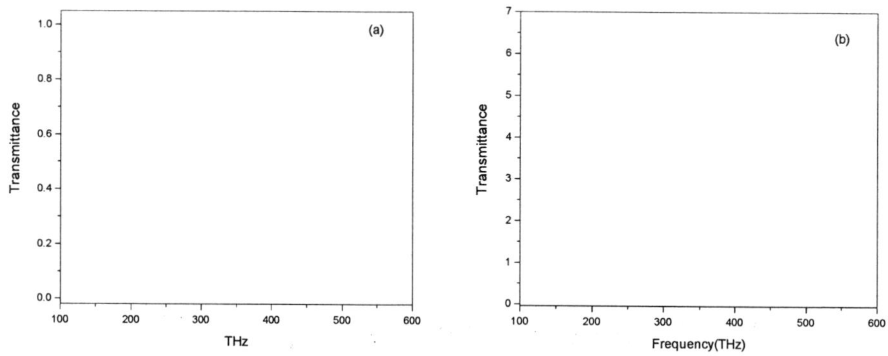

Fig. 15. Transmittance for S polarization as a function of frequency calculated for the crystal with dielectric constant of (a)11.4 and (b)11.4-0.05i with 261 dielectric rods. Where the side length of the square in the quasi-periodic photonic crystal is 330nm, and the radius of hole is 70nm

We proceed to the quantitative evaluation of the enhancement of stimulated emission for a 2-D dodecagonal quasi-periodic PhC composed of dielectric rods. Figure 14 shows the schematic structure of a 12-fold symmetric QPC, which was formed by placing dielectric rods with a circular cross section in the vertices of 2-D 12-fold symmetric quasiperiodic lattices. The pattern was tiled with squares and rhombuses (with an acute angle of 36°) with equal side lengths of 330 nm, and the crystal consisted of 189 dielectric rods with a rod radius of 70 nm. In the calculations, we assumed the imaginary part of the dielectric constant is no longer equal to zero, and has a negative value, which means that the impurity atoms with population inversion are uniformly distributed in the crystal. As mentioned in section 4 and 5, when there are impurity atoms (gain medium), the calculated transmission can be greater than unity because of the stimulated emission that takes place in the crystals [15]. Figure 2(a) plots the transmittance as a function of frequency for a transverse magnetic (TM) polarized (E field parallel to the cylindrical rods) incident wave for the QPC with a dielectric constant of 11.4, and where the incident light propagates perpendicularly to the 2-D QPC. Figure 15(b) shows the transmittance of the crystal with a dielectric constant of 11.4-0.08i, with the other parameters the same as that in Fig. 15(a). The negative imaginary part of the dielectric constant represents the inverted population of the impurity atoms [35]. The existence of calculated transmittance with a value greater than unity means that the stimulated emission occurred in the crystals. The sharp stimulated peaks at 471, 546, and 551 THz coincide with the frequencies in the photonic band, where the group velocities v_g are small [15]. This can also be explained by Eq.(56) and Eq.(57) approximately.

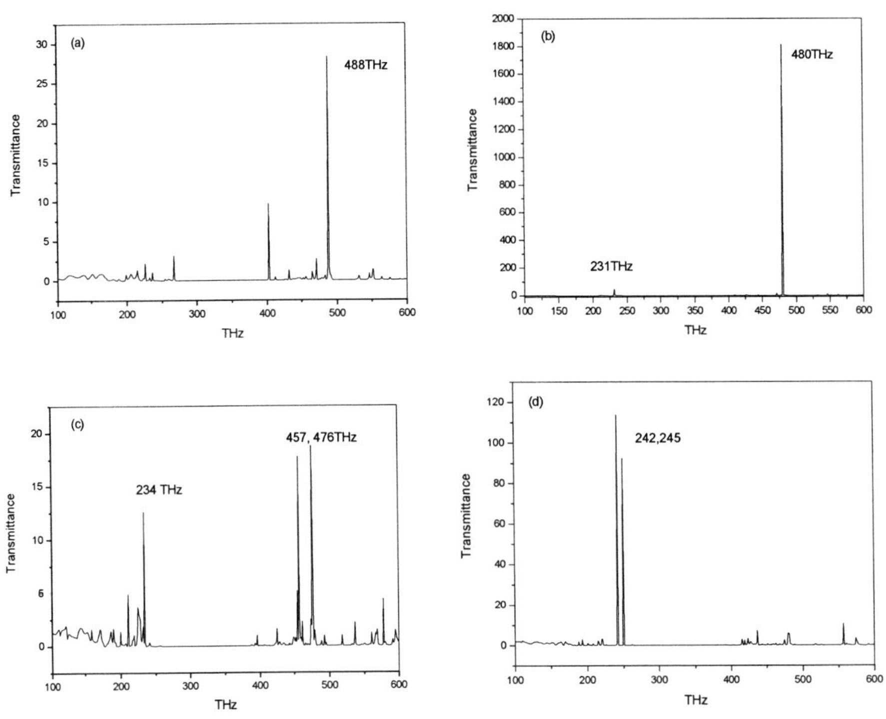

Fig. 16. Transmittance for the S polarization as a function of frequency calculated for a crystal with numbers of (a) 101, (b) 150, (c) 261, (d) 334 rods with a dielectric constant of 11.4-0.1i

We found that there were some interesting phenomena when the size of the PhC varies. Because the number of dielectric rods with a dielectric constant of 11.4-0.1i varies from 101 (Fig. 16(a)) to 189 (Fig. 16(b)), the intensity of the strongest stimulated emission rises from 27 to 1812. The frequency difference between these two peaks (at 488 THz in Fig. 3(a) and 480 THz in Fig. 16(b)) is only 8 THz. Both of these peaks coincide with the band gap edge where the group velocity is small. As shown in Figs. 16(c) and 16(d), the peaks in the crystals with 261 and 334 rods are not very strong. This means that the peak intensity is strong if the size of the crystal is adequate at a particular frequency. The enhancement factor, which was estimated as the ratio of the transmittance in a QPC with impurity atoms (Fig. 16(b)) and that in a QPC without impurity atoms (Fig. 15(a)) was as large as 51,994. Comparing the stimulated emissions in a QPC with those in a periodic crystal with similar sizes and parameters, we found that the strongest peak intensity in a square lattice periodic crystal (Fig. 17) is only 30, while that in a QPC is 1812 (see Fig. 16(b)), so the enhancement factor reaches 60. Compared with the periodic crystal, the quasiperiodic structure has a lower lasing threshold, which might be caused by the localization in a quasi-crystals that makes the optical paths of waves much longer than that in a periodic medium.[5] As the crystal size increases, some original peaks remain, other peaks disappear, and some new peaks appear, although most of them are weak. For example, three peaks appear at 234, 457, and 476 THz with respective intensities of 13, 17 and 18 in Fig. 16(c). In Fig. 16(d), two strong peaks appear at 242 THz with an intensity of 114 and 245 THz with an intensity of 92.5, and some weak peaks appeared at 437, 479, 481, and 557 THz with intensities of only about 10. This means that the number of lasing points satisfying the standing wave condition increases as the size of the quasi-crystals increases, and that some stimulated emission peaks come from the light oscillations between the two boundaries of the crystal, so the resonant frequency is determined by the size of the PhCs.

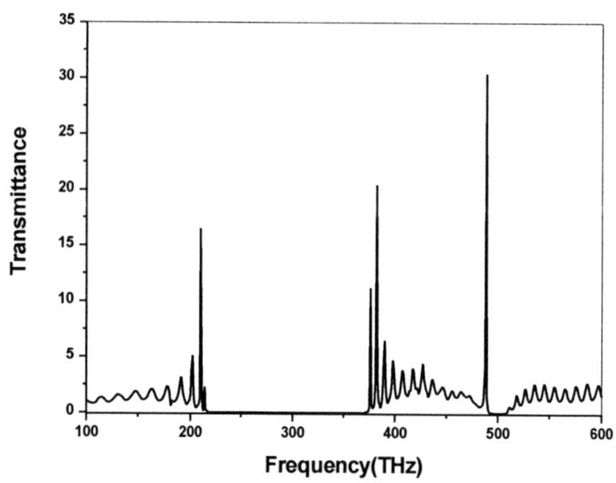

Fig. 17. Transmittance for a periodic crystal with 195 dielectric rods with a dielectric constant of 11.4-0.1i (similar parameters, similar size and same dielectric constant as those of quasiperiodic PhC in Fig.16(b)).

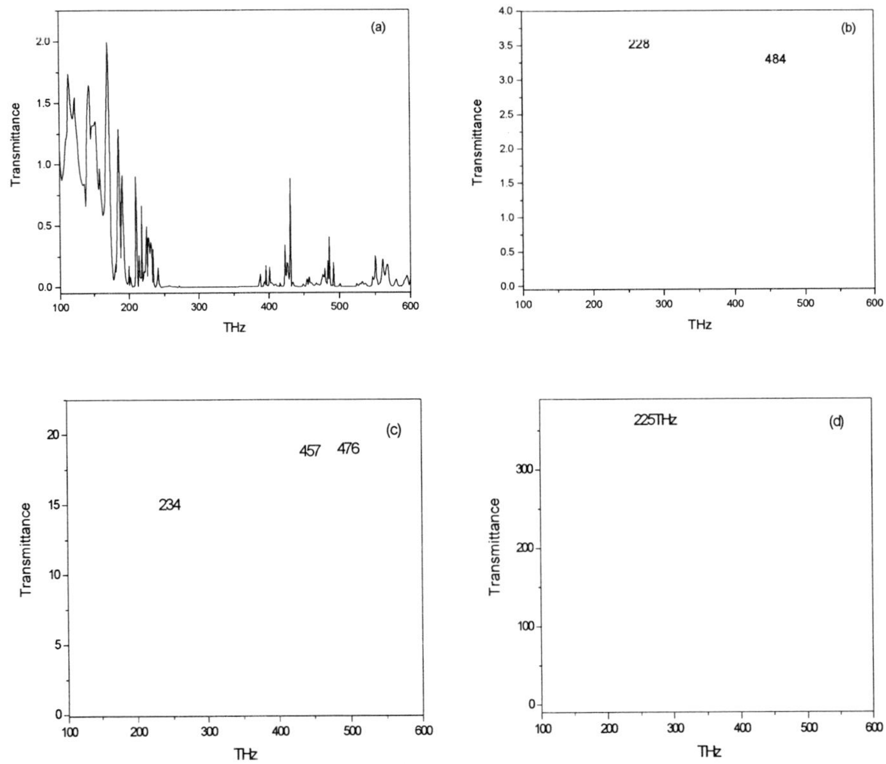

Fig. 18. The transmittance spectra of the quasi-periodic crystal with dielectric rods numbers of 261 with various dielectric constant, (a) 11.4-0.005i, (b) 11.4-0.02, (c) 11.4-0.1i, (d) 11.4-0.2i.

There is some new phenomenon as the imaginary part of the dielectric constant (demoted as ε') varies. The gain increases, more peaks appear (see Fig.18); moreover, the frequencies and the intensities of the stimulated emission peaks vary. When ε'=0.005, there is no strong peak, but two small peaks 228THz and 484THz with the intensities both only about 3.5. When ε'=0.1, there are some stronger peaks appear, such as 234THz with transmittance intensity 13, 457THz with intensity 17, 476THz with intensity 18. When ε' increases to 0.2, the intensity of the main peak of 225THz reach to 380.

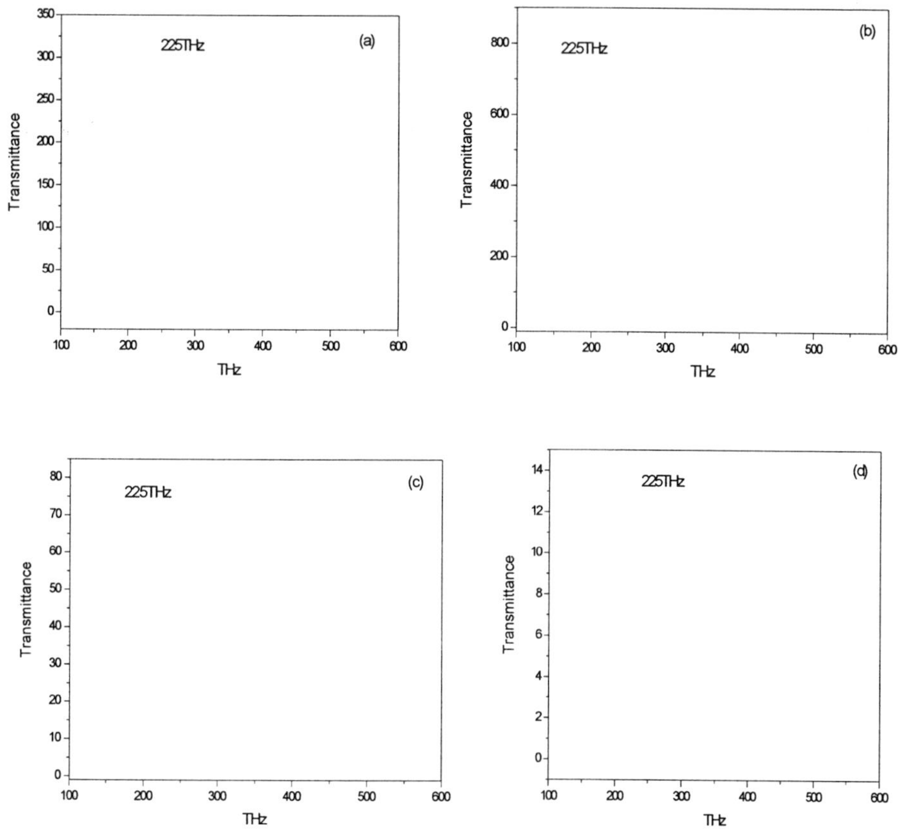

Fig. 19. The transmittance spectra of the quasi-periodic crystal with various incident angles (a) 5, (b) 10, (c) 15 and (d) 20 degrees, with rods number 261

Fig.19 shows the transmission spectra of the crystal with various incident angles. There are several peaks are independent on the direction of the incident angles of the light, especially the peak of 225THz. We think that these peaks occur from the localization of quasi-periodic structure of the PhCs. Localization from disorder will enhance the stimulated emission, because random systems are expected to possess a reduced gain threshold for lasing with enhanced optical paths from multiple scattering [21]. It is believed that the localized mode is determined by the quasi-periodic structure, so is the stimulated emission peak. It can be found that from Fig.14, the central region of the quasi-periodic PhC forming a high-refractive-indices ring, the electromagnetic waves can easily be localized there [44]. There may be three mechanisms for stimulated emission in the quasi-periodic crystal. One comes from small group velocity near the band edge or the anomaly velocity in the photonic band [15]; another is attributed to light oscillation between the PhC boundaries; the third comes from the localization. The physical process can sum up like this: Quasi-periodic PhC loses short-range order and translation symmetry, possesses long-range order and 12-fold rotational symmetry. This structure of quasi-periodic crystal will induce localization. The photonic band gap is independent on incident angle, so is the localized mode. The localized mode will contribute to the stimulated emission. Moreover, disorder can enhance the stimulated emission. So the lasing peaks are stronger and not dependent on incident direction of light.

To explain the stimulated emission described above, we calculated the local density of states in the 12-fold symmetric quasi-periodic crystals with gain medium by using the multiple-scattering method [45,46]. For some frequencies, the light is strongly localized in QPC, which can be clearly seen from the local density of states distribution in Fig. 20(a). Compared with periodic crystal, a QPC is a special regular structure without translation symmetry, so the localized mode may be determined by the structure of QPC without adding defects [44]. Localization induced by the structure of QPC will enhance the optical path due to multiple scattering and will then reduce the lasing threshold, which is similar to the case of random systems [21]. Similar to the explanation from M. Notomi *et al.*, these stimulated emissions occur due to the feedback of the quasiperiodic long-range order, which are different from those in both periodic crystal and random media. Furthermore, the stimulated emission peaks and corresponding local density of states are determined by standing wave modes in the quasiperiodic structure. The standing wave condition is satisfied only at symmetry points in the first Brillouin zone for periodic PhCs, but there are a large number of lasing points that satisfy the standing wave condition for quasiperiodic crystals [16], which explains why the lasing modes in QPC are more abundant than those in periodic crystals. It should be noted that the local density of states in a QPC with gain is stronger than that without gain. The largest local density of states is up to 21.3 for the crystal with gain in Fig. 20(a), while it is only 2.6 for the crystal without gain in Fig. 20(b). The local density of states in vacuum is only 0.25. The enhanced local density of states seems to be evidence that optical gain is beneficial to wave localization, which is similar to the case in which optical gain helps spatial confinement of light in a random medium [47].

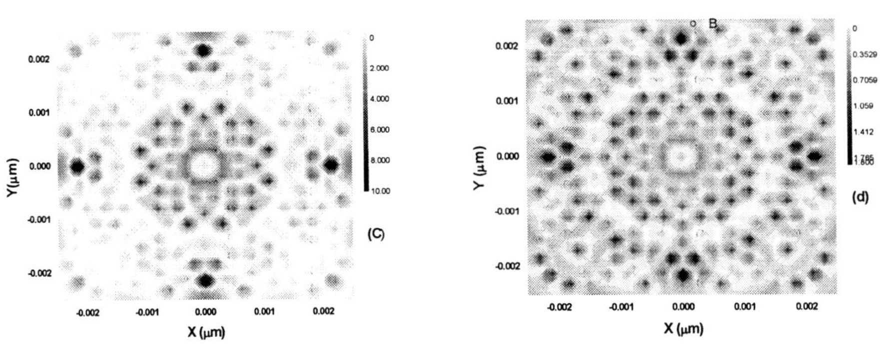

Fig.20 The calculated local density of states in a 12-fold quasi-periodic crystal (a) at 225THz with 261 dielectric rods diclectric constant 11.4-0.2i; (b) at 225THz with 261 dielectric rods with dielectric constant 11.4.

7. Summary and Conclusion

In conclusion, we have analyzed the stimulated emission and lasing threshold in arbitrary 2-D periodic photonic crystals and quasi-periodic photonic crystal by Maxwell equations, Bloch equations and numerical calculation. Both the transient coherent emission and steady emission were studied. We found that the transient emissions of the two-level atom were proportional to the density of states and oscillated with the Rabi frequency. The time evolution of the emission and the way that the steady state was reached depended crucially on the background decay rate and the Rabi frequency. A transient gain was found to exist if the Rabi frequency is larger than the background decay rate. When the transition frequency of the atom is tuned to the band gap edge frequency, the transient emission and its oscillation can be largely enhanced. The optical gain in the photonic crystals was proportional to the population inversion and inversely proportional to group velocity. It showed that the gain can be induced by population inversion and be enhanced by the small group velocity. The lasing peak was stronger than that in periodic, and some lasing peaks were independent on incident angle. The theory was demonstrated by numerical calculation. The enhancement factor of the stimulated emission in the periodic photonic crystals was calculated, it reached to 140 folds with proper parameters, while the stimulated emission in quasiperiodic crystal was 60 folds larger than that in periodic photonic crystals. It demonstrated that the results of numerical calculations were consistent with the analytical results.

Acknowledgement

The authors acknowledge the support of the NNSFC (Grant No. 60345008, 60377011, 60537010) and CNKBRSF (Grant NO.2001CB6104).

The authors acknowledges to Elsevier, Springer-Verlag and IOP Publishing Ltd to give the right to extend corresponding articles (Ref.[30], Ref.[36] and Ref.[42]) into this dissertation.

References

[1] D. Kleppner, *Phys. Rev. Lett.*, 27 (1981)233.

[2] *Photonic band gaps and localization*, edited by C. M. Soukoulis (Plenum Press, New York,1993).

[3] Z. Li, L. Lin, Z. Zhang, *Phys. Rev. Lett.* 84 (2000)4341.

[4] S. Y. Zhu, Y. P. Yang, H. Chen, H. Zheng, and M. S. Zubairy, *Phys. Rev. Lett.* 84 (2000)2136.

[5] Y. P. Yang, S. Y. Zhu, *Phys. Rev. A*, 61 (2000)043809.

[6] D. G. Angelakis, E. Paspalakis, and P. L. Knight, *Phys. Rev. A*, 61 (2000)055802.

[7] S. John, J. Wang: *Phys. Rev. lett.* 64 (1990)2418.

[8] M. Konopka, *Phys. Rev. A* 60 (1999)4183.

[9] V. I. Kopp, B. Fan and H. Vithana and A. Z. Genack, *Opt. Lett.* 23 (1998)1707.

[10] S. Nojima, *Jpn. J. Appl. Phys. Lett.* 37 (1998)L565.

[11] N. Sussa, *J. Appl. Phys*. 89 (2001)815.
[12] M. Centini, G.. D'Auguanno, M. Scalora, M. Bloemer, C. M. Bowden, C. Sibilia, N. Mattiucci, and M. Bertolotti, *Phys. Rev.* E 67 (2003)036617.
[13] L. Florescu, K. Busch, and S. John, *J. Opt. Soc. Am. B* 19 (2002)2215
[14] K. Sakoda and K. Ohtaka, *Phys. Rev. B* 54 (1996) 5732.
[15] K. Sakoda, K. Ohtaka, and T. Ueta, *Opt. Express* 4 (1999) 481.
[16] M. Notomi, H. Suzuki, T. Tamamura, and K. Edagawa, *Phys. Rev. Lett.* 92(2004)123906.
[17] F. Scheffold, R. Lenke, R. Tweer, and G. Maret, *Nature* 398 (1999)206.
[18] A. A. Chabanov, M. Stoytchev, and A. Z. Genack, *Nature* 404 (2000)850.
[19] H. Cao, Y. G. Zhao, S. T. Ho, E. W. Seelig, Q. H. Wang, and R. P. H. Chang, *Phys. Rev. Lett.* 82 (1999) 2278.
[20] X. Y. Jiang and C. M. Soukoulis, *Phys. Rev.* B 59 (1999)6159.
[21] X. Y. Jiang, Q. M. Li, and C. M. Soukoulis, *Phys. Rev.* B 59 (1999) R9007.
[22] L. Florescu, K. Busch, S. John, J. Opt. Soc. Am. B 19 (2002)2215.
[23] S. Zamith, J. Degert, S. Stock, B. Beauvoit, V. Blanchet, M. Aziz Bouchene, and B. Girad, *Phys. Rev. Lett.* 87 (2001)033001.
[24] P. Borri, W. Langbein, S. Schneider, and U. Woggon, *Phys. Rev. B* 66 (2002)081306®.
[25] Optical properties of photonic crystals, K. Sakoda (Springer Series in Optical Sciences, 2001) P.107.
[26] N. Vats, S. John: *Phys. Rev. A* 65 (2002)043808.
[27] H. Gieben, J. D. Berger, G. Mohs, and P. Meystre, *Phs. Rev. A* 53 (1996)2816.
[28] J. P. Dowling, M. Scalora, M. J. Bloemer, and C. M. Bowden: *J. Appl. Phys.* 75 (1994) 1896.
[29] T. Quang, M. Woldeyohannes, and S. John, *Phys. Rev. Lett*., 79 (1997)5238.
[30] X. S. Xu, B. Y. Cheng, D. Z. Zhang, *Appl. Phys. B* 77 (2003) 515.
[31] E. Paspalakis, N. J. Kylstra, and P. L. Knight, *Phys. Rev. A* 60 (1999) R33.
[32] L. Florescu, K. Busch, and S. John, *J. Opt. Soc. Am. B* 19 (2002) 2215.
[33] M. Centini, G. D'Auguanno, M. Scalora, M. Bloemer, C. M. Bowden, C. Sibilia, N. Mattiucci, and M. Bertolotti, *Phys. Rev. E* 67 (2003) 036617.
[34] D. G. Angelakis D. G., E. Paspalakis, and P. L. Knight, *Phys. Rev. A* 61 (2000)055802.
[35] K. Sakoda, *Opt. Express* 4 (1999)167.
[36] X. S. Xu, B. Y. Cheng, D. Z. Zhang, *J. Phys.: Condens. Matt.* 15 (2003) 7455-7461.
[37] K. Inoue, N. Kawai, Y. Sugimoto, N. Carlsson, N. Ikeda, and K. Asakawa, *Phys. Rev. B* 65 (2002)121308.
[38] J. M. Bendickson, J. P. Dowling and M. Scalra, *Phys. Rev. E* 53 (1996) 4107.
[39] M. Imada, S. Noda, A. Chutinan, T. Tokuda, M. Murata and G. Sasaki, *Appl. Phys. Lett*. 75 (1999) 316.
[40] M. Notomi, H. Suzuki and T. Tamamura, *Appl. Phys. Lett*. 78 (2001) 1325.
[41] M. Scalora, R. J. Flynn, S. B. Reinhardt, R. L. Fork, M. J. Bloemer, M. D. Tocci, C. M. Bowden, H. S. Ledbetter, J. M. Bendickson, J. P. Dowling and R. P. Leavitt, *Phys. Rev. E* 54 (1996)1078.
[42] X. S. Xu, B. Y. Cheng, D. Z. Zhang, *Optical Materials* 27 (2005)1149.
[43] C. M. Soukoulis, *Photonic Crystals and Light Localization in the 21st Century*, (USA: Nato Science Series, 2000).

[44] Y. Q. Wang, X. Y. Hu, X. S. Xu, B. Y. Cheng, and D. Z. Zhang, *Phys. Rev. B* 68 (2003)165106.
[45] C. J. Jin, B. Y. Cheng, B. Y. Man, Z. L. Li, and D. Z. Zhang, *Phys. Rev. B* 61 (2000)10762.
[46] A. A. Asatryan, K. Busch, R.C. McPhedran, L.C. Botten, C.Martijn de Sterke, and N. A. Nicorovici, Phys. Rev. E 63 (2001) 046612.
[47] H. Cao, J. Y. Xu, and D. Z. Zhang, S. H. Chang, S. T. Ho, E. W. Seelig, X. Liu, and R. P. H. Chang, Phys. Rev. Lett. 84 (2000)5584.

INDEX

A

absorption spectra, vii, viii, 2, 5, 7, 10, 11, 12, 19, 32, 33, 42, 43, 45, 48, 55, 56, 57, 69, 72, 75, 85, 86, 116, 208
accumulation, 132
accuracy, 23, 114
acetic acid, 95
acetone, 4, 95, 96, 116, 117, 120, 200
acid, 93, 95, 96, 195, 196, 198, 199, 200
acrylate, ix, 185, 188, 189, 190, 192, 202
acrylic acid, 215, 216
activation, 83
adsorption, 7, 208
affect, 138, 144, 164, 168
agent, 93
aggregates, viii, 11, 37, 45, 46, 55, 56, 63, 72, 81, 86, 87, 192
aggregation, 10, 13, 14, 48, 93, 102, 208
alcohol, 199
alternative, 138
aluminium, vii, 1, 3, 31, 148, 149
aluminum, 56, 68
ambient air, 118
amines, 200
ammonium, 208
amorphous polymers, 190
amplitude, 21, 24, 223, 225, 226, 232, 233, 238, 239, 244, 245
aniline, 188, 190, 191, 192, 194, 196, 197, 198
annealing, 105, 143, 144
annihilation, 81
arbitrary structure, x, 219, 222, 238
assessment, 166
assignment, 61
atomic force, 200, 201, 205
atomic force microscope, 200
atoms, 2, 6, 144, 169, 180, 220, 221, 222, 223, 225, 237, 238, 239, 243, 246, 247, 248, 250
attacks, 96
attention, 6, 7, 114, 195, 220, 225
availability, ix, 137, 161, 180, 198
azo groups, ix, 185, 204, 213
azo polymers, ix, 185, 187, 188, 191, 196, 198, 201, 203, 206
azobenzene units, ix, 185, 195, 202, 203, 204, 212, 213

B

background noise, 120
bandwidth, 120, 138
barium, ix, 137, 138, 161, 162, 168, 169, 171, 179, 180, 183
beams, ix, 6, 116, 149, 185, 186, 200, 201, 207, 211
behavior, ix, x, 63, 108, 127, 129, 185, 204, 213, 219, 220, 226, 227, 229, 230, 233, 234, 244
Beijing, 185, 219
benzene, 76, 81, 204
bias, 3
binding energy, 32
biological systems, 92
birefringence, 186
bleaching, 115, 125, 130, 131, 132
blueshift, 36
bonding, 2, 10, 79, 87
bonds, 139, 140, 150, 153, 155, 160, 161, 164, 166, 168, 169, 173, 174, 176, 179, 180
borophosphate glasses, ix, 137, 138, 169, 171, 179, 180
branching, 217
buffer, 6
building blocks, 140

C

Ca^{2+}, 164
calcium, ix, 137, 138, 161, 162, 168, 169, 180, 183
cancer, 7
candidates, ix, 7, 92, 137, 138, 181
carbon, 96, 147
carbon dioxide, 147
carrier, 35, 56, 198
cast, 143, 144
cation, 140, 153, 155, 161, 164, 166, 174, 180
cell, 4, 11
chain rigidity, x, 185
channels, 21, 44, 103, 138
chelates, 2, 31
chemical reactions, 195
chemical stability, 7
chemical structures, 78, 188, 189, 190, 193, 196, 197, 198, 199, 203, 209
China, 185, 219
chloroform, 3, 79
clusters, 101, 102, 180
CMC, 93
CO2, 147
coatings, 6
cobalt, 3
coherence, 120
colloid particles, 192
combined effect, 141, 149, 176
competition, viii, 55
complexity, 49
components, 2, 3, 8, 11, 14, 19, 20, 21, 23, 24, 37, 86, 138, 147
composition, ix, 97, 98, 138, 144, 150, 151, 152, 162, 164, 169, 171, 207, 213
compounds, 3, 5, 10, 69, 86, 87, 116, 118, 120, 133, 186, 196, 200, 203, 217
computing, 6
concentration, viii, 13, 32, 37, 38, 41, 42, 45, 56, 60, 65, 68, 72, 73, 76, 81, 83, 86, 87, 91, 92, 93, 97, 98, 100, 101, 102, 103, 105, 106, 107, 111, 116, 117, 120, 126, 147, 199, 208
configuration, 10, 35, 81, 194
confinement, 92, 129, 253
conformity, 166
consensus, 7
consumption, 2
contamination, 164
control, ix, 93, 137, 207, 212
conversion, 35, 43, 108, 149, 164, 194, 195
cooling, 144
copper, 3, 5
correlation, 28, 116, 153, 187
coupling, ix, 10, 11, 32, 35, 43, 49, 79, 80, 82, 99, 185, 187, 188, 189, 190, 191, 192, 193, 194, 196, 197, 198, 213, 215
covalent bonding, viii, 2, 55, 79, 87
critical value, 43
crystal growth, 145
crystal structure, viii, 91, 92, 97, 98, 100, 111, 222
crystallization, 11, 87, 98, 144, 145
crystals, vii, x, 2, 10, 42, 45, 46, 219, 220, 221, 230, 243, 247, 248, 249, 253, 254, 255
cycles, 186, 208, 209
cycling, 148

D

damage, 144, 201
decay, viii, x, 2, 92, 101, 102, 110, 113, 118, 120, 123, 124, 127, 128, 132, 219, 220, 223, 230, 231, 233, 234, 236, 254
decay times, 120
decomposition, 7, 32
decomposition temperature, 7
deconvolution, 9
deduction, 161, 202, 238, 247
defects, 105, 253
deformation, 11, 214
degenerate, 7, 81
demand, 114
density, x, 7, 10, 80, 81, 84, 110, 140, 149, 155, 161, 166, 169, 176, 180, 185, 198, 205, 207, 219, 220, 221, 226, 228, 233, 234, 236, 237, 239, 245, 247, 253, 254
density functional theory, 10
deposition, ix, 4, 6, 32, 185, 207, 208, 209, 216
deposition rate, 6
depression, 75, 82
derivatives, vii, 1, 2, 3, 7, 45, 48, 186, 190
desorption, 7
detection, 5, 120
deviation, 63, 120, 125, 132
DFT, 12
dielectric constant, vii, 2, 11, 29, 30, 31, 32, 69, 83, 84, 95, 220, 243, 246, 247, 248, 249, 250, 251, 253
diffraction, 95, 97, 98, 99, 200, 201, 202, 204, 205, 214
diffusion, 82, 114, 214
diffusion process, 114
dimethylformamide, 84, 190
diode laser, 92, 95
discs, 7
disorder, x, 220, 221, 252
disordered systems, 221

dispersion, 28, 93, 95, 223
displacement, 12
dissociation, 127
distilled water, 4
distribution, vii, 2, 30, 31, 56, 79, 84, 85, 86, 93, 116, 121, 237, 253
divergence, 129, 133
division, 180
DMF, 84, 190, 194, 196, 199, 200, 207, 208, 209, 210, 211, 212
domain, 56, 83, 88
donors, 204
dopants, viii, 6, 37, 49, 91
doping, 2, 97
drying, 194
DSC, 198
DTA curve, 145
durability, 138, 140, 149, 161, 169, 173
duration, 116, 120, 124, 125, 128, 221, 223, 227, 229, 234

E

earth, viii, 91, 92, 93, 98, 99, 106, 110, 111, 139, 149, 161, 168
effusion, 4
Egypt, 51
elasto-optic properties, vii
electric field, 6, 83, 187, 220, 221, 222, 223, 226, 229, 233, 235, 237, 238, 239, 241
electroluminescence, viii, 3, 55, 56, 68, 80
electromagnetic, 220, 222, 228, 237, 252
electromagnetic waves, 252
electron diffraction, 12, 99
electron microscopy, 95
electronic structure, vii, 1, 3, 10, 19, 48, 49
electrons, 2, 10, 32, 35, 56, 79, 81, 82, 92, 102, 122
emission, viii, x, 2, 6, 7, 12, 13, 14, 15, 17, 19, 35, 37, 43, 45, 46, 48, 49, 56, 57, 58, 59, 60, 61, 62, 63, 64, 65, 66, 67, 68, 69, 70, 71, 72, 73, 74, 75, 76, 77, 78, 79, 80, 81, 83, 84, 85, 86, 87, 88, 91, 99, 100, 101, 102, 103, 104, 105, 106, 107, 108, 109, 110, 111, 114, 115, 116, 123, 219, 220, 221, 222, 225, 226, 227, 228, 229, 230, 231, 232, 233, 234, 235, 236, 237, 239, 240, 241, 242, 243, 244, 245, 247, 248, 249, 251, 252, 253, 254
emission field, x, 219, 221, 226, 236
emulsifying agents, 93
emulsions, viii, 91, 93, 95
energy, vii, viii, 1, 2, 3, 7, 10, 11, 19, 21, 24, 32, 35, 37, 39, 41, 43, 45, 48, 49, 56, 57, 63, 69, 74, 79, 80, 81, 82, 83, 88, 91, 92, 101, 102, 103, 104, 106, 109, 144, 195, 217, 220, 227, 234
energy transfer, vii, viii, 1, 3, 32, 37, 49, 63, 79, 82, 91, 92, 101, 103, 106
environment, 68, 86, 220
equating, 142
equilibrium, 12, 84, 85, 144, 153, 237, 239, 240, 244
erbium, 95, 100, 101, 102, 111
ester, 96
etching, 229
ethanol, 3, 45, 84, 195, 199
evaporation, 3, 4, 87, 118, 143, 153
evidence, 125, 150, 166, 176, 253
evolution, 235, 236, 254
excitation, vii, viii, 1, 3, 5, 21, 32, 33, 35, 43, 47, 55, 57, 58, 59, 60, 63, 64, 66, 68, 69, 70, 71, 72, 73, 74, 75, 76, 77, 78, 81, 82, 88, 91, 92, 99, 101, 102, 103, 104, 105, 106, 108, 109, 110, 111, 113, 114, 115, 116, 117, 120, 122, 123, 124, 125, 126, 127, 128, 129, 130, 131, 132, 223, 227, 234
exciton, vii, 1, 10, 14, 15, 16, 19, 32, 35, 41, 43, 44, 46, 49
exploitation, 7
exposure, 207
expression, x, 110, 123, 219, 220, 237
extinction, vii, 2, 3, 11, 26, 27, 28, 29, 49
extraction, 3
extrapolation, 178

F

fabrication, 3, 87, 200
family, 138, 181
feedback, 53, 253
field theory, 214
film thickness, 201, 208
films, vii, viii, ix, 1, 3, 5, 10, 11, 26, 29, 32, 41, 43, 45, 48, 49, 55, 56, 57, 60, 61, 67, 69, 70, 72, 74, 76, 77, 78, 80, 81, 83, 84, 86, 87, 88, 114, 118, 185, 186, 202, 207, 208, 209, 211, 212, 213, 214, 215, 217
filters, ix, 116, 117, 137, 138, 180
filtration, 199, 200
financial support, 111, 134, 181, 213
flocculation, 96
fluorescence, viii, 2, 14, 19, 45, 46, 57, 95, 106, 110, 113, 114, 115, 116, 117, 118, 119, 120, 123, 124, 125, 126, 127, 128, 129, 130, 131, 132, 133
fluorescence decay, 116, 119, 120, 124
focusing, 117
food, 6
France, 1, 50, 52
free energy, 93
free volume, 187
friction, 87

FTIR, 146, 147
full width half maximum, 36
functionalization, ix, 185, 187, 188, 191, 192, 193, 194, 195, 198, 205, 206, 213, 214

G

gain threshold, 244
gas sensors, 2, 7
gases, vii, 7
gel, 93, 94, 95, 111, 197
gel permeation chromatography, 197
generation, 6, 7, 24, 92, 105, 132, 195, 217
glass transition temperature, 79, 138, 144, 145, 160, 165, 168, 174, 179, 180, 186, 190, 198
glasses, vii, ix, 137, 138, 139, 140, 142, 143, 144, 146, 147, 149, 150, 151, 152, 153, 154, 155, 156, 157, 158, 159, 160, 161, 162, 163, 164, 165, 166, 167, 168, 169, 170, 171, 172, 173, 174, 175, 176, 177, 178, 179, 180, 181, 183, 217
GPC, 197, 198
graph, 106, 124, 125, 126, 129, 130, 131
graphite, 143, 148
gratings, ix, 137, 138, 180, 186, 200, 201, 202, 211, 212, 213, 214, 215, 217
Greece, 113
groups, ix, 32, 85, 86, 170, 185, 190, 191, 192, 194, 195, 196, 197, 198, 201, 202, 203, 204, 212, 213, 216, 217
growth, 2, 37, 93, 105, 196, 197, 198

H

halogen, 5
harvesting, 195, 217
HE, 214
heating, 3, 96, 98, 105, 144, 148, 186
height, 58, 61, 63, 74, 76, 86, 145
helium, 5
homogeneity, 127
host, ix, 32, 45, 49, 56, 60, 63, 64, 79, 83, 87, 92, 101, 105, 111, 137
hydrogen bonds, 32
hydrolysis, 95, 96, 149, 164, 174, 194
hydroxide, 96
hydroxyl groups, 105, 192, 194
hyperbranched polymers, ix, 185, 195, 203

I

identification, ix, 138
identity, 32
illumination, 122, 123, 133
images, 57, 87, 114, 201, 206, 209, 211, 212
immersion, 117
India, 91
indication, 93, 126, 129, 133
indices, 252
indium, 56
induction, 223
industry, 138
infinite, 228, 236, 240
initial state, 227, 234
input, 117, 222, 229
integration, 224
intensity, vii, viii, 2, 6, 11, 15, 19, 32, 35, 43, 46, 47, 48, 49, 58, 59, 60, 61, 63, 64, 65, 75, 79, 81, 82, 83, 84, 86, 91, 95, 100, 101, 102, 103, 105, 106, 107, 108, 109, 110, 111, 114, 116, 123, 127, 129, 148, 201, 213, 221, 227, 229, 234, 236, 239, 240, 241, 242, 249, 251
interaction(s), vii, viii, 1, 2, 10, 11, 14, 15, 16, 19, 35, 41, 43, 44, 45, 46, 55, 56, 75, 81, 82, 83, 85, 86, 87, 88, 99, 102, 110, 146, 187, 220, 221, 223, 234, 236, 240, 244, 246
interest, 3, 6, 32, 113, 192, 230
interface, 6, 86, 93
interference, ix, 26, 38, 147, 148, 185, 186, 207, 211, 220, 227, 230, 234
intermolecular interactions, 3, 12, 32, 187
interpretation, 24
interval, 21, 226, 233
intervention, 126, 133
inversion, x, 219, 220, 221, 223, 237, 239, 240, 241, 243, 244, 245, 246, 247, 248, 254
ions, viii, 49, 79, 91, 92, 96, 97, 98, 99, 100, 101, 102, 103, 104, 105, 106, 109, 111, 138, 161, 173, 174
Ireland, 112
iridium, 59, 75
iris, 116
iron, 144
irradiation, viii, 6, 113, 117, 123, 124, 125, 127, 128, 129, 130, 131, 132, 186, 201, 202, 204, 205, 206, 207, 213
isolation, 195
isomerization, 186, 203, 204
isomers, 32, 84, 86, 186

J

Japan, 55, 89
Jordan, 49, 53

K

K^+, 155, 156, 160, 180
knowledge, 7, 49, 120, 196

L

lasers, vii, 92, 114, 220
lasing threshold, x, 219, 220, 221, 244, 245, 246
lattices, 248
layer-by-layer growth, 207
LED, 55
lens, 117
lifetime, 79, 81, 82, 101, 102, 103, 230
ligand, 10, 75, 79, 80, 81, 82, 84, 85
light beam, 110
light emitting diode, viii, 2, 55
limitation, 146
linear dependence, 127
linear polymers, ix, 185
links, 140
liquid crystals, 2
liquids, vii, 93, 114, 214
lithography, 229
localization, x, 79, 220, 221, 227, 247, 250, 252, 253, 254
location, 56, 201
low temperatures, 35, 59, 61
luminescence, vii, viii, 1, 5, 12, 32, 34, 35, 37, 43, 46, 79, 91, 92, 101, 103, 106, 107, 111
luminescence efficiency, 93
lying, 12, 15, 19

M

macromolecules, 194
magnesium, 3, 12
magnetic field, 191
mass, 186, 212, 214
matrix, viii, 91, 92, 111, 168, 169, 237
Maxwell equations, x, 219, 221, 254
measurement, ix, 21, 74, 117, 138, 143, 144, 145, 146, 147, 229
mechanical properties, vii, 198
media, 221, 253
melt(s), 143, 195
melting, 7, 139, 143, 144, 145, 151, 153, 162, 164, 170, 171
melting temperature, 143
memory, 125
metallophthalocyanines, vii, 1, 2, 3, 6, 7, 10, 11, 28, 48
metals, vii, 12
methanol, 95, 120, 199
methyl groups, 204
microfabrication, 114
microscope, 92, 117, 118
microscopic investigations, 214
microscopy, 87, 114, 201, 205, 212
microstructure, 144
migration, 186
miniaturization, 92
mixing, 10, 11, 80, 82
mobility, 6, 56
MOCVD, 229
mode, 85, 92, 116, 117, 147, 200, 209, 220, 252, 253
models, 3, 187
molar ratios, 190
molar volume, 142, 155
mole, 159, 160
molecular mobility, 145
molecular structure, 2, 4, 6, 33, 93
molecular weight distribution, 194, 195, 197, 198, 203
molecules, viii, 2, 3, 7, 10, 11, 13, 14, 32, 35, 37, 38, 41, 42, 43, 46, 48, 49, 55, 56, 57, 61, 62, 63, 64, 65, 67, 68, 69, 72, 79, 80, 81, 82, 83, 84, 85, 86, 87, 88, 93, 95, 96, 113, 114, 115, 116, 120, 122, 123, 129, 133, 187
monitoring, 97, 144
monolayer, 208
monomers, 187, 190, 195, 196
morphology, 5, 11, 56, 84, 87, 93, 95, 209
motion, 214
movement, 85, 187
multilayer films, ix, 185, 192, 207, 208, 209, 211, 212, 213

N

Na^+, 173, 176
nanocrystals, viii, 91, 92, 96, 97, 98, 99, 100, 101, 102, 103, 105, 107, 108, 109, 110, 111
nanomaterials, 92
nanometer scale, 92
nanoparticles, 92, 93, 94, 95, 96, 100, 103, 104, 105, 106, 107
Nd, 116
NdCl3, 108
neat films, viii, 55, 60, 69, 86, 88
needs, 220
neglect, 232
network, 138, 139, 140, 141, 152, 155, 160, 161, 163, 166, 168, 169, 170, 173, 176, 179, 180
nitrogen, 95, 143, 144

NMR, 115, 190, 191, 194, 197
noise, 123, 138
non-radiative transition, 43, 79
nucleation, 93, 145

O

observations, 122, 125, 128, 133, 155
oil, 93, 95, 117, 199
optical communications, 6
optical gain, x, 219, 221, 232, 240, 244, 253, 254
optical parameters, vii, 1
optical properties, vii, 3, 6, 7, 56, 92, 113, 195, 215
optical pulses, 237
optical systems, vii
optoelectronics, 6
orbit, 19, 75, 79, 80, 81, 82, 83, 99
organic compounds, 3, 79, 114
organic solvents, 93
orientation, 6, 11, 85, 86, 187
oscillation, x, 219, 220, 221, 223, 227, 236, 245, 252, 254
output, 92
oxidation, 199
oxide nanoparticles, 92, 93, 94
oxides, 139, 164
oxygen, 32, 98, 139, 140, 143, 150, 152, 153, 162, 170

P

parameter, 19, 21, 62, 82, 84, 114, 120, 220
particles, 7, 33, 93, 94, 95, 96, 98, 105, 144
passive, ix, 137
pathways, 126
periodicity, 222, 229
permittivity, 83
perylene, vii, 1, 2, 3, 41, 42, 45, 46, 48
pH, 95, 96, 192
phase diagram, 23, 24
phase shifts, 23
phenol, 216
phonons, 92
phosphate glasses, ix, 137, 140, 149, 161, 162, 169, 180, 181
phosphates, 140, 161
phosphorescence, 59, 82
phosphorus, 139, 150, 161
photoabsorption, vii, 1, 19, 21, 22, 49
photobleaching, viii, 113, 114, 115, 117, 118, 120, 122, 123, 124, 125, 126, 127, 128, 129, 130, 131, 132, 133
photoconductivity, 3
photodetectors, 5
photoluminescence, vii, viii, 1, 2, 3, 5, 7, 14, 15, 16, 17, 18, 19, 20, 24, 33, 37, 43, 45, 46, 47, 48, 55, 56, 58, 62, 65, 69, 75, 76, 79, 81, 82, 89
photonic crystals, x, 219, 220, 221, 254
photonics, viii, 113
photons, viii, 106, 113, 114, 123, 125, 126, 131, 132, 133, 195, 217, 237
photosensitivity, 138
photovoltaic cells, 2
physical properties, vii, 7, 138
physics, viii, 3, 113
PL spectrum, 12, 13, 14, 19, 34, 36, 37, 43, 46, 47, 58, 61, 62, 63, 67, 68, 76, 84, 85
plastics, 6
platinum, 67, 143
PM, 214
PMMA, 118
PMS, 116
Poland, 1
polarity, 93
polarizability, 84
polarization, 83, 116, 200, 214, 222, 223, 225, 232, 238, 243, 246, 247, 248, 249
polycarbonate, 69, 82, 87, 88
polycondensation, 190
polydispersity, 197
polyesters, 214, 216
polymer chains, 186
polymer films, ix, 114, 118, 185, 186, 200, 201, 214, 215
polymer materials, 187
polymer structure, 187, 201, 213
polymer synthesis, 187
polymeric chains, 194
polymerization, ix, 185, 187, 190, 194, 195, 196, 197, 198, 213, 216
polymers, vii, ix, 2, 183, 185, 186, 187, 188, 189, 190, 191, 192, 194, 195, 196, 197, 198, 199, 200, 201, 202, 203, 204, 205, 206, 207, 213, 214, 215, 216, 217
polyphosphates, 150, 155, 160
polystyrene, 60, 61, 64, 68, 70, 76, 77, 78, 86, 87, 88, 198, 199
polyurethane, 144
poor, 87, 140, 161, 207
population, x, 79, 81, 109, 110, 219, 220, 221, 223, 226, 230, 231, 235, 236, 237, 239, 240, 241, 243, 244, 245, 246, 247, 248, 254
population density, 109
population inversion, x, 219, 223, 237, 240, 254

postpolymerization, 187, 188, 189, 190, 191, 193, 213
potassium, ix, 137, 138, 149, 150, 151, 155, 160, 161, 180, 183
potassium aluminophosphate glasses, ix, 137, 149, 150, 160, 161, 180, 183
power, viii, 2, 6, 81, 91, 106, 107, 108, 109, 110, 114, 116, 117, 120, 122, 123, 124, 125, 126, 128, 129, 130, 131, 132, 138, 186, 200, 226, 229, 233, 234, 245
precipitation, 194
prediction, 166
pressure, 4, 186
principle, 57, 95
probability, 19, 43, 46, 79, 81, 82, 92, 110, 124
probe, 145, 222, 229
production, 6, 7
program, 134, 229
propagation, 222, 225, 232, 237
propane, 56
protons, 190, 191, 194
pulse, viii, x, 113, 116, 120, 125, 126, 128, 219, 220, 221, 222, 223, 227, 229, 234, 236
pumps, 101
purification, 195, 199
pyrophosphate, 140
pyrylium, viii, 113, 114, 115, 116, 117, 118, 119, 120, 122, 123, 124, 126, 127, 128, 129, 130, 131, 132, 133

Q

quantization, 92
quantum electrodynamics, 220
quantum optics, 108
quantum state, 57
quantum theory, 220
quantum well, 229
quantum yields, 114, 118, 120, 133
quartz, vii, 1, 4, 7, 10, 17, 27, 41, 42, 44, 46, 47, 48, 68, 116, 117, 208, 209, 211
quasi-periodic photonic crystals, x, 219, 221

R

Rabi frequency, x, 219, 220, 223, 224, 225, 226, 227, 228, 229, 230, 231, 233, 234, 254
Rabi oscillations, 223, 236
radiation, 27, 37, 81, 92, 95, 146, 222, 227, 229, 234, 236, 237, 240, 244
radical copolymerization, 216
radical polymerization, 187, 190, 192, 194
radius, 88, 246, 248
range, vii, 1, 2, 5, 6, 15, 17, 18, 19, 25, 35, 43, 45, 46, 49, 78, 85, 92, 114, 123, 125, 130, 131, 133, 138, 145, 147, 149, 162, 166, 168, 174, 176, 181, 194, 201, 229, 252, 253
raw materials, 143
reactive groups, 194
reactivity, 216
recombination, 35, 56
reconstruction, 24
recovery, 114
recrystallization, 32
red shift, 13, 14, 37, 76, 83
redistribution, 15, 16
reduction, 46, 82, 144, 176, 198, 199, 245
reflection, vii, 1, 2, 3, 7, 26, 27, 28, 38, 39, 40, 49, 147, 148, 244, 245
reflectivity, 245
refractive index, vii, ix, 2, 3, 26, 28, 29, 38, 40, 49, 92, 137, 138, 141, 142, 146, 149, 154, 155, 164, 171, 174, 179, 180, 244
refractive indices, 3, 27, 29, 160, 168, 174
relationship(s), 176, 187, 191, 194, 201, 204, 205
relaxation, x, 37, 43, 92, 101, 102, 105, 219, 221, 223, 228, 230, 237, 240, 244
relaxation coefficient, x, 219, 221, 240
relaxation process(es), 101, 102, 223, 228, 230
relaxation rate, 105
reliability, 133, 138
renormalization, 36
replacement, 92
resistance, 3
resolution, 64, 85, 116, 118, 147
returns, 145
rings, 10, 32, 140, 204
rods, 227, 243, 246, 248, 249, 250, 251, 252, 253
room temperature, vii, viii, 1, 4, 12, 19, 43, 47, 56, 61, 64, 67, 69, 70, 74, 79, 81, 82, 83, 84, 85, 87, 91, 95, 143, 144, 146, 199, 200
roughness, 144, 209
routing, 138

S

salts, viii, 113, 115, 117, 118, 119, 120, 122, 127, 128, 129, 131, 132, 133, 187, 188, 190, 197, 200
sample, 6, 97, 100, 103, 110, 116, 117, 125, 144, 145, 146, 147, 148, 154, 163, 164, 171, 172, 173, 201, 211, 223
sample design, 148
saturation, ix, 106, 108, 120, 132, 185, 213, 239
scattering, 7, 229, 252, 253
second generation, 7

selected area electron diffraction, 98, 99
selecting, 24, 138, 186, 191, 195
self, ix, 9, 43, 45, 48, 58, 61, 62, 63, 64, 65, 67, 75, 76, 93, 132, 185, 192, 207, 208, 211, 213, 215, 216
self-assembly, ix, 185, 207, 211, 213, 215, 216
semiconductor, 2, 53, 138, 230
semiconductor lasers, 53
semiconductors, 5, 19, 56
sensitivity, 19, 114, 180
sensors, 6, 186
separation, 77, 85, 95, 96, 163, 171
series, viii, 24, 35, 91, 93, 150, 152, 153, 154, 155, 156, 157, 158, 160, 170, 171, 173, 174, 176, 187, 191, 203, 204, 205, 208
shape, 7, 11, 20, 21, 23, 34, 43, 46, 56, 57, 60, 61, 62, 63, 71, 83, 84, 85, 86, 87, 88, 100, 120, 144
sharing, 168
shoulders, 7, 21, 43
sign, 142
signals, 21, 120, 132, 229
signs, 171
silica, 138, 143, 144
silicon, vii, 1, 4, 5, 148, 208, 210
similarity, 169
simulation, 220
SiO2, 139
sites, 105, 139, 144, 155, 161, 166, 168, 169, 180
sodium, ix, 137, 138, 146, 150, 166, 169, 170, 171, 179, 180, 183, 200
solar cells, 6
solid phase, 3, 43
solid state, viii, 6, 10, 11, 43, 45, 48, 55, 67, 69, 76, 86
solubility, 96, 190, 194, 195, 198, 207
solvation, viii, 55, 67, 69, 76, 83, 86, 87
solvents, 3, 4, 5, 7, 69, 190, 207, 208, 209, 210
spectroscopy, vii, viii, 1, 2, 3, 5, 7, 19, 24, 49, 53, 92, 113, 114, 115, 116, 133
spectrum, 6, 7, 8, 9, 11, 13, 14, 15, 16, 18, 19, 20, 21, 22, 23, 24, 25, 26, 32, 33, 36, 37, 39, 41, 43, 44, 45, 46, 56, 57, 58, 59, 60, 61, 62, 63, 64, 65, 67, 68, 69, 70, 71, 73, 74, 75, 76, 84, 85, 86, 88, 118, 120, 125, 126, 147, 186, 194, 197, 220
spin, ix, 3, 19, 56, 58, 59, 60, 61, 63, 64, 66, 67, 73, 75, 79, 80, 81, 82, 83, 85, 87, 88, 99, 185, 201, 207, 211, 212, 213
square lattice, 229, 243, 245, 246, 250
stability, ix, 6, 32, 137, 148, 198
stabilization, viii, 91, 93, 97
star polymers, 216
stars, 21
stimulated emission, x, 219, 220, 221, 228, 229, 234, 240, 241, 243, 246, 247, 248, 250, 252, 253, 254
stock, 187
stoichiometry, 140, 155, 161
storage, 6, 114, 186, 215
strain, 11, 148
strategies, 187
strength, 10, 12, 87, 153, 155, 156, 160, 164, 166, 173, 176, 180, 229, 236
stress, 143
stretching, 194
striae, 143
strong interaction, 56, 88
structural changes, 144
structural characteristics, 195
styrene, 199
substitutes, ix, 2, 185, 213
substitution, vii, 2, 11, 127, 132, 155, 196, 197, 198, 204
substitution effect, 132, 204
substitution reaction, 196, 197, 198
substrates, vii, 1, 2, 3, 4, 7, 41, 45, 186, 208
subtraction, 147
sulfuric acid, 199
Sun, 112, 134
supply, 187, 229
suppression, 24, 25
surface area, 93, 106
surface energy, 93
surface modification, 201, 212
surface-relief-grating (SRG), ix, 185, 211
surfactant, viii, 91, 93, 95, 96
survival, 114
susceptibility, 6
switching, 6, 186, 238
symmetry, viii, 2, 5, 10, 12, 19, 33, 43, 45, 49, 57, 81, 82, 98, 252, 253
synthesis, ix, 4, 5, 93, 111, 143, 153, 164, 170, 185, 200, 214, 216
systems, ix, 108, 137, 138, 181, 186, 203, 208, 230, 252, 253

T

TDI, 190
technology, 2, 6, 45, 138, 180, 229
temperature, vii, ix, 1, 2, 3, 4, 5, 7, 12, 15, 16, 17, 18, 19, 20, 32, 34, 35, 36, 43, 45, 46, 47, 48, 49, 56, 58, 59, 60, 61, 63, 64, 69, 70, 75, 79, 81, 82, 83, 84, 85, 87, 88, 96, 97, 98, 105, 118, 137, 138, 141, 142, 143, 144, 145, 148, 149, 180, 181, 186, 190, 192, 196, 199

temperature dependence, vii, 1, 7, 15, 17, 20, 34, 35, 49, 83, 138, 141
tension, 93
tetrahydrofuran, 3
theory, 81, 83, 220, 221, 247, 254
therapy, 7, 114
thermal activation energy, 35, 81, 82, 83
thermal evaporation, 4, 45, 56
thermal expansion, ix, 36, 137, 138, 140, 141, 142, 145, 149, 159, 160, 168, 179, 180, 181
thermal properties, vii, ix, 137, 138, 143, 149, 150, 152, 153, 160, 161, 162, 163, 169, 171, 180, 183
thermal quenching, vii, 1, 15, 17, 35, 37, 46, 47, 48
thermal stability, 32, 145, 195
thermolysis, 198
thin films, vii, viii, 1, 3, 4, 5, 6, 7, 10, 11, 32, 37, 43, 45, 46, 48, 49, 51, 55, 56, 58, 84, 85, 86, 207, 211, 213, 214, 217
threshold, x, 32, 39, 219, 220, 221, 239, 244, 245, 246, 247, 250, 252, 253, 254
time, vii, viii, 2, 10, 21, 23, 24, 43, 49, 80, 87, 101, 113, 114, 115, 116, 117, 118, 120, 123, 124, 132, 133, 144, 145, 148, 153, 201, 202, 204, 205, 206, 207, 213, 223, 225, 226, 227, 232, 233, 234, 235, 236, 237, 240, 244, 254
time resolution, 118, 132
tin, 56
toluene, 64, 69, 70, 79
total internal reflection, 146
toxicity, 6
TPA, viii, 113, 114, 116, 120, 122, 123, 125, 131, 133
transient emission, x, 219, 221, 227, 230, 236, 254
transistor, 6
transition(s), vii, x, 2, 7-12, 15, 18-25, 32, 33, 35, 41-46, 48, 49, 57, 59, 79, 82, 83, 84, 85, 92, 99, 101, 102, 103, 109, 110, 139, 140, 144, 156, 165, 166, 168, 169, 175, 176, 179, 180, 186, 197, 215, 219, 220, 221, 223, 227, 230, 231, 236, 237, 254
transition metal, 139
transition rate, 109
transition temperature, 140, 156, 165, 175, 176, 179, 180, 181
translation, 214, 252, 253
transmission, vii, 2, 26, 27, 38, 39, 40, 49, 138, 147, 153, 154, 164, 173, 174, 200, 248, 252
transmittance spectra, 251, 252
transparency, 198
transport, 2, 6, 32, 56, 65, 66, 67, 86, 186, 212, 214
trend, ix, 137, 168
tungsten, 5
tunneling, 82, 83

U

UG, 116
UK, 137, 181, 182, 183
uncertainty, 144, 149
uniform, 208, 225, 232, 241, 247
urethane, ix, 185, 188, 190, 202
UV, vii, 2, 5, 10, 18, 33, 139, 186, 195, 197, 208, 209
UV radiation, 33

V

vacancies, 98
vacuum, 3, 4, 32, 56, 196, 199, 200, 220, 253
valence, 10, 140, 153, 161
validity, 181
values, viii, 12, 19, 21, 23, 24, 28, 29, 35, 82, 84, 85, 91, 93, 100, 102, 103, 105, 113, 114, 120, 122, 123, 124, 125, 127, 129, 131, 132, 149, 158, 160, 166, 168, 169, 176, 178, 179, 180, 181, 226, 233, 241
variable, 5
variation, 12, 58, 60, 63, 99, 100, 223, 225, 238
vector, 21
velocity, x, 219, 221, 237, 240, 241, 242, 243, 244, 245, 246, 247, 249, 252, 254
versatility, 138
vibration, 85
viscosity, 145, 195, 207
volatility, 161

W

water, 4, 65, 93, 95, 139, 147, 153, 161, 163, 166, 168, 180, 199, 200, 207, 208
water vapor, 65
wave propagation, 221
wave vector, 222
wavelengths, 32, 37, 70, 71, 72, 73, 74, 114, 115, 120, 122, 123, 125, 139, 147
wealth, 113
weight ratio, 118
workers, 32
writing time, 201, 204

X

X-ray diffraction (XRD), 86, 95, 96, 97, 98

Y

yield, 92, 114, 120, 122, 131, 190, 198
yttrium, 102

Z

zinc, 3, 12, 17, 19
zirconia, 92, 95, 97, 102
zirconium, 102
ZnO, 92